Lecture Notes in Mathematics 1606

Springer
Berlin
Heidelberg
New York
Barcelona
Budapest
Hong Kong
London
Milan
Paris
Tokyo

Pei-Dong Liu   Min Qian

Smooth Ergodic Theory of Random Dynamical Systems

Springer

Authors

Pei-Dong Liu
Min Qian
Department of Mathematics and Institute of Mathematics
Peking University
Beijing 100871, P. R. China
E-mail: mathpu@bepc2.ihep.ac.cn

Library of Congress Cataloging-in-Publication Data

Liu, Pei-Dong, 1964-
Smooth ergodic theory of random dynamical systems / Pei-Dong Liu, Min Qian.
p. cm. -- (Lecture notes in mathematics ; 1606)
Includes bibliographical references and index.
ISBN 3-540-60004-3 (Berlin : acid-free). -- ISBN 0-387-60004-3 (New York : acid-free)
1. Ergodic theory. 2. Differentiable dynamical systems. 3. Stochastic differential equations. I. Ch'ien, Min. II. Title. III. Series: Lecture notes in mathematics (Springer-Verlag) ; 1606.
QA3.L28 no. 1605
[QA641.5]
510 s--dc20
[514'.74] 95-21996
CIP

Mathematics Subject Classification (1991): 58F11, 34F05, 34D08, 34C35, 60H10

ISBN 3-540-60004-3 Springer-Verlag Berlin Heidelberg New York

Printed in Great Britain

SPIN: 10130336 46/3142-543210 - Printed on acid-free paper

Table of Contents

Introduction

This book aims to present a systematic treatment of a series of results concerning invariant measures, entropy and Lyapunov exponents of smooth random dynamical systems. We first try to give a short account about this subject and the brief history leading to it.

Smooth ergodic theory of deterministic dynamical systems, i.e. the qualitative study of iterates of a single differentiable transformation on a smooth manifold is nowadays a well-developed theory. Among the major concepts of this theory are the notions of invariant measures, entropy and Lyapunov (characteristic) exponents which culminated in a theorem well known under the name of Oseledec, and there have been numerous relevant results interesting in theory itself as well as in applications. One of the most important classes of the results is Pesin's work on ergodic theory of differentiable dynamical systems possessing a smooth invariant measure. Another is related to the ergodic theory of Anosov diffeomorphisms or Axiom A attractors developed mainly by Sinai, Bowen and Ruelle. A brief review of these two classes of works is now given in the next two paragraphs.

In his paper $[\text{Pes}]_1$, Pesin proved some general theorems concerning the existence and absolute continuity of invariant families of stable and unstable manifolds of a smooth dynamical system, corresponding to the non-zero Lyapunov exponents. This set up the machinary for transferring the linear theory of Lyapunov exponents into non-linear results in neighbourhoods of typical trajectories. Using these tools Pesin then derived a series of deep results in ergodic theory of diffeomorphisms preserving a smooth measure ($[\text{Pes}]_2$). Among these results is the remarkable Pesin's entropy formula which expresses the entropy of a smooth dynamical system in terms of its Lyapunov exponents. Part of the work above has been extended and applied to dynamical systems preserving only a Borel measure ([Kat], [Fat] and $[\text{Rue}]_2$).

We now turn to some results related to the ergodic theory of Axiom A attractors. Recall that for a given Axiom A attractor there exists a unique invariant measure, called Sinai-Bowen-Ruelle (or simply SBR) measure, that is characterized by each of the following properties ([Sin], $[\text{Bow}]_2$ and $[\text{Rue}]_3$):

(1) Pesin's entropy formula holds true for the associated system.

(2) Its conditional measures on unstable manifolds are absolutely continuous with respect to Lebesgue measures on these manifolds.

(3) Lebesgue almost every point in an open neighbourhood of the attractor is generic with respect to this measure.

Each one of these properties has been shown to be significant in its own right, but it is also remarkable that they are equivalent to one another. More crucially, Ledrappier and Young proved later in their well-known paper $[\text{Led}]_2$ that the properties (1) and (2) above remain equivalent for all C^2 diffeomorphisms (That (2) implies (1) was proved by Ledrappier and Strelcyn in $[\text{Led}]_3$). All results mentioned above are fundamental and stand at the heart of smooth ergodic theory of deterministic dynamical systems.

In recent years the counterpart in random dynamical systems has also been investigated. For an introduction to the scope of random dynamical systems, one can hardly find better description than that given by Walter $[\mathrm{Wal}]_2$ in reviewing the pioneering book *Ergodic Theory of Random Transformations* by Kifer ($[\mathrm{Kif}]_1$):

> *"Traditionally ergodic theory has been the qualitative study of iterates of an individual transformation, of a one-parameter flow of transformations (such as that obtained from the solutions of an autonomous ordinary differential equation), and more generally of a group of transformations of some state space. Usually ergodic theory denotes that part of the theory obtained by considering a measure on the state space which is invariant or quasi-invariant under the group of transformations. However in 1945 Ulam and von Neumann pointed out the need to consider a more general situation when one applies in turn different transformations chosen at random from some space of transformations. Considerations along these lines have applications in the theory of products of random matrices, random Schrödinger operators, stochastic flows on manifolds, and differentiable dynamical systems".*

In his book $[\mathrm{Kif}]_1$, Kifer presented the first systematic treatment of ergodic theory of evolution processes generated by independent actions of transformations chosen at random from a certain class according to some probability distribution. Among the major contributions of this treatment are the introduction of the notions of invariant measures, entropy and Lyapunov exponents for such processes and a systematic exposition of some very useful properties of them. This pioneering book establishes a foundation for further study of this subject, especially for the purpose of the development of the present book.

In this book we are mainly concerned with ergodic theory of random dynamical systems generated by (discrete or continuous) stochastic flows of diffeomorphisms on a smooth manifold, which we sometimes call *smooth ergodic theory of random dynamical systems.* Our main purpose here is to exhibit a systematic generalization to the case of such flows of a major part of the fundamental deterministic results described above. Most generalizations presented in this book turn out to be non-trivial and some are in sharp contradistinction with the deterministic case. This is described in a more detailed way in the following paragraphs.

This book has the following structure. Chapter 0 consists of some necessary preliminaries. In this chapter we first present some basic concepts and theorems of measure theory. Proofs are only included when they cannot be found in standard references. Secondly, we give a quick review of the theory of measurable partitions of Lebesgue spaces and conditional entropies of such partitions.

Contents of this part come from Rohlin's fundamental papers $[\mathrm{Roh}]_{1,2}$. The major part of this chapter is devoted to developing a general theory of conditional entropies of measure-preserving transformations on Lebesgue spaces. The concept of conditional entropies of measure-preserving transformations was first introduced by Kifer (see Chapter II of $[\mathrm{Kif}]_1$), but his treatment was only justified for finite measurable partitions of a probability space. Here we deal with the concept in the case of general measurable partitions (maybe uncountable) of Lebesgue spaces and prove some associated properties mainly along the line of $[\mathrm{Roh}]_2$, though the paper of Rohlin only deals with the usual entropies of measure-preserving transformations. Results presented in this chapter serve as a basis of the later chapters.

The concepts in Chapter I are mainly adopted from Kifer's book $[\mathrm{Kif}]_1$. But for an adequate treatment of entropy formula (in Chapters II, IV, VI and VII) an extension of the notion of entropy to general measurable partitions is indispensible. So we have to formulate and prove the related theorems in this setting, which could be accomplished if the reader is familiar with the preliminaries in Chapter 0. In Section I.1 we first introduce the random dynamical system $\mathcal{X}^+(M,v)$ (see Section I.1 for its precise meaning). Then we discuss some properties of invariant measures of $\mathcal{X}^+(M,v)$. When associated with an invariant measure μ, $\mathcal{X}^+(M,v)$ will be referred to as $\mathcal{X}^+(M,v,\mu)$. Section I.2 consists of the concept of the (measure-theoretic) entropy $h_\mu(\mathcal{X}^+(M,v))$ of $\mathcal{X}^+(M,v,\mu)$ and of some useful properties of it deduced from its relationship with conditional entropies of (deterministic) measure-preserving transformations. In Section I.3 we introduce the notion of Lyapunov exponents of $\mathcal{X}^+(M,v,\mu)$ by adapting Oseledec multiplicative ergodic theorem to this random case.

In Chapter II we carry out the estimation of the entropy of $\mathcal{X}^+(M,v,\mu)$ from above through its Lyapunov exponents. We prove that for any given $\mathcal{X}^+(M,v,\mu)$ the following inequality holds true:

$$h_\mu(\mathcal{X}^+(M,v)) \le \int \sum_i \lambda^{(i)}(x)^+ m_i(x) d\mu,$$

where $\lambda^{(1)}(x) < \lambda^{(2)}(x) < \cdots < \lambda^{(r(x))}(x)$ are the Lyapunov exponents of $\mathcal{X}^+(M,v,\mu)$ at point $x \in M$ and $m_i(x)$ is the multiplicity of $\lambda^{(i)}(x)$. This is an extension to the present random case of the well-known Ruelle's (or Margulis-Ruelle) inequality in deterministic dynamical systems. As in the deterministic case the above inequality is sometimes also called Ruelle's (or Margulis-Ruelle) inequality. In the random case this type of inequality was first considered by Kifer in Chapter V of $[\mathrm{Kif}]_1$ (see Theorem 1.4 there), but the proof of his theorem contains a nontrivial mistake and this led the authors of the present book to an essentially different approach to this problem (see Chapter II for details). Our presentation here comes from $[\mathrm{Liu}]_1$. As compared with the deterministic case, it involves substantially new techniques (especially the introduction of relation numbers and the related estimates).

After the first version of this book was completed, the authors received a preprint [Bah] by J. Banmüller and T. Bogenschütz which gives an alternative treatment of Ruelle's inequality. Their argument shows that the mistake mentioned above is inessential and can be corrected with some careful modifications, and their argument is also carried out within a more general framework of "stationary" random dynamical systems. It turns out, then, that the correction of the mistake in the original Chapter II is at the expense of an extraneous hypothesis (see Remark 2.1 of Chapter II). However, the treatment in that chapter (for example, the argument about the C^2-norms and relation numbers) is besides its own right very useful for the later chapters. For this consideration and in order not to change drastically the original (carefully organized) sketch of the book, we retain here the original Chapter II and also introduce Bahnmüller and Bogenschütz's argument (with some modifications) in the Appendix (it involves some results in Chapter VI).

Chapter III deals with the theory of stable invariant manifolds of $\mathcal{X}^+(M, v, \mu)$. We present there an extension to the random case of Pesin's results concerning the existence and absolute continuity of invariant families of stable manifolds ([Pes]$_1$). Although some new technical approaches are employed, our treatment goes mainly along Pesin's line with some ideas being adopted from [Kat], [Fat] and [Bri]. Besides their own rights, results in this chapter serve as powerful tools for the treatment of entropy formula given in later chapters.

In Chapter IV we extend Pesin's entropy formula to the case of $\mathcal{X}^+(M, v, \mu)$, i.e. we prove that

$$h_\mu(\mathcal{X}^+(M, v)) = \int \sum_i \lambda^{(i)}(x)^+ m_i(x) d\mu$$

when μ is absolutely continuous with respect to the Lebesgue measure on M. This formula takes the same form as in the deterministic case, but now the meaning of the invariant measure μ is quite different since it is no longer necessarily invariant for individual sample diffeomorphisms; we also have to point out that the implication of this result exhibits a sharp contradistinction with that in the deterministic case (see the arguments in Section IV.1 and those at the end of Chapter V). This result was first proved by Ledrappier and Young ([Led]$_1$) in the setting of the two-sided random dynamical system $\mathcal{X}(M, v, \mu)$ (see Chapter VI for its meaning), and a more readable treatment of it was later given in [Liu]$_2$ within the present one-sided setting $\mathcal{X}^+(M, v, \mu)$. In this chapter we follow the latter paper. As compared with the deterministic case ([Pes]$_2$ and [Led]$_3$), the proof of the result given here involves the new ideas of employing the theory of conditional entropies and of applying stable manifolds instead of unstable manifolds. Aside from these points, the proof follows essentially the same line as in the deterministic case, although the technical details are much more complicated.

In Chapter V we apply our results obtained in the previous chapters to the case of stochastic flows of diffeomorphisms. Such flows arise essentially as solution flows of stochastic differential equations and all the assumptions made in the previous chapters can be automatically verified in this case. Thus we reach and finish with an important application to the theory of stochastic processes.

Chapter VI is devoted to an extension of the main result (Theorem A) of Ledrappier and Young's remarkable paper $[\mathrm{Led}]_2$ to the case of random diffeomorphisms. Roughly speaking, in the deterministic case one has Theorem A in $[\mathrm{Led}]_2$ which asserts that Pesin's entropy formula holds true if and only if the associated invariant measure has SBR (Sinai-Bowen-Ruelle) property, i.e. it has absolutely continuous conditional measures on unstable manifolds; for the case of random diffeomorphisms we prove in this chapter that Pesin's entropy formula holds true if and only if the associated family of sample measures, i.e. the natural invariant family of measures associated with individual realizations of the random process has SBR property. This result looks to be a natural generalization of the deterministic result to the random case, but it has a non-trivial consequence (Corollary VI.1.2) which looks unnatural and which seems hopeless to obtain if one follows a similar way as in the deterministic case (i.e. by using the absolute continuity of unstable foliations). This generalization was actually known first to Ledrappier and Young themselves, though not clearly stated. Here we present the first detailed treatment of this result. Although the technical details are rather different, our treatment follows the line in the deterministic case provided by $[\mathrm{Led}]_{2,3}$. The sources of this chapter are $[\mathrm{Led}]_{1,2,3}$ and $[\mathrm{Liu}]_3$.

In Chapter VII we study the case when a hyperbolic attractor is subjected to certain random perturbations. Based on our elaboration given in the previous chapters, a random version of the deterministic results mentioned above for Axiom A attractors is derived here. The idea of this chapter comes from [You] and $[\mathrm{Liu}]_5$.

Random dynamical systems, though only at an early stage of development by now, have been widely used and taken care of, especially in applications. In this book, our intention is to touch upon only a part of this subject which we can treat with mathematical rigor. For this reason, we naturally restrict ourselves to the finite dimensional case. Infinite dimensional dynamical systems with random effect should be more interesting from a physical point of view. Scientists from both probability theory and partial differential equations have already paid jointly sufficient attention to this new and important field (a conference was organized by P. L. Chow and Skorohod in 1994). We hope their efforts will lead to a substantially new mathematical theory which, we believe, could be considered as the core of the so-called "Nonlinear Science".

We would like to express our sincere thanks to Prof. Ludwig Arnold since conversations with him were very useful for the preparation of Chapter VII. Our gratitude also goes to Profs. Qian Min-Ping and Gong Guang-Lu for helpful discussions. During the elaboration of this book the first author is supported by the National Natural Science Foundation of China and also by the Peking University Science Foundation for Young Scientists. Finally, it is acknowledged that part of the work on this book was done while the first author was in the Institute of Mathematics, Academia Sinica as a postdoctor and he expresses here his gratitude for its hospitality.

Chapter 0 Preliminaries

In this chapter we first present some basic concepts and facts from measure theory. Then we give a quick review of the theory of measurable partitions of Lebesgue spaces and conditional entropies of such partitions. A detailed treatment of this theory is presented in Rohlin's fundamental papers $[\mathrm{Roh}]_{1,2}$. The rest of this chapter is devoted to developing, following the scheme of $[\mathrm{Roh}]_2$, a general theory of conditional entropies of measure-preserving transformations on Lebesgue spaces.

§1 Measure Theory

Let $(X, \mathcal{B}, \mu)$ be a measure space. Two sets $B_1, B_2 \in \mathcal{B}$ are said to be equivalent modulo zero, written $B_1 = B_2$ (mod 0), if the symmetric difference $B_1 \Delta B_2$ has measure zero. When we write $\mathcal{A}_1 = \mathcal{A}_2$ (mod 0) for two subsets $\mathcal{A}_1, \mathcal{A}_2$ of $\mathcal{B}$ we mean that for each $A_1 \in \mathcal{A}_1$ there exists $A_2 \in \mathcal{A}_2$ such that $A_1 = A_2$ (mod 0) and vice versa. Let $\mathcal{A}$ be a subset of $\mathcal{B}$, we say that $\mathcal{A}$ generates $\mathcal{B}$ (mod 0) if $\mathcal{B} = \mathcal{B}_0$ (mod 0), where $\mathcal{B}_0$ is the σ-algebra generated by $\mathcal{A}$. The following is the well-known approximation theorem (see, e.g. [Rud]):

Theorem 1.1. *If $(X, \mathcal{B}, \mu)$ is a probability space, a subalgebra $\mathcal{A} \subset \mathcal{B}$ generates $\mathcal{B}$ (mod 0) if and only if, for every $B \in \mathcal{B}$ and $\varepsilon > 0$, there exists $A \in \mathcal{A}$ such that $\mu(A \Delta B) \leq \varepsilon$.*

Before going further, let us first review some definitions and simple facts about function spaces on a measure space. Let $(X, \mathcal{B}, \mu)$ be a measure space and let $1 \leq p < +\infty$. We denote by $L^p(X, \mathcal{B}, \mu)$ the quotient of the set of functions $f : X \to \mathbf{C}$ such that $|f|^p$ is integrable, under the equivalence relation that identifies functions which coincide a.e. We endow $L^p(X, \mathcal{B}, \mu)$ with the norm $\|\cdot\|_p$ defined by

$$\|f\|_p = \left(\int_X |f|^p d\mu \right)^{\frac{1}{p}}$$

which makes $L^p(X, \mathcal{B}, \mu)$ a Banach space. For $p = 2$, the norm $\|\cdot\|_2$ comes from an inner product

$$< f, g > = \int_X f g d\mu$$

with respect to which $L^2(X, \mathcal{B}, \mu)$ is a Hilbert space.

Given a probability space $(X, \mathcal{B}, \mu)$, if there exists a countable subset of $\mathcal{B}$ which generates $\mathcal{B}$ (mod 0), then we say it is *separable*, several equivalent descriptions of this kind of separability are given by the following

Theorem 1.2. *For a probability space* $(X, \mathcal{B}, \mu)$, *the following properties are equivalent:*

1) $(X, \mathcal{B}, \mu)$ *is separable;*

2) $L^1(X, \mathcal{B}, \mu)$ *is separable;*

3) $L^p(X, \mathcal{B}, \mu)$ *is separable for every* $1 \leq p < +\infty$;

4) There exists a countable subalgebra $\mathcal{A} \subset \mathcal{B}$ *which generates* $\mathcal{B}$ *(mod 0).*

A complete separable metric space is known as a *Polish space*. This kind of spaces provide an important class of separable measure spaces by the following theorem (see, for instance, [Man]$_1$):

Theorem 1.3. *If* X *is a Polish space,* $\mathcal{B}$ *is the Borel* σ*-algebra of* X *and* μ *is a probability measure on* $\mathcal{B}$, *then* $(X, \mathcal{B}, \mu)$ *is separable.*

We formulate below a theorem concerning the regularity of finite Borel measures on Polish spaces (see [Coh]).

Theorem 1.4. *Let* $(X, \mathcal{B}, \mu)$ *be a probability space, where* X *is a Polish space and* $\mathcal{B}$ *is the Borel* σ*-algebra of* X. *Then for every Borel set* $B \in \mathcal{B}$ *and* $\varepsilon > 0$ *there exists a compact set* $K \subset B$ *such that* $\mu(B \backslash K) < \varepsilon$.

Let $(X, \mathcal{B})$ be a measurable space and let $\mu : \mathcal{B} \to [0, +\infty]$ and $\nu : \mathcal{B} \to [0, +\infty]$ be measures. We say that μ is *absolutely continuous* with respect to ν, and we write $\mu << \nu$, if $B \in \mathcal{B}$ and $\nu(B) = 0$ imply $\mu(B) = 0$. The following Radon-Nikodym theorem characterizes this kind of absolute continuity.

Theorem 1.5. *If* $(X, \mathcal{B}, \nu)$ *is* σ*-finite, then* $\mu << \nu$ *if and only if there exists* $f : X \to \mathbf{R}^+$, *integrable with respect to* ν *on all sets* $B \in \mathcal{B}$ *such that* $\nu(B) < +\infty$, *satisfying the following condition for every* $B \in \mathcal{B}$:

$$\mu(B) = \int_B f d\nu.$$

The function f *is unique a.e. (with respcct to* ν*), and is denoted by* $d\mu/d\nu$. *A function* $g : X \to \mathbf{C}$ *is in* $L^1(X, \mathcal{B}, \mu)$ *if and only if* gf *is in* $L^1(X, \mathcal{B}, \nu)$, *and in this case we have*

$$\int_X g d\mu = \int_X g f d\nu.$$

If $\mu(X) < +\infty$ *then* $f \in L^1(X, \mathcal{B}, \nu)$.

The function f is called the *Radon-Nikodym derivative* of μ with respect to ν.

Let $(X, \mathcal{B}, \mu)$ be a measure space and let $\mathcal{A}$ be a sub-σ-algebra of $\mathcal{B}$. For every $f \in L^1(X, \mathcal{B}, \mu)$, the Radon-Nikodym theorem allows us to prove easily that there exists a unique function, written $E(f|\mathcal{A})$, in $L^1(X, \mathcal{A}, \mu)$ such that $\int_A E(f|\mathcal{A}) d\mu = \int_A f d\mu$ for every $A \in \mathcal{A}$. This function $E(f|\mathcal{A})$ is called the *conditional expectation* of f with respect to $\mathcal{A}$. We now define the *conditional expectation operator* $E(\cdot|\mathcal{A}) : L^1(X,\mathcal{B},\mu) \to L^1(X,\mathcal{A},\mu)$, $f \mapsto E(f|\mathcal{A})$.

Theorem 1.6. *Let $(X,\mathcal{B},\mu)$ be a probability space and let $\mathcal{A}$ be a sub-σ-algebra of $\mathcal{B}$. The restriction of the conditional expectation operator $E(\cdot|\mathcal{A})$ to $L^2(X,\mathcal{B},\mu)$ is the orthogonal projection of $L^2(X,\mathcal{B},\mu)$ onto $L^2(X,\mathcal{A},\mu)$.*

Proof. For $f \in L^1(X,\mathcal{B},\mu)$ we know that $E(f|\mathcal{A})$ is the only $\mathcal{A}$-measurable function such that $\int_A E(f|\mathcal{A})d\mu = \int_A f d\mu$ for every $A \in \mathcal{A}$. Let P denote the orthogonal projection of $L^2(X,\mathcal{B},\mu)$ onto the closed subspace $L^2(X,\mathcal{A},\mu)$. If $f \in L^2(X,\mathcal{B},\mu)$, then Pf is $\mathcal{A}$-measurable and if $A \in \mathcal{A}$

$$\int_A f d\mu = < f, \chi_A > = < f, P\chi_A > = < Pf, \chi_A > = \int_A Pf d\mu.$$

Therefore $Pf = E(f|\mathcal{A})$. □

The Radon-Nikodym theorem also allows us to introduce the general definition of Jacobian of absolutely continuous maps between measure spaces.

Let $(X,\mathcal{B},\mu)$ and $(Y,\mathcal{A},\nu)$ be two σ-finite measure spaces, and let $T : X \to Y$ be a map. We say that T is ***absolutely continuous*** if the following three conditions hold: (i) T is injective; (ii) if $B \in \mathcal{B}$ then $TB \in \mathcal{A}$; (iii) $B \in \mathcal{B}$ and $\mu(B) = 0$ imply $\nu(TB) = 0$.

Assume that T is absolutely continuous. We now define on $\mathcal{B}$ a new measure μ_T by the formula $\mu_T(B) = \nu(TB)$. The measure μ_T is absolutely continuous with respect to the measure μ. Thus by the Radon-Nikodym theorem one can introduce the measurable function $J(T) = d\mu_T/d\mu$ defined on X, it is called the *Jacobian* of the map T.

It is easy to see that, when the absolutely continuous map T is bijective and T^{-1} is also absolutely continuous, we have

$$J(T) = \frac{1}{J(T^{-1}) \circ T}$$

for μ almost all points of X (we admit here $1/0 = +\infty$ and $1/+\infty = 0$).

When X and Y are two Riemannian manifolds without boundary and of the same finite dimension, $f : X \to Y$ is a C^1 diffeomorphism, and λ_X and λ_Y are the respective Lebesgue measures on X and Y induced by the Riemannian metrics, then in this particular case it is easy to see that for any $x \in X$ one has

$$J(f)(x) = \frac{d(\lambda_Y \circ f)}{d\lambda_X}(x) = |\det T_x f|,$$

where $T_x f$ is the derivative of f at x, and for any $h \in L^1(Y,\lambda_Y)$ one has

$$\int_X (h \circ f)(x) |\det T_x f| d\lambda_X(x) = \int_Y h(y) d\lambda_Y(y).$$

Next, we have the following Lebesgue-Vitali theorem on differentiation (see, e.g., [Shi]):

Theorem 1.7. *Let $A \subset \mathbf{R}^n$ be a Borel set, and $h : A \to \mathbf{C}$ an integrable function with respect to the Lebesgue measure λ of $\mathbf{R}^n$. Then the following holds true for λ-almost every $x \in A$:*

$$\lim_{r \to 0} \frac{1}{\lambda(B_r(x) \cap A)} \int_{B_r(x) \cap A} h d\lambda = h(x),$$

where $B_r(x) = \{y \in \mathbf{R}^n : \quad d(x,y) \leq r\}$.

A simple application of the Lebesgue-Vitali theorem yields the following:

Theorem 1.8. *Let $T : (X, \mathcal{B}, \mu) \to (Y, \mathcal{A}, \nu)$ be an absolutely continuous map, where X is a Borel subset of $\mathbf{R}^n$ with $\lambda(X) < +\infty$, $\mathcal{B}$ is the σ-algebra of Borel subsets of X and μ is absolutely continuous with respect to $\lambda|_X$. Then there holds the following formula for μ-almost every $x \in X$:*

$$\lim_{r \to 0} \frac{\mu_T(B_r(x) \cap X)}{\mu(B_r(x) \cap X)} = J(T)(x).$$

Proof. Let $h = d\mu/d\lambda$, then

$$\begin{aligned} &\lim_{r \to 0} \frac{\mu_T(B_r(x) \cap X)}{\mu(B_r(x) \cap X)} \\ &= \lim_{r \to 0} \frac{\lambda(B_r(x) \cap X)^{-1} \int_{B_r(x) \cap X} J(T) h d\lambda}{\lambda(B_r(x) \cap X)^{-1} \int_{B_r(x) \cap X} h d\lambda} \\ &= J(T)(x), \quad \mu - a.e. x \in X. \end{aligned}$$

This completes the proof. □

An easy application of Theorem 1.7 also gives

Theorem 1.9. *If $A \subset \mathbf{R}^n$ is a Borel set and μ is a Borel measure on $\mathbf{R}^n$ which is absolutely continuous with respect to λ, then the limit*

$$\lim_{r \to 0} \frac{\mu(B_r(x) \cap A)}{\mu(B_r(x))}$$

exists μ-almost everywhere in $\mathbf{R}^n$, and is equal to 0 if $x \notin A$ and to 1 if $x \in A$.

When the above limit exists for $x \in A$ and is equal to 1, we call x a *density point* of A with respect to μ.

We conclude this section with the notion of Lebesgue spaces. A map between two measure spaces is called an invertible measure-preserving transformation if it is bijective and measure-preserving and so is its inversion. Two measure spaces $(X_i, \mathcal{B}_i, \mu_i), i = 1, 2$ are said to be isomorphic mod 0 if there exist $Y_1 \in \mathcal{B}_1, Y_2 \in \mathcal{B}_2$ with $\mu_1(X_1 \backslash Y_1) = 0 = \mu_2(X_2 \backslash Y_2)$ and there exists an invertible measure-preserving transformation $\phi : (Y_1, \mathcal{B}_1|_{Y_1}, \mu_1|_{Y_1}) \to (Y_2, \mathcal{B}_2|_{Y_2}, \mu_2|_{Y_2})$. Given a probability space $(X, \mathcal{B}, \mu)$, let $X_0 = X \backslash \{x : x \in X$ with $\{x\} \in \mathcal{B}$ and $\mu(\{x\}) > 0\}$ and $s = \mu(X_0)$. We call $(X, \mathcal{B}, \mu)$ a *Lebesgue space* if $(X_0, \mathcal{B}|_{X_0}, \mu|_{X_0})$ is isomorphic mod 0 to the space $([0, s], \mathcal{L}([0, s]), l)$, where $\mathcal{L}([0, s])$ is the σ-algebra of Lebesgue measurable subsets of $[0, s]$ and l is the usual Lebesgue measure. There is now the following important theorem (see [Roy]):

Theorem 1.10. *Let X be a Polish space, μ a Borel probability measure on X, and $\mathcal{B}_\mu(X)$ the completion of the Borel σ-algebra of X with respect to μ. Then $(X, \mathcal{B}_\mu(X), \mu)$ is a Lebesgue space.*

Throughout the remaining sections of this chapter it is always assumed that $(X, \mathcal{B}, \mu)$ is a Lebesgue space.

§2 Measurable Partitions

Let $(X, \mathcal{B}, \mu)$ be a Lebesgue space.

Any collection of non-empty disjoint sets that covers X is said to be a partition of X. Subsets of X that are unions of elements of a partition ξ are called ξ-sets.

A countable system $\{B_\alpha : \alpha \in \mathcal{A}\}$ of measurable ξ-sets is said to be a basis of ξ if, for any two elements C and C' of ξ, there exists an $\alpha \in \mathcal{A}$ such that either $C \subset B_\alpha, C' \not\subset B_\alpha$ or $C \not\subset B_\alpha, C' \subset B_\alpha$. A partition with a basis is said to be *measurable*. Obviously, every element of a measurable partition is a measurable set.

For $x \in X$ we will denote by $\xi(x)$ the element of a partition ξ which contains x. If ξ, ξ' are measurable partitions of X, we write $\xi \le \xi'$ if $\xi'(x) \subset \xi(x)$ for μ-almost every $x \in X$, $\xi = \xi'$ is also considered up to mod 0.

For any system of measurable partitions $\{\xi_\alpha\}$ of X there exists a product $\bigvee_\alpha \xi_\alpha$, defined as the measurable partition ξ satisfying the following two conditions : 1) $\xi_\alpha \le \xi$ for all α; 2) if $\xi_\alpha \le \xi'$ for all α, then $\xi \le \xi'$.

For any system of measurable partitions $\{\xi_\alpha\}$ of X there exists an intersection $\bigwedge_\alpha \xi_\alpha$, defined as the measurable partition ξ satisfying the conditions : 1) $\xi_\alpha \ge \xi$ for all α; 2) if $\xi_\alpha \ge \xi'$ for all α, then $\xi \ge \xi'$.

For measurable partitions ξ_n, $n \in \mathbf{N}$ and ξ of X, the symbol $\xi_n \nearrow \xi$ indicates that $\xi_1 \le \xi_2 \le \cdots$ and $\bigvee_{n=1}^{+\infty} \xi_n = \xi$, the symbol $\xi_n \searrow \xi$ indicates that $\xi_1 \ge \xi_2 \ge \cdots$ and $\bigwedge_{n=1}^{+\infty} \xi_n = \xi$.

If $\{B_1, B_2, \cdots\}$ is a basis for the partition ξ and β_n is the partition of X into the sets B_n and $X \backslash B_n$, then the partitions $\xi_n = \bigvee_{i=1}^n \beta_i$ form an increasing sequence and $\bigvee_{n=1}^{+\infty} \xi_n = \xi$. Thus, for any measurable partition ξ there exists a sequence of finite partitions ξ_n such that $\xi_n \nearrow \xi$.

One can define an equivalence relation on $\mathcal{B}$ by saying that A and B are equivalent ($A \sim B$) if and only if $\mu(A \Delta B) = 0$. We denote the set of equivalence classes by $\widetilde{\mathcal{B}}$. The operations of countable union, countable intersection and subtraction of sets in $\mathcal{B}$ can be carried over to the same operations on classes, making $\widetilde{\mathcal{B}}$ a σ-algebra. Any part of $\widetilde{\mathcal{B}}$ that is closed with respect to these operations is said to be a sub-σ-algebra of $\widetilde{\mathcal{B}}$.

It is clear that the intersection $\bigcap_\alpha \widetilde{\mathcal{B}}_\alpha$ of any system of sub-σ-algebras $\{\widetilde{\mathcal{B}}_\alpha\}$ is a sub-σ-algebra of $\widetilde{\mathcal{B}}$. The sum $\bigvee_\alpha \widetilde{\mathcal{B}}_\alpha$ of a system of sub-σ-algebras $\{\widetilde{\mathcal{B}}_\alpha\}$ is defined to be the intersection of all sub-σ-algebras each of which contains all the $\widetilde{\mathcal{B}}_\alpha$. If $\widetilde{\mathcal{B}}_1 \subset \widetilde{\mathcal{B}}_2 \subset \cdots$ and $\bigvee_{n=1}^{+\infty} \widetilde{\mathcal{B}}_n = \widetilde{\mathcal{B}}'$, then we write $\widetilde{\mathcal{B}}_n \nearrow \widetilde{\mathcal{B}}'$. If $\widetilde{\mathcal{B}}_1 \supset \widetilde{\mathcal{B}}_2 \supset \cdots$ and $\bigcap_{n=1}^{+\infty} \widetilde{\mathcal{B}}_n = \widetilde{\mathcal{B}}'$, then we write $\widetilde{\mathcal{B}}_n \searrow \widetilde{\mathcal{B}}'$.

For any measurable partition ξ, we denote by $\widetilde{\mathcal{B}}(\xi)$ the sub-σ-algebra of $\widetilde{\mathcal{B}}$ consisting of classes of measurable ξ-sets. If $\widetilde{\mathcal{B}}(\xi) = \widetilde{\mathcal{B}}(\xi')$, then $\xi = \xi'$, and for any sub-σ-algebra of $\widetilde{\mathcal{B}}$ there exists a measurable partition ξ such that $\widetilde{\mathcal{B}}(\xi)$ is this sub-σ-algebra. Thus, the sub-σ-algebras of $\widetilde{\mathcal{B}}$ are in one-to-one correspondence with the classes of mod 0-equal measurable partitions. Here $\widetilde{\mathcal{B}}(\xi) \subset \widetilde{\mathcal{B}}(\xi')$ if and only if $\xi \leq \xi'$. We also have

$$\widetilde{\mathcal{B}}(\bigvee_\alpha \xi_\alpha) = \bigvee_\alpha \widetilde{\mathcal{B}}(\xi_\alpha), \widetilde{\mathcal{B}}(\bigwedge_\alpha \xi_\alpha) = \bigcap_\alpha \widetilde{\mathcal{B}}(\xi_\alpha).$$

In particular, $\widetilde{\mathcal{B}}(\xi_n) \nearrow \widetilde{\mathcal{B}}(\xi)$ if and only if $\xi_n \nearrow \xi$ and $\widetilde{\mathcal{B}}(\xi_n) \searrow \widetilde{\mathcal{B}}(\xi)$ if and only if $\xi_n \searrow \xi$.

The *factor-space* of X with respect to a partition ξ is the measure space whose points are the elements of ξ and whose measurable structure and measure μ_ξ are defined as follows: Let p be the map taking each point $x \in X$ to $\xi(x)$, a set Z is considered to be measurable if $p^{-1}(Z) \in \mathcal{B}$, and we define $\mu_\xi(Z) = \mu(p^{-1}(Z))$. This factor-space will be denoted by X/ξ. When ξ is a measurable partition X/ξ is a Legesgue space.

A very important property of measurable partitions of a Lebesgue space $(X, \mathcal{B}, \mu)$ is that associated with every such partition ξ there exists a unique system of measures $\{\mu_C\}_{C\in\xi}$ satisfying the following two conditions:

i) $(C, \mathcal{B}|_C, \mu_C)$ is a Lebesgue space for μ_ξ-a.e. $C \in X/\xi$;

ii) for every $A \in \mathcal{B}, \mu_C(A \cap C)$ is measurable on X/ξ and

$$\mu(A) = \int_{X/\xi} \mu_C(A \cap C) d\mu_\xi.$$

Such a system of measures $\{\mu_C\}_{C\in\xi}$ is called a *canonical system of conditional measures* of μ associated with ξ. Its uniqueness implies that any two systems $\{\mu_C\}_{C\in\xi}$ and $\{\mu'_C\}_{C\in\xi}$ satisfying i) and ii) above are mod 0-equal (that is, $\mu_C = \mu'_C$ for μ_ξ-a.e. $C \in X/\xi$).

Let ξ and η be measurable partitions such that $\xi \leq \eta$ and let A be an element of η and C an element of ξ containing A. As an element of the partition

η of X, $(A, \mathcal{B}|_A, \mu_A)$ is a Lebesque space. On the other hand, as an element of the partition $\eta|_C$ of C with the σ-algebra $\mathcal{B}|_C$ and the measure μ_C, $(A, \mathcal{B}|_A, (\mu_C)_A)$ is also a Lebesgue space. The uniqueness of canonical systems of conditional measures implies that $(\mu_C)_A = \mu_A$ for μ_η-a.e. $A \in X/\eta$. This property is called the *transitivity* of canonical systems of conditional measures.

Let $\{\mu_C\}_{C\in\xi}$ be a canonical system of conditional measures of μ associated with a measurable partition ξ. By the standard method of measure theory we know that for any $g \in L^1(X, \mathcal{B}, \mu)$, the section g_C defined as

$$g_C(x) = g(x) \qquad \text{if} \quad x \in C$$

is integrable on $(C, \mathcal{B}|_C, \mu_C)$ for μ_ξ-a.e. $C \in X/\xi$, $\int_C g_C d\mu_C$ is measurable on X/ξ and

$$\int_X g(x) d\mu = \int_{X/\xi} \left(\int_C g_C d\mu_C \right) d\mu_\xi . \tag{2.1}$$

§3 Conditional Entropies of Measurable Partitions

In this section we give a brief review of several notions about entropy for measurable partitions of Lebesgue spaces. Here we assume again that $(X, \mathcal{B}, \mu)$ is a Lebesgue space.

Let ξ be a measurable partition of X and let $C_1, C_2, \cdots$ be the elements of ξ of positive μ measure. We put

$$H_\mu(\xi) = \begin{cases} -\sum_k \mu(C_k) \log \mu(C_k) & \text{if} \quad \mu(X \setminus \cup_k C_k) = 0 \\ +\infty & \text{if} \quad \mu(X \setminus \cup_k C_k) > 0 \end{cases} \tag{3.1}$$

The sum in the first part of (3.1) can be finite or infinite. $H_\mu(\xi)$ is called the *entropy* of ξ.

If ξ and η are two measurable partitions of X, then almost every partition ξ_B, defined as the restriction $\xi|_B$ of ξ to B, $B \in X/\eta$, has a well-defined entropy $H_{\mu_B}(\xi_B)$. This is a non-negative measurable function on the factor-space X/η, called the *conditional entropy* of ξ with respect to η. Put

$$H_\mu(\xi|\eta) = \int_{X/\eta} H_{\mu_B}(\xi_B) d\mu_\eta . \tag{3.2}$$

This integral can be finite or infinite. We call it the *mean conditional entropy* of ξ with respect to η. Obviously, when η is the trivial partition whose single element is X, $H_\mu(\xi|\eta)$ coincides with the entropy $H_\mu(\xi)$.

We put $\log 0 = -\infty, 0 \log 0 = 0$ and define for each $x \in X$

$$m(x, \xi) = \mu(\xi(x)),$$
$$m(x, \xi|\eta) = \mu_{\eta(x)}(\xi(x) \cap \eta(x)),$$

then they are measurable functions on X and (3.1), (3.2) can be written respectively in the forms:

$$H_\mu(\xi) = -\int_M \log m(x,\xi)d\mu, \tag{3.3}$$

$$H_\mu(\xi|\eta) = -\int_M \log m(x,\xi|\eta)d\mu. \tag{3.4}$$

The mean conditional entropy has, in particular, the following properties:

1) If $\xi_n \nearrow \xi$, then for any measurable partition η

$$H_\mu(\xi_n|\eta) \nearrow H_\mu(\xi|\eta); \tag{3.5}$$

2) If $\xi_n \searrow \xi$, then for any measurable partition η with $H_\mu(\xi_1|\eta) < +\infty$,

$$H_\mu(\xi_n|\eta) \searrow H_\mu(\xi|\eta); \tag{3.6}$$

3) For any measurable partitions ξ, η and ζ,

$$H_\mu(\xi \vee \eta|\zeta) = H_\mu(\xi|\zeta) + H_\mu(\eta|\xi \vee \zeta); \tag{3.7}$$

4) If $\eta_n \nearrow \eta$ and ξ is a measurable partition such that $H_\mu(\xi|\eta_1) < +\infty$, then

$$H_\mu(\xi|\eta_n) \searrow H_\mu(\xi|\eta); \tag{3.8}$$

5) If $\eta_n \searrow \eta$, then for any measurable partition ξ

$$H_\mu(\xi|\eta_n) \nearrow H_\mu(\xi|\eta); \tag{3.9}$$

6) If $(X_i, \mathcal{B}_i, \mu_i), i = 1,2$ are two Lebesgue spaces and T is a measure-preserving transformation from $(X_1, \mathcal{B}_1, \mu_1)$ to $(X_2, \mathcal{B}_2, \mu_2)$, then for any measurable partitions ξ and η of X_2,

$$H_{\mu_1}(T^{-1}\xi|T^{-1}\eta) = H_{\mu_2}(\xi|\eta). \tag{3.10}$$

The proofs of 1)-5) can be found in [Roh]$_2$. Now we give a proof of 6).

When ξ and η are finite partitions, it is easy to see that

$$m(x, T^{-1}\xi|T^{-1}\eta) = m(Tx, \xi|\eta), \quad \mu_1 - \text{a.e.} \quad x \in X_1.$$

Since $T\mu_1 = \mu_2$, then (3.10) follows from (3.4). In addition, it is obvious that $H_{\mu_2}(\xi|\eta) < +\infty$.

In general, let ξ and η be any two measurable partitions of X_2. Choose two sequences of finite measurable partitions $\{\xi_n\}_{n=1}^{+\infty}$ and $\{\eta_n\}_{n=1}^{+\infty}$ such that $\xi_n \nearrow \xi$ and $\eta_n \nearrow \eta$. Since for any $k, n \in \mathbf{N}$

$$H_{\mu_1}(T^{-1}\xi_k|T^{-1}\eta_n) = H_{\mu_2}(\xi_k|\eta_n),$$

letting $n \to +\infty$, by (3.8) we have for every $k \in \mathbf{N}$

$$H_{\mu_1}(T^{-1}\xi_k|T^{-1}\eta) = H_{\mu_2}(\xi_k|\eta),$$

then (3.10) is deduced from (3.5) when $k \to +\infty$.

§4 Conditional Entropies of Measure-Preserving Transformations: I

Now we begin to deal with the concept of conditional entropies of measure-preserving transformations. Let T be a measure-preserving transformation on a Lebesgue space $(X, \mathcal{B}, \mu)$ and $\mathcal{A}$ a sub-σ-algebra of $\mathcal{B}$ satisfying

$$T^{-1}\mathcal{A} \subset \mathcal{A}. \tag{4.1}$$

Then there exists a unique (mod 0) measurable partition ζ_0 of X such that $\widetilde{\mathcal{B}}(\zeta_0)$ is the σ-algebra $\widetilde{\mathcal{A}}$ consisting of classes of sets in $\mathcal{A}$, and from (4.1) it follows that

$$T^{-1}\zeta_0 \leq \zeta_0. \tag{4.2}$$

Let ζ_0 be as obtained above. We denote by $\mathcal{Z}(\zeta_0)$ the set of measurable partitions ξ of X satisfying $H_\mu(\xi|\zeta_0) < +\infty$. For $\xi, \eta \in \mathcal{Z}(\zeta_0)$, put

$$\rho(\xi, \eta) = H_\mu(\xi|\eta \vee \zeta_0) + H_\mu(\eta|\xi \vee \zeta_0). \tag{4.3}$$

Then it can be easily shown that for any $\xi, \eta, \zeta \in \mathcal{Z}(\zeta_0)$

$$|H_\mu(\xi|\zeta \vee \zeta_0) - H_\mu(\eta|\zeta \vee \zeta_0)| \leq \rho(\xi, \eta), \tag{4.4}$$

$$|H_\mu(\xi|\eta \vee \zeta_0) - H_\mu(\xi|\zeta \vee \zeta_0)| \leq \rho(\eta, \zeta), \tag{4.5}$$

and if $\xi_n, \xi \in \mathcal{Z}(\zeta_0)$ with $\xi_n \nearrow \xi$ or $\xi_n \searrow \xi$ then

$$\rho(\xi_n, \xi) \to 0 \quad \text{as} \quad n \to +\infty. \tag{4.6}$$

In fact, (4.4) and (4.5) follows respectively from

$$\begin{aligned} H_\mu(\xi|\zeta \vee \zeta_0) - H_\mu(\eta|\zeta \vee \zeta_0) &\leq H_\mu(\xi \vee \eta|\zeta \vee \zeta_0) - H_\mu(\eta|\zeta \vee \zeta_0) \\ &= H_\mu(\xi|\eta \vee \zeta \vee \zeta_0) \leq H_\mu(\xi|\eta \vee \zeta_0) \end{aligned}$$

and

$$\begin{aligned} H_\mu(\xi|\eta \vee \zeta_0) - H_\mu(\xi|\zeta \vee \zeta_0) &\leq H_\mu(\xi \vee \zeta|\eta \vee \zeta_0) - H_\mu(\xi|\eta \vee \zeta \vee \zeta_0) \\ &= H_\mu(\zeta|\eta \vee \zeta_0), \end{aligned}$$

and (4.6) follows from (3.5), (3.6), (3.8) and (3.9).

For any measurable partition ξ of X, we use the following notations:

$$\xi_T^n = \bigvee_{k=0}^{n-1} T^{-k}\xi, \quad \xi_T^- = \bigvee_{k=0}^{+\infty} T^{-k}\xi, \quad \xi_T^0 = \{X\},$$

and by ε we denote the partition of X into single points. Sometimes we use the simpler notations ξ^n, ξ^- and ξ^0 to denote ξ_T^n, ξ_T^- and ξ_T^0 respectively.

Theorem 4.1. *Let $T, \mathcal{A}$ and ζ_0 be as given above. Then for any $\xi \in \mathcal{Z}(\zeta_0)$*

$$h_\mu^{\mathcal{A}}(T, \xi) = \lim_{n \to +\infty} \frac{1}{n} H_\mu(\xi_T^n|\zeta_0) \tag{4.7}$$

exists and equals $\inf_{n\geq 1}\frac{1}{n}H_\mu(\xi_T^n|\zeta_0)$. *This limit is called the* $\mathcal{A}$*-conditional entropy of* T *with respect to* ξ.

Proof. Let $a_n = H_\mu(\xi_T^n|\zeta_0)$, then by (3.7), (3.9) and (3.10)

$$\begin{aligned} a_{n+m} &\leq H_\mu(\xi_T^n|\zeta_0) + H_\mu\left(\bigvee_{k=n}^{n+m-1} T^{-k}\xi|\zeta_0\right) \\ &\leq a_n + H_\mu(T^{-n}\xi_T^m|T^{-n}\zeta_0) \\ &= a_n + a_m \end{aligned}$$

for any $m, n \in \mathbf{N}$. Fix $p \in \mathbf{N}$ arbitrarily. Each $n \in \mathbf{N}$ can be written as $n = kp+i$ with $0 \leq i < p$. Then

$$\frac{a_n}{n} = \frac{a_{i+kp}}{i+kp} \leq \frac{a_i}{kp} + \frac{a_{kp}}{kp} \leq \frac{a_i}{kp} + \frac{ka_p}{kp} = \frac{a_i}{kp} + \frac{a_p}{p}.$$

Letting $n \to +\infty$, we have

$$\limsup_{n\to+\infty}\frac{a_n}{n} \leq \frac{a_p}{p}$$

and therefore

$$\limsup_{n\to+\infty}\frac{a_n}{n} \leq \inf_{p\geq 1}\frac{a_p}{p}.$$

But

$$\inf_{p\geq 1}\frac{a_p}{p} \leq \liminf_{n\to+\infty}\frac{a_n}{n},$$

so $\lim_{n\to+\infty}\frac{1}{n}a_n$ exists and equals $\inf_{p\geq 1}\frac{1}{p}a_p$. □

Definition 4.1. *Let* T *and* $\mathcal{A}$ *be as given above. Then*

$$h_\mu^{\mathcal{A}}(T) = \sup_{\xi\in\mathcal{Z}(\zeta_0)} h_\mu^{\mathcal{A}}(T,\xi) \tag{4.8}$$

is called the $\mathcal{A}$*-conditional entropy of* T.

Remark 4.1. The right-hand side of (4.8) does not change if the supremum is taken only over the set of finite measurable partitions.

In fact, this remark can be deduced from the following lemma and (4.6).

Lemma 4.1. *Let* $h_\mu^{\mathcal{A}}(T,\xi)$ *be as defined by (4.7). Then for any* $\xi,\eta \in \mathcal{Z}(\zeta_0)$

$$|h_\mu^{\mathcal{A}}(T,\xi) - h_\mu^{\mathcal{A}}(T,\eta)| \leq \rho(\xi,\eta). \tag{4.9}$$

Proof. From (3.7) it follows that

$$\begin{aligned} H_\mu(\xi^n|\zeta_0) + H_\mu(\eta^n|\xi^n \vee \zeta_0) &= H_\mu(\xi^n \vee \eta^n|\zeta_0) \\ &= H_\mu(\eta^n|\zeta_0) + H_\mu(\xi^n|\eta^n \vee \zeta_0) \end{aligned} \tag{4.10}$$

which yields

$$|H_\mu(\xi^n|\zeta_0) - H_\mu(\eta^n|\zeta_0)| \le H_\mu(\xi^n|\eta^n \vee \zeta_0) + H_\mu(\eta^n|\xi^n \vee \zeta_0). \tag{4.11}$$

But

$$\begin{aligned} H_\mu(\xi^n|\eta^n \vee \zeta_0) &\le \sum_{k=0}^{n-1} H_\mu(T^{-k}\xi|\eta^n \vee \zeta_0) \\ &\le \sum_{k=0}^{n-1} H_\mu(T^{-k}\xi|T^{-k}\eta \vee T^{-k}\zeta_0) \\ &= nH_\mu(\xi|\eta \vee \zeta_0), \end{aligned} \tag{4.12}$$

and similarly

$$H_\mu(\eta^n|\xi^n \vee \zeta_0) \le nH_\mu(\eta|\xi \vee \zeta_0), \tag{4.13}$$

then (4.9) follows from (4.7) and (4.11)-(4.13). □

Now we give some further properties of the $\mathcal{A}$-conditional entropy.

Theorem 4.2. *Let $T, \mathcal{A}$ and ζ_0 be as given before. Then for any $\xi, \eta \in \mathcal{Z}(\zeta_0)$ the following hold true:*

1) $h_\mu^{\mathcal{A}}(T,\xi) \le H_\mu(\xi|\zeta_0)$.
2) $h_\mu^{\mathcal{A}}(T,\xi \vee \eta) \le h_\mu^{\mathcal{A}}(T,\xi) + h_\mu^{\mathcal{A}}(T,\eta)$.
3) $\xi \le \eta$ implies $h_\mu^{\mathcal{A}}(T,\xi) \le h_\mu^{\mathcal{A}}(T,\eta)$.
4) $h_\mu^{\mathcal{A}}(T,\xi) \le h_\mu^{\mathcal{A}}(T,\eta) + H_\mu(\xi|\eta \vee \zeta_0)$.
5) $h_\mu^{\mathcal{A}}(T,T^{-1}\xi) \le h_\mu^{\mathcal{A}}(T,\xi)$.
6) $h_\mu^{\mathcal{A}}(T,\xi) = h_\mu^{\mathcal{A}}(T,\xi_T^k)$ for any $k \in \mathbf{N}$.
7) $h_\mu^{\mathcal{A}}(T^k,\xi_T^k) = kh_\mu^{\mathcal{A}}(T,\xi)$ for any $k \in \mathbf{N}$.
8) If $\xi \in \mathcal{Z}(T^{-1}\zeta_0)$, then $h_\mu^{\mathcal{A}}(T,\xi) = h_\mu^{T^{-1}\mathcal{A}}(T,\xi)$.
9) If T is invertible and $T^{-1}\mathcal{A} = \mathcal{A}$, then

$$h_\mu^{\mathcal{A}}(T^{-1},\xi) = h_\mu^{\mathcal{A}}(T,\xi).$$

Proof. 1) From

$$\begin{aligned} \frac{1}{n}H_\mu(\xi^n|\zeta_0) &\le \frac{1}{n}\sum_{k=0}^{n-1} H_\mu(T^{-k}\xi|\zeta_0) \\ &\le \frac{1}{n}\sum_{k=0}^{n-1} H_\mu(T^{-k}\xi|T^{-k}\zeta_0) = H_\mu(\xi|\zeta_0) \end{aligned}$$

and (4.7) we obtain 1).

2) Since

$$H_\mu((\xi \vee \eta)^n|\zeta_0) = H_\mu(\xi^n \vee \eta^n|\zeta_0) \leq H_\mu(\xi^n|\zeta_0) + H_\mu(\eta^n|\zeta_0),$$

then 2) holds obviously.

3) $\xi \leq \eta$ implies that $\xi^n \leq \eta^n$ for all $n \in \mathbf{N}$. Then 3) follows from (4.7) and (3.5).

4) Since

$$H_\mu(\xi^n|\zeta_0) \leq H_\mu(\xi^n \vee \eta^n|\zeta_0) = H_\mu(\eta^n|\zeta_0) + H_\mu(\xi^n|\eta^n \vee \zeta_0)$$

and

$$\begin{aligned} H_\mu(\xi^n|\eta^n \vee \zeta_0) &\leq \sum_{k=0}^{n-1} H_\mu(T^{-k}\xi|\eta^n \vee \zeta_0) \\ &\leq \sum_{k=0}^{n-1} H_\mu(T^{-k}\xi|T^{-k}(\eta \vee \zeta_0)) \\ &= nH_\mu(\xi|\eta \vee \zeta_0), \end{aligned}$$

this together with (4.7) yields 4).

$$\begin{aligned} 5) \quad h_\mu^{\mathcal{A}}(T, T^{-1}\xi) &= \lim_{n\to+\infty} \frac{1}{n} H_\mu\left(\bigvee_{k=1}^{n} T^{-k}\xi|\zeta_0\right) \\ &\leq \lim_{n\to+\infty} \frac{1}{n} H_\mu(T^{-1}\xi^n|T^{-1}\zeta_0) = h_\mu^{\mathcal{A}}(T,\xi). \end{aligned}$$

$$\begin{aligned} 6) \quad h_\mu^{\mathcal{A}}(T, \xi^k) &= \lim_{n\to+\infty} \frac{1}{n} H_\mu(\xi^{n+k-1}|\zeta_0) \\ &= \lim_{n\to+\infty} \frac{1}{n+k-1} H_\mu(\xi^{n+k-1}|\zeta_0) = h_\mu^{\mathcal{A}}(T,\xi). \end{aligned}$$

$$\begin{aligned} 7) \quad h_\mu^{\mathcal{A}}(T^k, \xi^k) &= \lim_{n\to+\infty} \frac{1}{n} H_\mu\left(\bigvee_{j=0}^{n-1} T^{-kj}\left(\bigvee_{i=0}^{k-1} T^{-i}\xi\right)\middle|\zeta_0\right) \\ &= \lim_{n\to+\infty} k\frac{1}{kn} H_\mu(\xi^{kn}|\zeta_0) = kh_\mu^{\mathcal{A}}(T,\xi). \end{aligned}$$

$$\begin{aligned} 8) \quad h_\mu^{T^{-1}\mathcal{A}}(T, \xi) &= \lim_{n\to+\infty} \frac{1}{n} H_\mu(\xi^n|T^{-1}\zeta_0) \\ &= \lim_{n\to+\infty} \frac{1}{n}\left[H_\mu\left(\bigvee_{k=1}^{n-1} T^{-k}\xi|T^{-1}\zeta_0\right) + H_\mu\left(\xi\middle|\bigvee_{k=1}^{n-1} T^{-k}\xi \vee T^{-1}\zeta_0\right)\right] \\ &= \lim_{n\to+\infty} \frac{1}{n} H_\mu(\xi^{n-1}|\zeta_0) = h_\mu^{\mathcal{A}}(T,\xi). \end{aligned}$$

9) If T is invertible and $T^{-1}\mathcal{A} = \mathcal{A}$, then

$$h_\mu^{\mathcal{A}}(T,\xi) = \lim_{n\to+\infty} \frac{1}{n} H_\mu(\xi_T^n|\zeta_0) = \lim_{n\to+\infty} \frac{1}{n} H_\mu(T^{n-1}\xi_T^n|T^{n-1}\zeta_0)$$
$$= \lim_{n\to+\infty} \frac{1}{n} H_\mu(\xi_{T^{-1}}^n|\zeta_0) = h_\mu^{\mathcal{A}}(T^{-1},\xi).$$

This completes the proof. □

Theorem 4.3. *Let T and $\mathcal{A}$ be as given before. Then for any $k \in \mathbf{N}$*

$$h_\mu^{\mathcal{A}}(T^k) = k h_\mu^{\mathcal{A}}(T). \tag{4.14}$$

If T is invertible and $T^{-1}\mathcal{A} = \mathcal{A}$, then

$$h_\mu^{\mathcal{A}}(T^{-1}) = h_\mu^{\mathcal{A}}(T). \tag{4.15}$$

Proof. From 7) of Theorem 4.2 it follows obviously that

$$h_\mu^{\mathcal{A}}(T^k) \geq k h_\mu^{\mathcal{A}}(T). \tag{4.16}$$

On the other hand, for any $\xi \in \mathcal{Z}(\zeta_0)$

$$h_\mu^{\mathcal{A}}(T^k,\xi) \leq h_\mu^{\mathcal{A}}(T^k,\xi_T^k) = k h_\mu^{\mathcal{A}}(T,\xi).$$

So

$$h_\mu^{\mathcal{A}}(T^k) \leq k h_\mu^{\mathcal{A}}(T)$$

which together with (4.16) proves (4.14). (4.15) follows from 9) of Theorem 4.2. □

Theorem 4.4. *Let T and $\mathcal{A}$ be as given before. If T is invertible and we let $\mathcal{A}^+ = \bigvee_{k=0}^{+\infty} T^k\mathcal{A}$, then*

$$h_\mu^{\mathcal{A}}(T) = h_\mu^{\mathcal{A}^+}(T).$$

Proof. From 8) of Theorem 4.2 we know that for any finite measurable partition ξ and any $k \in \mathbf{N}$

$$h_\mu^{\mathcal{A}}(T,\xi) = h_\mu^{T^k\mathcal{A}}(T,\xi). \tag{4.17}$$

Let $\zeta_0^+ = \bigvee_{k=0}^{+\infty} T^k\zeta_0$. Since for every $n \in \mathbf{N}$

$$\frac{1}{n} H_\mu(\xi^n|T^k\zeta_0) \searrow \frac{1}{n} H_\mu(\xi^n|\zeta_0^+) \tag{4.18}$$

as $k \to +\infty$, by Theorem 4.1 we can easily show that for any $\varepsilon > 0$ there exists $k \in \mathbf{N}$ such that

$$h_\mu^{\mathcal{A}^+}(T,\xi) \geq h_\mu^{T^k\mathcal{A}}(T,\xi) - \varepsilon = h_\mu^{\mathcal{A}}(T,\xi) - \varepsilon.$$

Hence

$$h_\mu^{\mathcal{A}^+}(T,\xi) \geq h_\mu^{\mathcal{A}}(T,\xi).$$

On the other hand, by (3.8)

$$h_\mu^{\mathcal{A}^+}(T,\xi) \leq h_\mu^{\mathcal{A}}(T,\xi).$$

Thus, for any finite measurable partition ξ

$$h_\mu^{\mathcal{A}^+}(T,\xi) = h_\mu^{\mathcal{A}}(T,\xi).$$

This together with Remark 4.1 yields

$$h_\mu^{\mathcal{A}^+}(T) = h_\mu^{\mathcal{A}}(T).$$

We complete the proof. □

Theorem 4.5. *Let T and $\mathcal{A}$ be as given before. If $\xi_1 \leq \xi_2 \leq \cdots$ is an increasing sequence of partitions in $\mathcal{Z}(\zeta_0)$ such that $(\bigvee_{n=1}^{+\infty} \xi_n) \vee \zeta_0 = \varepsilon$, then*

$$h_\mu^{\mathcal{A}}(T,\xi_n) \nearrow h_\mu^{\mathcal{A}}(T) \qquad \text{as} \quad n \to +\infty.$$

Proof. For any $\eta \in \mathcal{Z}(\zeta_0)$, by 4) of Theorem 4.2

$$h_\mu^{\mathcal{A}}(T,\eta) \leq h_\mu^{\mathcal{A}}(T,\xi_n) + H_\mu(\eta|\xi_n \vee \zeta_0).$$

Since, according to (3.8),

$$H_\mu(\eta|\xi_n \vee \zeta_0) \searrow 0 \qquad \text{as} \quad n \to +\infty,$$

and by 3) of Theorem 4.2 the sequence $h_\mu^{\mathcal{A}}(T,\xi_1), h_\mu^{\mathcal{A}}(T,\xi_2), \cdots$ is increasing, we have

$$h_\mu^{\mathcal{A}}(T,\eta) \leq \lim_{n\to+\infty} h_\mu^{\mathcal{A}}(T,\xi_n).$$

Since η is arbitrary in $\mathcal{Z}(\zeta_0)$ we see that

$$h_\mu^{\mathcal{A}}(T,\xi_n) \nearrow h_\mu^{\mathcal{A}}(T) \qquad \text{as} \quad n \to +\infty,$$

completing the proof. □

Theorem 4.6. *Let T and $\mathcal{A}$ be as given before. If $\xi \in \mathcal{Z}(\zeta_0)$ satisfies $\xi_T^- \vee \zeta_0 = \varepsilon$, then*

$$h_\mu^{\mathcal{A}}(T,\xi) = h_\mu^{\mathcal{A}}(T).$$

Proof. This theorem follows from Theorem 4.5 and 6) of Theorem 4.2. □

Theorem 4.7. *Assume that X is a Borel subset of a Polish space Y, μ is a Borel probability measure on X and $\mathcal{B}$ is the completion of the Borel σ-algebra of*

X with respect to μ. Let T be a measure-preserving transformation on $(X, \mathcal{B}, \mu)$, $\mathcal{A}$ a sub-σ-algebra of $\mathcal{B}$ satisfying $T^{-1}\mathcal{A} \subset \mathcal{A}$ and ζ_0 the measurable partition of X with $\widetilde{\mathcal{B}}(\zeta_0) = \widetilde{\mathcal{A}}$. If $\xi_1, \xi_2, \cdots$ is a sequence of countable partitions in $\mathcal{Z}(\zeta_0)$ such that

$$\lim_{n\to+\infty} diam(\xi_n) = 0$$

where $diam(\xi_n) = \sup_{C\in\xi_n} diam(C)$, then

$$h_\mu^{\mathcal{A}}(T) = \lim_{n\to+\infty} h_\mu^{\mathcal{A}}(T, \xi_n).$$

To prove this theorem we need the following

Lemma 4.2. *Let $r \geq 1$ be a fixed integer. Then for each $\varepsilon > 0$ there exists $\delta > 0$ such that if $\xi = \{A_1, \cdots, A_r\}$, $\eta = \{C_1, \cdots, C_r\}$ are two measurable partitions of X with $\sum_{i=1}^r \mu(A_i \Delta C_i) < \delta$ then $H_\mu(\xi|\eta) + H_\mu(\eta|\xi) < \varepsilon$.*

Proof. Let $\varepsilon > 0$ be given. Choose $\delta > 0$ so that $\delta < \frac{1}{4}$ and $-r(r-1)\delta\log\delta - (1-\delta)\log(1-\delta) < \frac{\varepsilon}{2}$. Let ζ be the partition into the sets $A_i \cap C_j (i \neq j)$ and $\cup_{n=1}^r (A_n \cap C_n)$. Then $\xi \vee \eta = \zeta \vee \eta$ and since $A_i \cap C_j \subset \cup_{n=1}^r (A_n \Delta C_n)(i \neq j)$ we have

$$\mu(A_i \cap C_j) < \delta(i \neq j) \quad \text{and} \quad \mu\left(\bigcup_{n=1}^r (A_n \cap C_n)\right) > 1 - \delta.$$

Hence $H_\mu(\zeta) < -r(r-1)\delta\log\delta - (1-\delta)\log(1-\delta) < \frac{\varepsilon}{2}$. Therefore $H_\mu(\eta) + H_\mu(\xi|\eta) = H_\mu(\xi \vee \eta) = H_\mu(\eta \vee \zeta) \leq H_\mu(\eta) + H_\mu(\zeta) < H_\mu(\eta) + \frac{\varepsilon}{2}$, and so $H_\mu(\xi|\eta) < \frac{\varepsilon}{2}$. Similarly we have $H_\mu(\eta|\xi) < \frac{\varepsilon}{2}$. Then the proof is completed. □

Proof of Theorem 4.7. Note that μ can be regarded as a Borel probability measure on Y and then $(X, \mathcal{B}, \mu)$ is a Lebesgue space. Fix now $\varepsilon > 0$ arbitrarily. We can choose a finite partition $\xi = \{A_1, \cdots, A_r\}$ of X such that its elements are Borel sets and $h_\mu^{\mathcal{A}}(T, \xi) > h_\mu^{\mathcal{A}}(T) - \varepsilon$ if $h_\mu^{\mathcal{A}}(T) < +\infty$,or $h_\mu^{\mathcal{A}}(T, \xi) > \varepsilon^{-1}$ if $h_\mu^{\mathcal{A}}(T) = +\infty$. Take $\delta > 0$ as corresponding to ε and r in the sense explained in Lemma 4.2. Since every finite Borel measure on a Polish space is regular (see Theorem 1.4) and μ can be regarded as a Borel measure on Y supported by X, we can choose compact sets $K_i \subset A_i$ with $\mu(A_i \backslash K_i) < \delta(2r)^{-1}, 1 \leq i \leq r$. Let $\delta' = \inf_{i\neq j} d(K_i, K_j)$ and choose n such that $\text{diam}(\xi_n) < 2^{-1}\delta'$.

For $1 \leq i < r$ let $E_n^{(i)}$ be the union of all the elements of ξ_n that intersect K_i and let $E_n^{(r)}$ be the union of the remaining elements of ξ_n. Since $\text{diam}(\xi_n) < 2^{-1}\delta'$ each $C \in \xi_n$ can intersect at most one K_i. Then $\xi_n' = \{E_n^{(1)}, \cdots, E_n^{(r)}\}$ is a measurable partition of X such that $\xi_n' \leq \xi_n$ and $\sum_{i=1}^r \mu(E_n^{(i)} \Delta A_i) = \sum_{i=1}^r \mu(E_n^{(i)} \backslash A_i) + \sum_{i=1}^r \mu(A_i \backslash E_n^{(i)}) \leq \mu(X \backslash \cup_{i=1}^r K_i) + \sum_{i=1}^r \mu(A_i \backslash K_i) < \delta$.

From Lemma 4.2 we have $H_\mu(\xi|\xi'_n) + H_\mu(\xi'_n|\xi) < \varepsilon$. Therefore, if n is such that $\mathrm{diam}(\xi_n) < \frac{\delta'}{2}$ then

$$h_\mu^{\mathcal{A}}(T,\xi) \le h_\mu^{\mathcal{A}}(T,\xi'_n) + H_\mu(\xi|\xi'_n \vee \zeta_0) \le h_\mu^{\mathcal{A}}(T,\xi_n) + \varepsilon.$$

Thus $\mathrm{diam}(\xi_n) < \frac{\delta'}{2}$ implies $h_\mu^{\mathcal{A}}(T,\xi_n) > h_\mu^{\mathcal{A}}(T) - 2\varepsilon$ if $h_\mu^{\mathcal{A}}(T) < +\infty$ or $h_\mu^{\mathcal{A}}(T,\xi_n) > \varepsilon^{-1} - \varepsilon$ if $h_\mu^{\mathcal{A}}(T) = +\infty$. So $\lim_{n\to+\infty} h_\mu^{\mathcal{A}}(T,\xi_n)$ exists and equals $h_\mu^{\mathcal{A}}(T)$. □

§5 Conditional Entropies of Measure-Preserving Transformations: II

Let T and $\mathcal{A}$ be as in Section 4. In this section we generalize the definition of the entropy function $h_\mu^{\mathcal{A}}(T,\xi)$, which was justified in Section 4 only for partitions in $\mathcal{Z}(\zeta_0)$, so that it is defined on the set of all measurable partitions of X. In order to do this we have to confine us to the case when $T^{-1}\mathcal{A} = \mathcal{A}$.

Definition 5.1. *Let $T, \mathcal{A}$ and ζ_0 be as in Section 4 and assume that $T^{-1}\mathcal{A} = \mathcal{A}$. For every measurable partition ξ of X, put*

$$h_\mu^{\mathcal{A}}(T,\xi) = H_\mu(\xi|T^{-1}\xi_T^- \vee \zeta_0), \tag{5.1}$$

we call this the $\mathcal{A}$-conditional entropy of T with respect to ξ.

Remark 5.1. If $\xi \in \mathcal{Z}(\zeta_0)$, then $h_\mu^{\mathcal{A}}(T,\xi)$ defined by (5.1) coincides with that defined by Theorem 4.1. Indeed, if $\xi \in \mathcal{Z}(\zeta_0)$, from (3.7), (3.10) and $T^{-1}\mathcal{A} = \mathcal{A}$ it follows that for every $n \in \mathbf{N}$

$$\begin{aligned} H_\mu(\xi^n|\zeta_0) &= H_\mu\left(\bigvee_{k=1}^{n-1} T^{-k}\xi|\zeta_0\right) + H_\mu\left(\xi|\bigvee_{k=1}^{n-1} T^{-k}\xi \vee \zeta_0\right) \\ &= H_\mu(T^{-1}\xi^{n-1}|T^{-1}\zeta_0) + H_\mu(\xi|T^{-1}\xi^{n-1} \vee \zeta_0) \\ &= H_\mu(\xi^{n-1}|\zeta_0) + H_\mu(\xi|T^{-1}\xi^{n-1} \vee \zeta_0), \end{aligned}$$

then an easy induction shows

$$H_\mu(\xi^n|\zeta_0) = H_\mu(\xi|\zeta_0) + \sum_{l=1}^{n-1} H_\mu(\xi|T^{-1}\xi^l \vee \zeta_0). \tag{5.2}$$

Since

$$H_\mu(\xi|T^{-1}\xi^l \vee \zeta_0) \searrow H_\mu(\xi|T^{-1}\xi^- \vee \zeta_0)$$

as $l \to +\infty$, by (5.2) we have

$$\frac{1}{n}H_\mu(\xi^n|\zeta_0) \searrow H_\mu(\xi|T^{-1}\xi^- \vee \zeta_0) \quad \text{as} \quad n \to +\infty. \tag{5.3}$$

This shows that the conclusion of Remark 5.1 is true. □

Theorem 5.1. *Let T and $\mathcal{A}$ be as given before and assume that $T^{-1}\mathcal{A} = \mathcal{A}$. Then*

$$h_\mu^{\mathcal{A}}(T) = \sup h_\mu^{\mathcal{A}}(T, \xi)$$

where the supremum is taken over the set of all measurable partitions of X.

Proof. It is enough to show that for any measurable partition ξ and any number $a < h_\mu^{\mathcal{A}}(T, \xi)$ there exists a finite measurable partition η such that $h_\mu^{\mathcal{A}}(T, \eta) > a$.

Let $\xi_1, \xi_2, \cdots$ be a sequence of finite measurable partitions such that $\xi_n \nearrow \xi$. Since $\xi_n \le \xi$ we have

$$h_\mu^{\mathcal{A}}(T, \xi_n) = H_\mu(\xi_n | T^{-1}(\xi_n)^- \vee \zeta_0) \ge H_\mu(\xi_n | T^{-1}\xi^- \vee \zeta_0).$$

From

$$H_\mu(\xi_n | T^{-1}\xi^- \vee \zeta_0) \longrightarrow H_\mu(\xi | T^{-1}\xi^- \vee \zeta_0) \qquad \text{as} \quad n \to +\infty$$

it follows that $h_\mu^{\mathcal{A}}(T, \xi_n) > a$ for sufficiently large n and we can take $\eta = \xi_n$. □

Before discussing further properties of $h_\mu^{\mathcal{A}}(T, \xi)$ given by Definition 5.1, we first give some preliminary results (Lemmas 5.1-5.3).

Lemma 5.1. *Let $T, \mathcal{A}$ and ζ_0 be as given before and assume that $T^{-1}\mathcal{A} = \mathcal{A}$. If ξ, η are two measurable partitions with $\eta \le \xi$, then*

$$H_\mu(\xi^n | T^{-n}\eta^- \vee \zeta_0) = \sum_{k=0}^{n-1} H_\mu(\xi | T^{-1}(\eta^- \vee \xi^k) \vee \zeta_0). \tag{5.4}$$

In particular,

$$H_\mu(\xi^n | T^{-n}\xi^- \vee \zeta_0) = n h_\mu^{\mathcal{A}}(T, \xi).$$

Proof. Since $\xi^k = \xi \vee T^{-1}\xi^{k-1}$ and $\eta \le \xi$, we have

$$\begin{aligned} &H_\mu(\xi^k | T^{-k}\eta^- \vee \zeta_0) \\ =&H_\mu(T^{-1}\xi^{k-1} | T^{-k}\eta^- \vee \zeta_0) + H_\mu(\xi | T^{-k}\eta^- \vee \zeta_0 \vee T^{-1}\xi^{k-1}) \\ =&H_\mu(\xi^{k-1} | T^{-(k-1)}\eta^- \vee \zeta_0) + H_\mu(\xi | T^{-1}(\eta^- \vee \xi^{k-1}) \vee \zeta_0), \end{aligned}$$

(5.4) is then derived from this equation by an obvious induction. □

Lemma 5.2. *Let $T, \mathcal{A}$ and ζ_0 be as given before and assume that $T^{-1}\mathcal{A} = \mathcal{A}$. If ξ, η are two measurable partitions with $\eta \le \xi$ and $H_\mu(\xi | T^{-1}\eta^- \vee \zeta_0) < +\infty$, then*

$$\frac{1}{n} H_\mu(\xi^n | T^{-n}\eta^- \vee \zeta_0) \searrow h_\mu^{\mathcal{A}}(T, \xi) \qquad \textit{as} \quad n \to +\infty. \tag{5.5}$$

Proof. Since $\xi^n \nearrow \xi^-$ and $\eta \leq \xi$, we have $\eta^- \vee \xi^n \nearrow \xi^-$ and by (3.8)

$$H_\mu(\xi|T^{-1}(\eta^- \vee \xi^n) \vee \zeta_0) \searrow H_\mu(\xi|T^{-1}\xi^- \vee \zeta_0).$$

Then (5.5) follows from (5.4). □

Lemma 5.3. *Let $T, \mathcal{A}$ and ζ_0 be as given before and assume that $T^{-1}\mathcal{A} = \mathcal{A}$. If ξ, η are two measurable partitions with $\xi \leq \eta$ and $H_\mu(\eta|T^{-1}\xi^- \vee \zeta_0) < +\infty$, then*

$$\lim_{n\to+\infty} \frac{1}{n} H_\mu(\xi^n|T^{-n}\eta^- \vee \zeta_0) = h_\mu^{\mathcal{A}}(T,\xi).$$

Proof. Let δ be a positive number. Since by Lemma 5.1

$$\frac{1}{n} H_\mu(\xi^n|T^{-n}\eta^- \vee \zeta_0) \leq \frac{1}{n} H_\mu(\xi^n|T^{-n}\xi^- \vee \zeta_0) = h_\mu^{\mathcal{A}}(T,\xi)$$

and by Lemma 5.2

$$\frac{1}{n} H_\mu(\eta^n|T^{-n}\xi^- \vee \zeta_0) \searrow h_\mu^{\mathcal{A}}(T,\eta) \qquad \text{as} \quad n \to +\infty,$$

it suffices to show that the inequality

$$\frac{1}{n} H_\mu(\xi^n|T^{-n}\eta^- \vee \zeta_0) > h_\mu^{\mathcal{A}}(T,\xi) - \delta$$

holds true for all n for which

$$\frac{1}{n} H_\mu(\eta^n|T^{-n}\xi^- \vee \zeta_0) < h_\mu^{\mathcal{A}}(T,\eta) + \delta.$$

This is clear from the following chain of relations:

$$\begin{aligned}
&\frac{1}{n} H_\mu(\xi^n|T^{-n}\eta^- \vee \zeta_0) \\
=&\frac{1}{n} H_\mu(\eta^n|T^{-n}\eta^- \vee \zeta_0) - \frac{1}{n} H_\mu(\eta^n|\xi^n \vee T^{-n}\eta^- \vee \zeta_0) \\
\geq& h_\mu^{\mathcal{A}}(T,\eta) - \frac{1}{n} H_\mu(\eta^n|\xi^n \vee T^{-n}\xi^- \vee \zeta_0) \\
\geq&\frac{1}{n} H_\mu(\eta^n|T^{-n}\xi^- \vee \zeta_0) - \delta - \frac{1}{n} H_\mu(\eta^n|\xi^n \vee T^{-n}\xi^- \vee \zeta_0) \\
=&\frac{1}{n} H_\mu(\xi^n|T^{-n}\xi^- \vee \zeta_0) - \delta \\
=&h_\mu^{\mathcal{A}}(T,\xi) - \delta.
\end{aligned}$$

The proof is completed. □

We are now prepared to deduce some properties of $h_\mu^{\mathcal{A}}(T,\xi)$ given by Definition 5.1.

Theorem 5.2. *Let $T, \mathcal{A}$ and ζ_0 be as given before and assume that $T^{-1}\mathcal{A} = \mathcal{A}$. Then for any measurable partitions ξ and η*

1) $h_\mu^{\mathcal{A}}(T,\xi) \le H_\mu(\xi|\zeta_0)$.

2) $h_\mu^{\mathcal{A}}(T,\xi \vee \eta) \le h_\mu^{\mathcal{A}}(T,\xi) + h_\mu^{\mathcal{A}}(T,\eta)$.

3) $h_\mu^{\mathcal{A}}(T,\xi) \le h_\mu^{\mathcal{A}}(T,\xi_T^n)$ for all $n \in \mathbf{N}$.

4) $h_\mu^{\mathcal{A}}(T,T^{-1}\xi) = h_\mu^{\mathcal{A}}(T,\xi)$.

5) $h_\mu^{\mathcal{A}}(T^n,\xi_T^n) = nh_\mu^{\mathcal{A}}(T,\xi)$.

6) If $\xi \le \eta$ and $H_\mu(\eta|T^{-1}\xi_T^- \vee \zeta_0) < +\infty$, then $h_\mu^{\mathcal{A}}(T,\xi) \le h_\mu^{\mathcal{A}}(T,\eta)$.

Proof. 1) $h_\mu^{\mathcal{A}}(T,\xi) = H_\mu(\xi|T^{-1}\xi^- \vee \zeta_0) \le H_\mu(\xi|\zeta_0)$.

2)
$$\begin{aligned} h_\mu^{\mathcal{A}}(T,\xi \vee \eta) &= H_\mu(\xi \vee \eta|T^{-1}(\xi \vee \eta)^- \vee \zeta_0) \\ &\le H_\mu(\xi|T^{-1}\xi^- \vee \zeta_0) + H_\mu(\eta|T^{-1}\eta^- \vee \zeta_0) \\ &= h_\mu^{\mathcal{A}}(T,\xi) + h_\mu^{\mathcal{A}}(T,\eta). \end{aligned}$$

3) Since $(\xi^n)^- = \xi^-$ we have
$$h_\mu^{\mathcal{A}}(T,\xi^n) = H_\mu(\xi^n|T^{-1}\xi^- \vee \zeta_0) \ge H_\mu(\xi|T^{-1}\xi^- \vee \zeta_0) = h_\mu^{\mathcal{A}}(T,\xi).$$

4)
$$\begin{aligned} h_\mu^{\mathcal{A}}(T,T^{-1}\xi) &= H_\mu(T^{-1}\xi|T^{-1}(T^{-1}\xi)^- \vee \zeta_0) \\ &= H_\mu(\xi|T^{-1}\xi^- \vee \zeta_0) = h_\mu^{\mathcal{A}}(T,\xi). \end{aligned}$$

5) Since $(\xi_T^n)_{T^n}^- = \xi_T^-$, we have
$$h_\mu^{\mathcal{A}}(T^n,\xi_T^n) = H_\mu(\xi_T^n|T^{-n}(\xi_T^n)_{T^n}^- \vee \zeta_0) = H_\mu(\xi_T^n|T^{-n}\xi_T^- \vee \zeta_0),$$
then we get the desired result by applying Lemma 5.1.

6) Since $\xi \le \eta$ we have
$$\frac{1}{n}H_\mu(\xi^n|T^{-n}\eta^- \vee \zeta_0) \le \frac{1}{n}H_\mu(\eta^n|T^{-n}\eta^- \vee \zeta_0).$$
By Lemma 5.1 the right-hand side is equal to $h_\mu^{\mathcal{A}}(T,\eta)$ and by Lemma 5.3 the limit of the left-hand side as $n \to +\infty$ is $h_\mu^{\mathcal{A}}(T,\xi)$, then 6) is proved. □

A measurable partition η of X is said to be *fixed* under T if every element C of η satisfies $T^{-1}C = C$. Let η be such a partition of X, we denote by T_C the measure-preserving transformation on $(C, \mathcal{B}|_C, \mu_C)$ induced by T for μ_η-a.e. C.

Theorem 5.3. *Let $T, \mathcal{A}$ and ζ_0 be as given before and assume that $T^{-1}\mathcal{A} = \mathcal{A}$. If $\xi \in \mathcal{Z}(\zeta_0)$ and η is a measurable partition fixed under T, then*
$$h_\mu^{\mathcal{A}}(T,\xi \vee \eta) = h_\mu^{\mathcal{A}}(T,\xi).$$

Proof. First we assume that η is a finite measurable partition. Since

$$\begin{aligned}
&\frac{1}{n}[H_\mu(\xi^n \vee \eta^n | T^{-n}(\xi^- \vee \eta^-) \vee \zeta_0) - H_\mu(\xi^n | T^{-n}(\xi^- \vee \eta^-) \vee \zeta_0)] \\
=&\frac{1}{n} H_\mu(\eta^n | \xi^n \vee T^{-n}\xi^- \vee T^{-n}\eta^- \vee \zeta_0) \\
=&\frac{1}{n} H_\mu(\eta^n | \xi^- \vee T^{-n}\eta^- \vee \zeta_0), \qquad (5.6)
\end{aligned}$$

and since $T^{-1}\eta = \eta$, the last term of (5.6) is zero. The first term of (5.6), by Lemma 5.1 and Lemma 5.2, converges as $n \to +\infty$ to $h_\mu^{\mathcal{A}}(T, \xi \vee \eta) - h_\mu^{\mathcal{A}}(T, \xi)$. Thus $h_\mu^{\mathcal{A}}(T, \xi \vee \eta) = h_\mu^{\mathcal{A}}(T, \xi)$.

If η is an arbitrary measurable partition fixed under T, then there exists a sequence of finite measurable partitions $\eta_1, \eta_2, \cdots$ such that $\eta_n \nearrow \eta$ and every η_n is fixed under T. For every $n \in \mathbf{N}$ we first have

$$h_\mu^{\mathcal{A}}(T, \xi \vee \eta_n) = h_\mu^{\mathcal{A}}(T, \xi). \qquad (5.7)$$

Then, from

$$h_\mu^{\mathcal{A}}(T, \xi \vee \eta_n) = H_\mu(\xi | \eta_n \vee T^{-1}\xi^- \vee \zeta_0),$$

$$h_\mu^{\mathcal{A}}(T, \xi \vee \eta) = H_\mu(\xi | \eta \vee T^{-1}\xi^- \vee \zeta_0)$$

and (3.8) and (5.7) we obtain

$$h_\mu^{\mathcal{A}}(T, \xi \vee \eta) = h_\mu^{\mathcal{A}}(T, \xi)$$

by letting $n \to +\infty$. □

Theorem 5.4. *Let $T, \mathcal{A}$ and ζ_0 be as given before and assume that $T^{-1}\mathcal{A} = \mathcal{A}$. If $\xi \in \mathcal{Z}(\zeta_0)$ and η is a measurable partition fixed under T, then*

$$h_\mu^{\mathcal{A}}(T, \xi) = \int_{X/\eta} h_{\mu_B}^{\mathcal{A}_B}(T_B, \xi_B) d\mu_\eta \qquad (5.8)$$

where $\mathcal{A}_B = \mathcal{A}|_B$.

Proof. Since

$$\begin{aligned}
h_{\mu_B}^{\mathcal{A}_B}(T_B, \xi_B) &= H_{\mu_B}(\xi_B | T_B^{-1}(\xi_B)_{T_B}^- \vee (\zeta_0)_B) \\
&= H_{\mu_B}(\xi_B | T_B^{-1}(\xi_T^-)_B \vee (\zeta_0)_B) \\
&= -\int_B \log m(x, \xi_B | T_B^{-1}(\xi_T^-)_B \vee (\zeta_0)_B) d\mu_B,
\end{aligned}$$

and since the function under the integral sign is the restriction to B of $\log m(x, \xi | \eta \vee T^{-1}\xi^- \vee \zeta_0)$, we have

$$\begin{aligned}&\int_{X/\eta} h_{\mu_B}^{\mathcal{A}_B}(T_B,\xi_B)d\mu_\eta\\&=-\int_{X/\eta}d\mu_\eta\int_B \log m(x,\xi|\eta\vee T^{-1}\xi^-\vee\zeta_0)d\mu_B\\&=-\int_X \log m(x,\xi|\eta\vee T^{-1}\xi^-\vee\zeta_0)d\mu\\&=H_\mu(\xi|\eta\vee T^{-1}\xi^-\vee\zeta_0)=h_\mu^{\mathcal{A}}(T,\xi\vee\eta).\end{aligned}$$

Then by Theorem 5.3 we get (5.8). □

Theorem 5.5. *Let $T,\mathcal{A}$ be as given before and assume that $T^{-1}\mathcal{A}=\mathcal{A}$. If η is a measurable partition fixed under T, then*

$$h_\mu^{\mathcal{A}}(T)=\int_{X/\eta} h_{\mu_B}^{\mathcal{A}_B}(T_B)d\mu_\eta. \tag{5.9}$$

Proof. Let $\xi_1,\xi_2,\cdots$ be a sequence of finite measurable partitions such that $\xi_n\nearrow\varepsilon$. According to Theorem 5.4, for every $n\in\mathbf{N}$ we have

$$h_\mu^{\mathcal{A}}(T,\xi_n)=\int_{X/\eta} h_{\mu_B}^{\mathcal{A}_B}(T_B,(\xi_n)_B)d\mu_\eta. \tag{5.10}$$

By Theorem 4.5, for μ_η-a.e. $B\in X/\eta$

$$h_\mu^{\mathcal{A}}(T,\xi_n)\nearrow h_\mu^{\mathcal{A}}(T),\quad h_{\mu_B}^{\mathcal{A}_B}(T_B,(\xi_n)_B)\nearrow h_{\mu_B}^{\mathcal{A}_B}(T_B) \tag{5.11}$$

as $n\to+\infty$. Then (5.9) follows from (5.10), (5.11) and the monotone convergence theorem. □

We end this section with the notion of *ergodic decompositions* of measure -preserving transformations. Let T be a measure-preserving transformation on the Lebesgue space $(X,\mathcal{B},\mu)$. T is called *ergodic* if every measurable set A satisfying $T^{-1}A=A$ has either measure 1 or measure 0. If T is not ergodic, then it can be decomposed into ergodic components in the following sense (see [Roh]$_3$): There exists a unique (mod 0) measurable partition η of X fixed under T such that $T_C:(C,\mathcal{B}|_C,\mu_C)\hookleftarrow$ is ergodic for μ_η-a.e. $C\in X/\eta$.

Chapter I Entropy and Lyapunov Exponents of Random Diffeomorphisms

The purpose of this chapter is to introduce some basic notions, especially entropy and Lyapunov exponents, of ergodic theory of random diffeomorphisms and to present some basic properties of them. This chapter serves as the foundation of this book. It is assumed throughout that the reader is familiar with the contents of Chapter I of Kifer's book $[\mathrm{Kif}]_1$.

§1 The Basic Measure Spaces and Invariant Measures

Throughout the remaining part of this book M is always (except mentioned otherwise) a C^∞ compact and connected Riemannian manifold without boundary, $\mathrm{Diff}^2(M)$ denotes the space of C^2 diffeomorphisms on M equipped with the C^2-topology (see, e.g., $[\mathrm{Hir}]_1$). To simplify notations we shall denote $\mathrm{Diff}^2(M)$ by Ω. Let v be a Borel probability measure on $(\Omega, \mathcal{B}(\Omega))$, where we denote by $\mathcal{B}(X)$ the Borel σ-algebra of a topological space X.

We are mainly concerned with in this book the ergodic theory of the evolution process generated by the successive applications of randomly chosen maps in Ω. These maps will be assumed to be independent and identically distributed with law v. There are now two cases to be distinguished when the process evolutes noninvertibly or invertibly in time. The former case seems to be more natural than the latter since many important physical processes such as diffusions are in general noninvertible in time. We are first mainly concerned with the former one and in Chapter VI we shall touch on the latter and give an important result about entropy formula and Sinai–Bowen–Ruelle measures for this case. Now let us first describe the former case in a more precise way.

Let

$$(\Omega^{\mathbf{N}}, \mathcal{B}(\Omega)^{\mathbf{N}}, v^{\mathbf{N}}) = \prod_{1}^{+\infty} (\Omega, \mathcal{B}(\Omega), v)$$

be the infinite product of copies of the measure space $(\Omega, \mathcal{B}(\Omega), v)$. For each $\omega = (f_0(\omega), f_1(\omega), \cdots) \in \Omega^{\mathbf{N}}$ and $n \geq 0$, define

$$f_\omega^0 = id, \quad f_\omega^n = f_{n-1}(\omega) \circ f_{n-2}(\omega) \circ \cdots \circ f_0(\omega).$$

Our purpose is to study the dynamical behaviour of these composed maps as $n \to +\infty$ for $v^{\mathbf{N}}$-a.e. ω, and the random system generated by $\{f_\omega^n : n \geq 0, \omega \in (\Omega^{\mathbf{N}}, \mathcal{B}(\Omega)^{\mathbf{N}}, v^{\mathbf{N}})\}$ will be referred to as $\mathcal{X}^+(M, v)$. Throughout this book, associated with any system $\mathcal{X}^+(M, v)$ it is always assumed that v satisfies

$$\int \log^+ |f|_{C^2} dv(f) < +\infty \tag{1.1}$$

where $\log^+ x = \max\{\log x, 0\}$ and $|f|_{C^2}$ is the usual C^2–norm of $f \in \Omega$ whose definition is given in the following paragraph for the convenience of the reader.

We shall always denote $\dim M$ by m_0. Let $h : M \to \mathbf{R}^{2m_0+1}$ be an C^∞ embedding. Choose a system of charts $\{(U_i, \varphi_i)\}_{i=1}^l$ which covers M and let $\{V_i\}_{i=1}^l$ be an open cover of M such that $\overline{V}_i \subset U_i, 1 \le i \le l$. Then for every $f \in \Omega$ define

$$|f|_{C^2} = \max_{1\le i\le l} \sup_{u\in\varphi_i(V_i)} \max\left\{\left|\frac{\partial f_j^{(i)}}{\partial u_k}(u)\right|, \left|\frac{\partial^2 f_j^{(i)}}{\partial u_k \partial u_r}(u)\right| : 1 \le k, r \le m_0, \right.$$
$$\left. 1 \le j \le 2m_0 + 1\right\}$$

where $(f_1^{(i)}, f_2^{(i)}, \cdots, f_{2m_0+1}^{(i)})^\top = h \circ f \circ \varphi_i^{-1}$ and $u = (u_1, \cdots, u_{m_0})^\top$. It is clear that $|\cdot|_{C^2}$ depends on the choice of $\{(U_i, \varphi_i), V_i\}_{i=1}^l$ and h , but if $\{(U_i', \varphi_i'), V_i'\}_{i=1}^{l'}$ and h' is another choice and let $|\cdot|_{C^2}'$ is the norm corresponding to it, then $|\cdot|_{C^2}$ and $|\cdot|_{C^2}'$ are equivalent, i.e., there exists a number $C > 0$ such that for all $f \in \Omega$

$$C^{-1}|f|_{C^2} \le |f|_{C^2}' \le C|f|_{C^2}.$$

Since we do not pay any attention to the difference between $|\cdot|_{C^2}$ and $|\cdot|_{C^2}'$, such a norm $|\cdot|_{C^2}$ will be called the C^2*-norm* on Ω. Another equivalent definition will be given in Chapter II without using charts.

Condition (1.1) will play a fundamental role in the discussions about entropy, Lyapunov exponents and stable manifolds of $\mathcal{X}^+(M, v)$ carried out in the later chapters. And it can actually be varified in the case when $\mathcal{X}^+(M, v)$ arises from a diffusion process (see Chapter V).

We remark that, when v is supported by one point of Ω, $\mathcal{X}^+(M, v)$ reduces to the deterministic case.

Now we introduce two important spaces $\Omega^{\mathbf{N}} \times M$ and $\Omega^{\mathbf{Z}} \times M$ and let them equipped respectively with the product σ–algebras $\mathcal{B}(\Omega)^{\mathbf{N}} \times \mathcal{B}(M)$ and $\mathcal{B}(\Omega)^{\mathbf{Z}} \times \mathcal{B}(M)$ and with the product topologies. The standard knowledge of differential topology (see, for instance, [Hir]$_1$) tells us that Ω is an open subset of $C^2(M, M)$, which is the set of all C^2 maps on M endowed with the C^2–topology, and $C^2(M, M)$ can be metrized to a Polish space, i.e., a complete separable metric space. Therefore, Ω can be metrized to a separable metric space. Hence we have

$$\mathcal{B}(\Omega)^{\mathbf{N}} \times \mathcal{B}(M) = \mathcal{B}(\Omega^{\mathbf{N}} \times M),$$

$$\mathcal{B}(\Omega)^{\mathbf{Z}} \times \mathcal{B}(M) = \mathcal{B}(\Omega^{\mathbf{Z}} \times M).$$

Let τ denote both the left shift operators on $\Omega^{\mathbf{N}}$ and $\Omega^{\mathbf{Z}}$, namely,

$$f_n(\tau\omega) = f_{n+1}(\omega)$$

for all $\omega = (f_0(\omega), f_1(\omega), \cdots) \in \Omega^{\mathbf{N}}$ and $n \ge 0$, and

$$f_n(\tau w) = f_{n+1}(w)$$

for all $w = (\cdots, f_{-1}(w), f_0(w), f_1(w), \cdots) \in \Omega^{\mathbf{Z}}$ and $n \in \mathbf{Z}$. Define then

$$F : \Omega^{\mathbf{N}} \times M \to \Omega^{\mathbf{N}} \times M, (\omega, x) \mapsto (\tau\omega, f_0(\omega)x);$$

$$G : \Omega^{\mathbf{Z}} \times M \to \Omega^{\mathbf{Z}} \times M, (w, x) \mapsto (\tau w, f_0(w)x).$$

The two systems $(\Omega^{\mathbf{N}} \times M, F)$ and $(\Omega^{\mathbf{Z}} \times M, G)$ will act as important bridges between ergodic theory of random dynamical systems and that of deterministic dynamical systems.

We now begin to discuss the invariant measures of $\mathcal{X}^+(M, v)$.

Definition 1.1. *A Borel probability measure μ on M is called an invariant measure of $\mathcal{X}^+(M, v)$ if*

$$\int_\Omega f\mu dv(f) = \mu \tag{1.2}$$

where $(f\mu)(E) = \mu(f^{-1}E)$ for all $E \in \mathcal{B}(M)$ and $f \in \Omega$.

We denote by $\mathcal{M}(\mathcal{X}^+(M, v))$ the set of all invariant measures of $\mathcal{X}^+(M, v)$. Since M is compact and v is a probability measure on Ω, by Lemma I.2.2 of [Kif]$_1$ we know that $\mathcal{M}(\mathcal{X}^+(M, v)) \neq \phi$. Furthermore, $\mathcal{M}(\mathcal{X}^+(M, v))$ can be proved to be a compact convex set with respect to the weak topology. In fact, for any $\mu_1, \mu_2 \in \mathcal{M}(\mathcal{X}^+(M, v))$ and $0 \leq t < 1$, it is clear that $t\mu_1 + (1-t)\mu_2 \in \mathcal{M}(\mathcal{X}^+(M, v))$, so $\mathcal{M}(\mathcal{X}^+(M, v))$ is a convex set. What remains is to show the compactness of it. Let $\mathcal{M}(M)$ be the space of Borel probability measures on M equipped with the weak topology (see, e.g., Chapter VI of [Wal]$_1$), and assume that $\{\mu_n\}_{n=1}^{+\infty}$ is a sequence in $\mathcal{M}(\mathcal{X}^+(M, v))$ such that $\mu_n \to \mu$ in $\mathcal{M}(M)$ as $n \to +\infty$. Since for every $g \in C^0(M)$,

$$\int g(x) d\mu_n(x) = \int \int g(x) d(f\mu_n)(x) dv(f), \quad \forall n \in \mathbf{N},$$

and as $n \to +\infty$

$$\int g(x) d\mu_n(x) \to \int g(x) d\mu(x),$$

$$\int g(x) d(f\mu_n)(x) \to \int g(x) d(f\mu)(x), \quad \forall f \in \Omega,$$

by the dominated convergence theorem we have

$$\int g(x) d\mu(x) = \int \int g(x) d(f\mu)(x) dv(f). \tag{1.3}$$

By the standard methods of measure theory (1.3) implies that

$$\mu = \int f\mu dv(f),$$

that is,

$$\mu \in \mathcal{M}(\mathcal{X}^+(M,v)).$$

Hence $\mathcal{M}(\mathcal{X}^+(M,v))$ is a closed subset of $\mathcal{M}(M)$. This together with the compactness of $\mathcal{M}(M)$ shows that $\mathcal{M}(\mathcal{X}^+(M,v))$ is compact.

Proposition 1.1. *Let $\mu \in \mathcal{M}(M)$. Then $\mu \in \mathcal{M}(\mathcal{X}^+(M,v))$ if and only if $v^{\mathbf{N}} \times \mu$ is F-invariant.*

See Lemma I.2.3 of [Kif]$_1$ for a proof.

Define now projection operators

$$P_1 : \Omega^{\mathbf{Z}} \times M \to \Omega^{\mathbf{Z}}, \quad (w,x) \mapsto w,$$

$$P_2 : \Omega^{\mathbf{Z}} \times M \to M, \quad (w,x) \mapsto x,$$

$$P : \Omega^{\mathbf{Z}} \times M \to \Omega^{\mathbf{N}} \times M, (w,x) \mapsto (w^+, x)$$

where $w^+ = (f_0(w), f_1(w), \cdots)$ for $w \in \Omega^{\mathbf{Z}}$. Then we have

Proposition 1.2. *For every $\mu \in \mathcal{M}(\mathcal{X}^+(M,v))$, there exists a unique Borel probability measure μ^* on $\Omega^{\mathbf{Z}} \times M$ such that $G\mu^* = \mu^*$ and $P\mu^* = v^{\mathbf{N}} \times \mu$. Moreover, $P_1\mu^* = v^{\mathbf{Z}}, P_2\mu^* = \mu$, and $G^n(v^{\mathbf{Z}} \times \mu)$ converges weakly to μ^* as $n \to +\infty$.*

Proof. For every $n \geq 0$ let $\mathcal{B}_n$ denote the σ-algebra $\left\{\prod_{-\infty}^{-n-1} \Omega \times W : W \in \mathcal{B}\left(\prod_{-n}^{+\infty} \Omega \times M\right)\right\}$ and μ_n^* the probability measure $G^n(v^{\mathbf{Z}} \times \mu)|_{\mathcal{B}_n}$. Clearly $\mathcal{B}_n \subset \mathcal{B}_{n+1}$ for all $n \geq 0$. Now we show that for every $n \geq 0$

$$\mu_{n+1}^*|_{\mathcal{B}_n} = \mu_n^*. \tag{1.4}$$

Since $\mathcal{B}_n$ is generated by the semi-algebra $\mathcal{A}_n = \left\{\left(\prod_{-\infty}^{-n-1} \Omega \times \prod_{-n}^{j} \Gamma_i \times \prod_{j+1}^{+\infty} \Omega\right) \times B : \Gamma_i \in \mathcal{B}(\Omega), -n \leq i \leq j, j \geq -n, B \in \mathcal{B}(M)\right\}$, by the extention theorems of measure theory (see, for example, Theorems 0.2, 0.3 and 0.4 of [Wal]$_1$), in order to prove (1.4) it is enough to show that

$$\mu_{n+1}^*(E) = \mu_n^*(E), \quad \forall E \in \mathcal{A}_n. \tag{1.5}$$

Now for every $E = \left(\prod_{-\infty}^{-n-1} \Omega \times \prod_{-n}^{j} \Gamma_i \times \prod_{j+1}^{+\infty} \Omega\right) \times B \in \mathcal{A}_n$, since $\mu \in \mathcal{M}(\mathcal{X}^+$

$(M, v))$ we have

$$
\begin{aligned}
& \mu_{n+1}^*(E) \\
= \; & (v^{\mathbf{Z}} \times \mu)(G^{-n-1}E) \\
= \; & \int_{\tau^{-n-1}\left(\prod_{-\infty}^{-n-1} \Omega \times \prod_{-n}^{j} \Gamma_i \times \prod_{j+1}^{+\infty} \Omega\right)} \mu(f_0(w)^{-1} \circ \cdots \circ f_n(w)^{-1} B) dv^{\mathbf{Z}} \\
= \; & \int_{\prod_{-\infty}^{-n-1} \Omega \times \prod_{-n}^{j} \Gamma_i \times \prod_{j+1}^{+\infty} \Omega} \mu(f_{-n-1}(w)^{-1} \circ \cdots \circ f_{-1}(w)^{-1} B) dv^{\mathbf{Z}} \\
= \; & \int_{\prod_{-n}^{j} \Gamma_i} \left\{ \int_{\Omega} \mu(f_{-n-1}^{-1} \circ \cdots \circ f_{-1}^{-1} B) dv(f_{-n-1}) \right\} \prod_{-n}^{j} dv(f_i) \\
= \; & \int_{\prod_{-n}^{j} \Gamma_i} \mu(f_{-n}^{-1} \circ \cdots \circ f_{-1}^{-1} B) \prod_{-n}^{j} dv(f_i).
\end{aligned}
\tag{1.6}
$$

In the same way as above we easily see that the last term of (1.6) equals $\mu_n^*(E)$, (1.5) is then proved.

Let $\mathcal{B}_\infty = \bigcup_{n=0}^{+\infty} \mathcal{B}_n$. Obviously, $\mathcal{B}_\infty$ is a sub–algebra of $\mathcal{B}(\Omega^{\mathbf{Z}} \times M)$ and it generates $\mathcal{B}(\Omega^{\mathbf{Z}} \times M)$. (1.4) allows one to define a probability measure μ_∞^* on $\mathcal{B}_\infty$ by letting

$$\mu_\infty^*|_{\mathcal{B}_n} = \mu_n^*, \quad \forall n \geq 0.$$

Now we show that it is indeed a measure on $\mathcal{B}_\infty$. In order to do this it is sufficient to confirm that for every decreasing sequence $E_0 \supset E_1 \supset \cdots$ of members of $\mathcal{B}_\infty$ with $\bigcap_{n=0}^{+\infty} E_n = \phi$ we have $\mu_\infty^*(E_n) \to 0$ as $n \to +\infty$.

Suppose that there exists a decreasing sequence $E_0 \supset E_1 \supset \cdots$ of members of $\mathcal{B}_\infty$ such that $\bigcap_{n=0}^{+\infty} E_n = \phi$ and $\mu_\infty^*(E_n) \geq \varepsilon_0$ for some $\varepsilon_0 > 0, \forall n \geq 0$. Then one can choose an increasing sequence $\{l_n\}_{n=0}^{+\infty} \subset \mathbf{Z}^+$ such that $E_n \in \mathcal{B}_{l_n}$ for all $n \geq 0$. Since Ω is an open subset of $C^2(M, M)$, $\prod_{-l_n}^{+\infty} \Omega \times M$ is a Borel subset of $\prod_{-l_n}^{+\infty} C^2(M, M) \times M$, which is equipped with the product topology and is then also a Polish space. Then by the regularity of Borel probability measures on Polish spaces (see Theorem 0.1.4), for every $n \geq 0$ we can find a compact subset W_n of $\prod_{-l_n}^{+\infty} \Omega \times M$ satisfying

$$\prod_{-\infty}^{-l_n-1} \Omega \times W_n \subset E_n$$

and

$$\mu_{l_n}^*\left(E_n \backslash \left(\prod_{-\infty}^{-l_n-1} \Omega \times W_n\right)\right) < 3^{-(n+1)}\varepsilon_0.$$

Put

$$F_n = \bigcap_{i=0}^{n} \left(\prod_{-\infty}^{-l_i-1} \Omega \times W_i\right)$$

for every $n \geq 0$, by a simple calculation we have

$$\mu_{l_n}^*(F_n) \geq \frac{\varepsilon_0}{3},$$

hence $F_n \neq \phi$. Then by using the diagonalization process we can choose a sequence $\{(w_n, x_n)\}_{n=1}^{+\infty}$ such that $(w_n, x_n) \in F_n, \forall n \geq 0$ and $(w_n, x_n) \to (w_0, x_0)$ as $n \to +\infty$. Clearly $(w_0, x_0) \in \bigcap_{n=0}^{+\infty} F_n \subset \bigcap_{n=0}^{+\infty} E_n$. This contradicts the fact that $\bigcap_{n=0}^{+\infty} E_n = \phi$. Therefore, μ_∞^* is indeed a measure on $\mathcal{B}_\infty$.

By the Hahn–Kolmogorov extension theorem μ_∞^* can be uniquely extended to a Borel probability measure on $\Omega^{\mathbf{Z}} \times M$ and we denote it by μ^*. It is obvious that

$$P\mu^* = v^{\mathbf{N}} \times \mu, P_1\mu^* = v^{\mathbf{Z}}, P_2\mu^* = \mu.$$

Moreover, since for all $E \in \mathcal{B}_\infty$

$$(G\mu^*)(E) = \mu^*(E),$$

and $\mathcal{B}_\infty$ generates $\mathcal{B}(\Omega^{\mathbf{Z}} \times M)$, we have

$$G\mu^* = \mu^*.$$

From the discussion above it also follows clearly that $G^n(v^{\mathbf{Z}} \times \mu)$ weakly converges to μ^* as $n \to +\infty$.

On the other hand, from the above construction of μ^* we know that it is the unique probability measure on $\mathcal{B}(\Omega^{\mathbf{Z}} \times M)$ such that $G\mu^* = \mu^*$ and $P\mu^* = v^{\mathbf{N}} \times \mu$. The proof of the proposition is then completed. □

From now on, $\mathcal{X}^+(M, v)$ associated with $\mu \in \mathcal{M}(\mathcal{X}^+(M, v))$ will be referred to as $\mathcal{X}^+(M, v, \mu)$. Next we shall discuss the ergodicity of the invariant measures of $\mathcal{X}^+(M, v)$. Let us first give a definition of this in a more intuitive way.

A Borel set $A \in \mathcal{B}(M)$ is said to be $\mathcal{X}^+(M, v)$*–invariant* if for μ–a.e. $x \in M$,

$$x \in A \Longrightarrow f(x) \in A \quad \text{for} \quad v - a.e. f \in \Omega,$$

$$x \in A^C \Longrightarrow f(x) \in A^C \quad \text{for} \quad v - a.e. f \in \Omega.$$

More generally, a bounded measurable function g on $(M, \mathcal{B}(M))$ is said to be $\mathcal{X}^+(M, v)$–invariant if

$$\int g(f(x)) dv(f) = g(x), \quad \mu - a.e. x.$$

Definition 1.2. *An invariant measure μ of $\mathcal{X}^+(M,v)$ is said to be ergodic if every $\mathcal{X}^+(M,v)$-invariant set has either μ-measure 1 or μ-measure 0 .*

We also call $\mathcal{X}^+(M,v,\mu)$ ergodic when μ is ergodic.

Proposition 1.3. *Let $\mathcal{X}^+(M,v,\mu)$ be given. Then the following conditions are equivalent:*

1) μ is ergodic;

2) If g is an $\mathcal{X}^+(M,v,\mu)$-invariant function, then

$$g = const. \quad \mu - a.e.;$$

3) $F : (\Omega^{\mathbf{N}} \times M, v^{\mathbf{N}} \times \mu) \hookleftarrow$ is ergodic;

4) $G : (\Omega^{\mathbf{Z}} \times M, \mu^*) \hookleftarrow$ is ergodic;

5) μ is an extreme point of $\mathcal{M}(\mathcal{X}^+(M,v))$.

Proof. See [Kif]$_1$ for a detailed proof of the equivalence of 1), 2) and 3). We complete the proof of this proposition by showing that 3) $\Longleftrightarrow$ 4), 3) $\Longrightarrow$ 5) and 5) $\Longrightarrow$ 1).

3) $\Longrightarrow$ 4). Let $E \in \mathcal{B}(\Omega^{\mathbf{Z}} \times M)$ be a set satisfying $G^{-1}E = E$. By the approximation theorem of measure theory (see Theorem 0.7 of [Wal$_1$]), for any given $\varepsilon > 0$ there exists $W \in \mathcal{B}(\Omega^{\mathbf{N}} \times M)$ such that

$$\mu^*(E \triangle P^{-1}W) < \varepsilon \tag{1.7}$$

and

$$v^{\mathbf{N}} \times \mu(W \triangle F^{-n}W) < \varepsilon, \quad \forall n \geq 0. \tag{1.8}$$

Since F is ergodic, by Birkhoff Ergodic Theorem

$$\lim_{n \to +\infty} \frac{1}{n} \sum_{k=0}^{n-1} \chi_W \circ F^k = v^{\mathbf{N}} \times \mu(W) \quad v^{\mathbf{N}} \times \mu - a.e. \tag{1.9}$$

Then from (1.8) and (1.9) it follows that

$$\int \left| \chi_W - v^{\mathbf{N}} \times \mu(W) \right| dv^{\mathbf{N}} \times \mu < \varepsilon,$$

thus

$$v^{\mathbf{N}} \times \mu(W) > \frac{1}{2}(1 + \sqrt{1 - 2\varepsilon})$$

or

$$v^{\mathbf{N}} \times \mu(W) < \frac{1}{2}(1 - \sqrt{1 - 2\varepsilon}).$$

The arbitrariness of $\varepsilon > 0$ together with (1.7) yields that $\mu^*(E) = 1$ or 0. Hence G is ergodic.

4) $\Longrightarrow$ 3) . This follows easily from $P\mu^* = v^{\mathbf{N}} \times \mu$ and $P \circ G = F \circ P$.
3) $\Longrightarrow$ 5) . Assume that $\mu = t\mu_1 + (1-t)\mu_2$ with $\mu_1, \mu_2 \in \mathcal{M}(\mathcal{X}^+(M,v))$ and $0 < t < 1$. We have $\mu_1 << \mu$ since $t \neq 0$. Let $g = d\mu_1/d\mu$ be the Radon–Nikodym derivative. Obviously $v^{\mathbf{N}} \times \mu_1 << v^{\mathbf{N}} \times \mu$ and $\tilde{g} = d(v^{\mathbf{N}} \times \mu_1)/d(v^{\mathbf{N}} \times \mu)$ where $\tilde{g} : \Omega^{\mathbf{N}} \times M \to \mathbf{R}^+$ is defined by $\tilde{g}(\omega, x) = g(x)$. Since $v^{\mathbf{N}} \times \mu_1$ and $v^{\mathbf{N}} \times \mu$ are both F–invariant, we have

$$\tilde{g} \circ F = F \quad v^{\mathbf{N}} \times \mu - a.e.$$

and then

$$\tilde{g} = \text{const.} \quad v^{\mathbf{N}} \times \mu - a.e.,$$

thus

$$g = \text{const.} \quad \mu - a.e.$$

So $\mu = \mu_1 = \mu_2$. The proof of 3) $\Longrightarrow$ 5) is completed.
5) $\Longrightarrow$ 1). Suppose that μ is not ergodic, then there exists an $\mathcal{X}^+(M,v,\mu)$–invariant set A_0 with $0 < \mu(A_0) < 1$. Define measures μ_1 and μ_2 by

$$\mu_1(A) = \frac{1}{\mu(A_0)}\mu(A \cap A_0),$$

$$\mu_2(A) = \frac{1}{\mu(A_0^C)}\mu(A \cap A_0^C)$$

for all $A \in \mathcal{B}(M)$. Since for every $A \in \mathcal{B}(M)$

$$\begin{aligned}
\int f\mu_1(A)dv(f) &= \frac{1}{\mu(A_0)} \int \mu(f^{-1}A \cap A_0)dv(f) \\
&= \frac{1}{\mu(A_0)} \int\int \chi_A(f(x))\chi_{A_0}(x)d\mu(x)dv(f) \\
&= \frac{1}{\mu(A_0)} \int\int \chi_A(f(x))\chi_{A_0}(f(x))d\mu(x)dv(f) \\
&= \frac{1}{\mu(A_0)} \int f\mu(A \cap A_0)dv(f) \\
&= \frac{1}{\mu(A_0)}\mu(A \cap A_0) = \mu_1(A)
\end{aligned}$$

and similarly

$$\int f\mu_2(A)dv(f) = \mu_2(A),$$

we know that $\mu_1, \mu_2 \in \mathcal{M}(\mathcal{X}^+(M,v))$. Obviously

$$\mu = \mu(A_0)\mu_1 + \mu(A_0^C)\mu_2,$$

this shows that μ is not extremal. Thus 5) implies 1). □

Finally we touch briefly on the ergodic decomposition of an invariant measure of $\mathcal{X}^+(M,v)$. Let $\mathcal{M}(\mathcal{X}^+(M,v))$ have the smallest measurable structure among those with respect to which every function of the form $\Phi_g(\mu) = \int g d\mu$ on $\mathcal{M}(\mathcal{X}^+(M,v))$ is measurable, where g is a bounded function on M measurable with respect to any $\mathcal{B}_\mu(M)$ which denotes the completion of $\mathcal{B}(M)$ for $\mu \in \mathcal{M}(\mathcal{X}^+(M,v))$. The following theorem is adopted from Appendix A.1 of [Kif]$_1$.

Theorem 1.1. *The set $\mathcal{M}_e$ of all ergodic measures is a measurable subset of $\mathcal{M}(\mathcal{X}^+(M,v))$ and each measure μ from $\mathcal{M}(\mathcal{X}^+(M,v))$ can be uniquely represented as an integral*

$$\mu = \int_{\mathcal{M}_e} \rho d\gamma_\mu \tag{1.9}$$

where γ_μ is a probability measure on $\mathcal{M}(\mathcal{X}^+(M,v))$ concentrated on $\mathcal{M}_e$.

(1.9) means that

$$\mu(A) = \int_{\mathcal{M}_e} \rho(A) d\gamma_\mu(\rho)$$

for all $A \in \mathcal{B}(M)$, or equivalently

$$\int_M g(x) d\mu(x) = \int_{\mathcal{M}_e} \int_M g(x) d\rho(x) d\gamma_\mu(\rho) \tag{1.10}$$

for any bounded measurable function g on $(M, \mathcal{B}(M))$.

Corollary 1.1. *Let $\mu \in \mathcal{M}(\mathcal{X}^+(M,v))$. Then for any Borel function h on $\Omega^{\mathbf{N}} \times M$ with $h^+ \in L^1(\Omega^{\mathbf{N}} \times M, v^{\mathbf{N}} \times \mu)$ and*

$$h \circ F = h \quad v^{\mathbf{N}} \times \mu - a.e. \tag{1.11}$$

we have

$$h(\omega, x) = \int h(\omega, x) dv^{\mathbf{N}}(\omega) \quad v^{\mathbf{N}} \times \mu - a.e.. \tag{1.12}$$

Proof. Let $W_h = \{(\omega, x) : h(\omega, x) \neq \int h(\omega, x) dv^{\mathbf{N}}(\omega)\}$ and $\mu = \int_{\mathcal{M}_e} \rho d\gamma_\mu$ be the ergodic decomposition of μ. By the standard method of measure theory it is easy to show that for any $W \in \mathcal{B}(\Omega^{\mathbf{N}} \times M)$, $v^{\mathbf{N}} \times \rho(W)$ is a measurable function on $\mathcal{M}_e$ and

$$v^{\mathbf{N}} \times \mu(W) = \int_{\mathcal{M}_e} v^{\mathbf{N}} \times \rho(W) d\gamma_\mu(\rho). \tag{1.13}$$

(1.11) and (1.13) yield that for $\gamma_\mu-$ a.e. $\rho \in \mathcal{M}_e$

$$h \circ F = h \quad v^{\mathbf{N}} \times \rho - a.e.$$

and then by the ergodicness of $F : (\Omega^{\mathbf{N}} \times M, v^{\mathbf{N}} \times \rho) \hookleftarrow$

$$v^{\mathbf{N}} \times \rho(W_h) = 0.$$

Hence

$$v^{\mathbf{N}} \times \mu(W_h) = \int_{\mathcal{M}_e} v^{\mathbf{N}} \times \rho(W_h) d\gamma_\mu(\rho) = 0$$

which implies (1.12). □

§2 Measure-Theoretic Entropies of Random Diffeomorphisms

In this section we discuss entropies (measure-theoretic entropies) of random diffeomorphisms. As in ergodic theory of deterministic dynamical systems, the entropy of a system generated by random diffeomorphisms should be a number which describes to some extent the "complexity" of the associated system. In the random case there may be different choices for the definition of the entropy, but in this book we are concerned with the entropy defined by Y. Kifer in [Kif]$_1$ for random transformations. We shall first give a review of this concept and then prove some useful properties of it.

Let $\mathcal{X}^+(M, v, \mu)$ be a system as defined in Section 1.

Theorem 2.1. *For any finite measurable partition ξ of M the limit*

$$h_\mu(\mathcal{X}^+(M,v),\xi) = \lim_{n\to\infty} \frac{1}{n} \int H_\mu\left(\bigvee_{k=0}^{n-1} (f_\omega^k)^{-1}\xi\right) dv^{\mathbf{N}}(\omega) \tag{2.1}$$

exists. This limit is called the entropy of $\mathcal{X}^+(M,v,\mu)$ with respect to ξ.

Proof. Define $g(x) = \begin{cases} 0 & \text{if } \quad x = 0 \\ -x\log x & \text{if } \quad x > 0 \end{cases}$. Let

$$a_n = \int H_\mu\left(\bigvee_{k=0}^{n-1} (f_\omega^k)^{-1}\xi\right) dv^{\mathbf{N}}(\omega),$$

then

$$a_{n+m} = \int H_\mu\left(\bigvee_{k=0}^{n+m-1} (f_\omega^k)^{-1}\xi\right) dv^{\mathbf{N}}(\omega)$$

$$\le \int H_\mu\left(\bigvee_{k=0}^{n-1}(f_\omega^k)^{-1}\xi\right)dv^{\mathbf{N}}(\omega)$$
$$+\int H_\mu\left((f_\omega^n)^{-1}\bigvee_{k=0}^{m-1}(f_{\tau^n\omega}^k)^{-1}\xi\right)dv^{\mathbf{N}}(\omega)$$
$$= a_n+\int\int H_\mu\left((f_\omega^n)^{-1}\bigvee_{k=0}^{m-1}(f_{\omega'}^k)^{-1}\xi\right)dv^{\mathbf{N}}(\omega)dv^{\mathbf{N}}(\omega')$$
$$= a_n+\int\int\sum_{C\in\xi_m(\omega')}g(\mu((f_\omega^n)^{-1}C))dv^{\mathbf{N}}(\omega)dv^{\mathbf{N}}(\omega')$$

where $\xi_m(\omega')=\vee_{k=0}^{m-1}(f_{\omega'}^k)^{-1}\xi$. Since g is concave on $[0,+\infty)$, one has

$$\begin{aligned} a_{n+m} &\le a_n+\int\sum_{C\in\xi_m(\omega')}g\left(\int\mu((f_\omega^n)^{-1}C)dv^{\mathbf{N}}(\omega)\right)dv^{\mathbf{N}}(\omega') \\ &= a_n+\int\sum_{C\in\xi_m(\omega')}g(\mu(C))dv^{\mathbf{N}}(\omega') \\ &= a_n+\int H_\mu\left(\bigvee_{k=0}^{m-1}(f_{\omega'}^k)^{-1}\xi\right)dv^{\mathbf{N}}(\omega') \\ &= a_n+a_m. \end{aligned}$$

Then from the proof of Theorem 0.4.1 we know that $\lim_{n\to+\infty}\frac{1}{n}a_n$ exists and equals $\inf_{n\ge1}\frac{1}{n}a_n$. □

Definition 2.1. $h_\mu(\mathcal{X}^+(M,v))=\sup h_\mu(\mathcal{X}^+(M,v),\xi)$ *is called the entropy of* $\mathcal{X}^+(M,v,\mu)$, *where the supremum is taken over the set of all finite measurable partitions of* M.

From this definition we see that $h_\mu(\mathcal{X}^+(M,v))$ describes to a certain extent the mean "complexity" of the dynamical behaviour of the systems $\{f_\omega^0, f_\omega^1,\cdots\}$, $\omega\in\Omega^{\mathbf{N}}$ and it is a generalization to the random case of the entropy of a measure-preserving transformation in the deterministic ergodic theory.

Now we proceed to prove some properties of the entropy $h_\mu(\mathcal{X}^+(M,v))$. Henceforth in this book, for a given system $\mathcal{X}^+(M,v,\mu)$ the associated σ-algebra on $\Omega^{\mathbf{N}}\times M$ will be the completion of $\mathcal{B}(\Omega^{\mathbf{N}}\times M)$ with respect to $v^{\mathbf{N}}\times\mu$ and that on $\Omega^{\mathbf{Z}}\times M$ will be the completion of $\mathcal{B}(\Omega^{\mathbf{Z}}\times M)$ with respect to μ^*, except mentioned otherwise. We introduce now the space $C^2(M,M)^{\mathbf{N}}\times M$ with the product topology. The general topology tells that it can be metrized to a Polish space. Since $\Omega^{\mathbf{N}}\times M$ is a Borel subset of $C^2(M,M)^{\mathbf{N}}\times M$ and $v^{\mathbf{N}}\times\mu$ can be regarded as a Borel probability measure on $C^2(M,M)^{\mathbf{N}}\times M$ which is supported by $\Omega^{\mathbf{N}}\times M$, then we know that $(\Omega^{\mathbf{N}}\times M, v^{\mathbf{N}}\times\mu)$ is a Lebesgue space (see Theorem 0.1.10). Similarly, $(\Omega^{\mathbf{Z}}\times M,\mu^*)$ is a Lebesgue space as well.

Let $F:\Omega^{\mathbf{N}}\times M\hookleftarrow$ and $G:\Omega^{\mathbf{Z}}\times M\hookleftarrow$ be as defined in Section 1 and let σ_0,σ^+ and σ denote respectively the σ-algebras $\{\Gamma\times M:\Gamma\in\mathcal{B}(\Omega^{\mathbf{N}})\}$,

$\left\{\prod_{-\infty}^{-1} \Omega \times \Gamma \times M : \Gamma \in \mathcal{B}(\prod_0^{+\infty} \Omega)\right\}$ and $\{\Gamma' \times M : \Gamma' \in \mathcal{B}(\Omega^{\mathbf{Z}})\}$. Obviously, in the sense as explained in Section 0.2 σ_0, σ^+ and σ correspond respectively to the measurable partitions $\{\{\omega\} \times M : \omega \in \Omega^{\mathbf{N}}\}$ of $\Omega^{\mathbf{N}} \times M$ and $\left\{\prod_{-\infty}^{-1} \Omega \times \{\omega\} \times M : \omega \in \prod_0^{+\infty} \Omega\right\}$ and $\{\{w\} \times M : w \in \Omega^{\mathbf{Z}}\}$ of $\Omega^{\mathbf{Z}} \times M$. We shall denote these partitions still by the corresponding σ_0, σ^+ and σ.

Theorem 2.2. *The following hold true for* $\mathcal{X}^+(M, v, \mu)$:

1) If $\xi = \{A_1, \cdots, A_r\}$ *is a finite measurable partition of* M *and* $\zeta = \{\Gamma_1, \Gamma_2, \cdots\}$ *is a countable (or finite) measurable partition of* $\Omega^{\mathbf{N}}$, *then*

$$h_\mu(\mathcal{X}^+(M, v), \xi) = h^{\sigma_0}_{v^{\mathbf{N}} \times \mu}(F, \zeta \times \xi) \tag{2.2}$$

where $\zeta \times \xi = \{\Gamma_i \times A_j : \Gamma_i \in \zeta, A_j \in \xi\}$.

2)

$$h_\mu(\mathcal{X}^+(M, v)) = h^{\sigma_0}_{v^{\mathbf{N}} \times \mu}(F). \tag{2.3}$$

Proof. Since F preserves $v^{\mathbf{N}} \times \mu$, $F^{-1}\sigma_0 \subset \sigma_0$ and $H_{v^{\mathbf{N}} \times \mu}(\zeta \times \xi | \sigma_0) = H_\mu(\xi) < +\infty$, the right–hand sides of (2.2) and (2.3) are well–defined (see Chapter 0).

1). Since for each $\omega \in \Omega^{\mathbf{N}}$

$$(v^{\mathbf{N}} \times \mu)_{\{\omega\} \times M} = \mu$$

and

$$(\zeta \times \xi)^n_F|_{\{\omega\} \times M} = \bigvee_{k=0}^{n-1} (f^k_\omega)^{-1}\xi, \quad \forall n \geq 0,$$

where we regard $\{\omega\} \times M$ as M, then

$$H_{v^{\mathbf{N}} \times \mu}((\zeta \times \xi)^n_F | \sigma_0) = \int H_\mu \left(\bigvee_{k=0}^{n-1} (f^k_\omega)^{-1}\xi \right) dv^{\mathbf{N}}(\omega), \forall n \geq 0.$$

This together with (2.1) proves (2.2).

2) From (2.2) it follows obviously that

$$h_\mu(\mathcal{X}^+(M, v)) \leq h^{\sigma_0}_{v^{\mathbf{N}} \times \mu}(F).$$

Then what remains is to show that

$$h_\mu(\mathcal{X}^+(M, v)) \geq h^{\sigma_0}_{v^{\mathbf{N}} \times \mu}(F),$$

and, by (2.2) and Remark 0.4.1, it is sufficient to prove that for every finite measurable partition α of $\Omega^{\mathbf{N}} \times M$ and every $\varepsilon > 0$ there exists a measurable partition β of the type $\zeta \times \xi$ as explained above of $\Omega^{\mathbf{N}} \times M$ such that

$$h^{\sigma_0}_{v^{\mathbf{N}} \times \mu}(F, \alpha) \leq h^{\sigma_0}_{v^{\mathbf{N}} \times \mu}(F, \beta) + \varepsilon.$$

Since M is a compact metric space (with respect to the metric $d(\ ,\ )$ induced by the Riemannian metric on M), one can easily find an increasing sequence of finite measurable partitions $\{\xi_n\}_{n=1}^{+\infty}$ of M such that $\vee_{n=1}^{+\infty}\xi_n$ is the partition of M into single points. Define $\beta_n = \{\Omega^{\mathbf{N}}\} \times \xi_n, n \geq 1$. By (3.8) of Chapter 0 one has as $n \to +\infty$

$$H_{v^{\mathbf{N}}\times\mu}(\alpha|\beta_n \vee \sigma_0) \quad \to 0.$$

This together with 4) of Theorem 0.4.2 yields that when n is large enough

$$\begin{aligned} h^{\sigma_0}_{v^{\mathbf{N}}\times\mu}(F,\alpha) &\leq h^{\sigma_0}_{v^{\mathbf{N}}\times\mu}(F,\beta_n) + H_{v^{\mathbf{N}}\times\mu}(\alpha|\beta_n \vee \sigma_0) \\ &\leq h^{\sigma_0}_{v^{\mathbf{N}}\times\mu}(F,\beta_n) + \varepsilon. \end{aligned}$$

The proof is completed. □

Theorem 2.3. *For $\mathcal{X}^+(M, v, \mu)$ it holds that*

$$h^{\sigma^+}_{\mu^*}(G) = h^{\sigma}_{\mu^*}(G) \tag{2.4}$$

and

$$h^{\sigma_0}_{v^{\mathbf{N}}\times\mu}(F) = h^{\sigma^+}_{\mu^*}(G). \tag{2.5}$$

Proof. Since G is invertible, $G^{-1}\sigma^+ \subset \sigma^+$ and $\bigvee_{n=0}^{+\infty} G^n\sigma^+ = \sigma$, by Theorem 0.4.4 we know that (2.4) holds true.

Let ξ be a finite measurable partition of $\Omega^{\mathbf{N}} \times M$, from the relation $P \circ G = F \circ P$ it follows that for every $n \in \mathbf{N}$

$$\begin{aligned} H_{v^{\mathbf{N}}\times\mu}\left(\bigvee_{k=0}^{n-1} F^{-k}\xi|\sigma_0\right) &= H_{\mu^*}\left(P^{-1}\bigvee_{k=0}^{n-1} F^{-k}\xi|P^{-1}\sigma_0\right) \\ &= H_{\mu^*}\left(\bigvee_{k=0}^{n-1} G^{-k}(P^{-1}\xi)|\sigma^+\right) \end{aligned}$$

and then

$$h^{\sigma_0}_{v^{\mathbf{N}}\times\mu}(F,\xi) = h^{\sigma^+}_{\mu^*}(G, P^{-1}\xi).$$

Therefore

$$h^{\sigma_0}_{v^{\mathbf{N}}\times\mu}(F) \leq h^{\sigma^+}_{\mu^*}(G). \tag{2.6}$$

Let $\{\xi_n\}_{n=1}^{+\infty}$ be an increasing sequence of finite measurable partitions of M such that $\vee_{n=1}^{+\infty}\xi_n$ is the partition of M into single points. Define $\widetilde{\beta}_n = \{\Omega^{\mathbf{Z}}\} \times \xi_n, n \geq 1$ and let $\varepsilon > 0$ be fixed arbitrarily. Since for every finite measurable partition $\widetilde{\alpha}$ of $\Omega^{\mathbf{Z}} \times M$, by 4) of Theorem 0.4.2,

$$h^{\sigma}_{\mu^*}(G,\widetilde{\alpha}) \leq h^{\sigma}_{\mu^*}(G,\widetilde{\beta}_n) + H_{\mu^*}(\widetilde{\alpha}|\widetilde{\beta}_n \vee \sigma)$$

and by (3.8) of Chapter 0

$$H_{\mu^*}(\widetilde{\alpha}|\widetilde{\beta}_n \vee \sigma) \longrightarrow 0 \quad \text{as} \quad n \longrightarrow +\infty,$$

then one has for sufficiently large n

$$h^{\sigma}_{\mu^*}(G) \leq h^{\sigma}_{\mu^*}(G, \widetilde{\beta}_n) + \varepsilon.$$

Let $\beta_n = \{\Omega^{\mathbf{N}}\} \times \xi_n, n \geq 1$. Since $P^{-1}\beta_n = \widetilde{\beta}_n$, one has for every $k \in \mathbf{N}$

$$\begin{aligned} H_{{}_v\mathbf{N}\times\mu}\left(\bigvee_{i=0}^{k-1} F^{-i}\beta_n|\sigma_0\right) &= H_{\mu^*}\left(P^{-1}\bigvee_{i=0}^{k-1} F^{-i}\beta_n|P^{-1}\sigma_0\right) \\ &= H_{\mu^*}\left(\bigvee_{i=0}^{k-1} G^{-i}\widetilde{\beta}_n|\sigma^+\right) \\ &\geq H_{\mu^*}\left(\bigvee_{i=0}^{k-1} G^{-i}\widetilde{\beta}_n|\sigma\right), \end{aligned}$$

hence for sufficiently large n

$$\begin{aligned} h^{\sigma_0}_{{}_v\mathbf{N}\times\mu}(F) &\geq h^{\sigma_0}_{{}_v\mathbf{N}\times\mu}(F, \beta_n) \geq h^{\sigma}_{\mu^*}(G, \widetilde{\beta}_n) \\ &\geq h^{\sigma}_{\mu^*}(G) - \varepsilon. \end{aligned}$$

Since ε is arbitrary, then

$$h^{\sigma_0}_{{}_v\mathbf{N}\times\mu}(F) \geq h^{\sigma}_{\mu^*}(G)$$

which together with (2.4) and (2.6) yields (2.5). □

Theorem 2.4. *If $\xi_1, \xi_2, \cdots$ is a sequence of countable measurable partitions of $\Omega^{\mathbf{N}} \times M$ such that $H_{{}_v\mathbf{N}\times\mu}(\xi_n|\sigma_0) < +\infty$ for all $n \in \mathbf{N}$ and*

$$\lim_{n\to+\infty} \mathit{diam}\ (\xi_n) = 0,$$

then

$$h^{\sigma_0}_{{}_v\mathbf{N}\times\mu}(F) = \lim_{n\to+\infty} h^{\sigma_0}_{{}_v\mathbf{N}\times\mu}(F, \xi_n).$$

Proof. This theorem follows from Theorem 0.4.7. □

Theorem 2.5. *If $\xi_1, \xi_2, \cdots$ is a sequence of finite measurable partitions of M with*

$$\lim_{n\to+\infty} \mathit{diam}(\xi_n) = 0,$$

then

$$h_\mu(\mathcal{X}^+(M,v)) = \lim_{n\to+\infty} h_\mu(\mathcal{X}^+(M,v),\xi_n). \tag{2.7}$$

Proof. Since $\Omega^{\mathbf{N}}$ is a separable metric space, for each ξ_n we can choose a countable measurable partition ζ_n of $\Omega^{\mathbf{N}}$ such that $\text{diam}(\zeta_n) \leq \text{diam}(\xi_n)$. (2.7) follows then from Theorems 2.2 and 2.4 since

$$\lim_{n\to+\infty} \text{diam}(\zeta_n \times \xi_n) = 0$$

and

$$H_{v^{\mathbf{N}}\times\mu}(\zeta_n \times \xi_n|\sigma_0) = H_\mu(\xi_n) < +\infty, \quad \forall n \in \mathbf{N}.$$

The proof is completed. □

Theorem 2.6. *For $\mathcal{X}^+(M,v,\mu)$, let $\mu = \int_{\mathcal{M}_e} \rho d\gamma_\mu(\rho)$ be the ergodic decomposition of μ, then*

$$h^\sigma_{\mu^*}(G) = \int_{\mathcal{M}_e} h^\sigma_{\rho^*}(G) d\gamma_\mu(\rho). \tag{2.8}$$

Proof. According to Appendix A.1 of [Kif]$_1$ and [Roh]$_3$ and (1.13), we know that, if $\mu = \int_{\mathcal{M}_e} \rho d\gamma_\mu(\rho)$ is the ergodic decomposition of μ, then

$$v^{\mathbf{N}} \times \mu = \int_{\mathcal{M}_e} v^{\mathbf{N}} \times \rho d\gamma_\mu(\rho) \tag{2.9}$$

is the ergodic decomposition of $(F, v^{\mathbf{N}} \times \mu)$ and there exists a measurable partition η of $\Omega^{\mathbf{N}} \times M$ fixed under F such that

$$\{v^{\mathbf{N}} \times \rho : \rho \in \mathcal{M}_e\} = \{(v^{\mathbf{N}} \times \mu)_C : C \in \Omega^{\mathbf{N}} \times M/\eta\} \mod 0,$$

where "mod 0" means that there exist a measurable set $\mathcal{M}'_e \subset \mathcal{M}_e$ with $\gamma_\mu(\mathcal{M}'_e) = 1$ and a measurable set $\mathcal{F} \subset \Omega^{\mathbf{N}} \times M/\eta$ with $(v^{\mathbf{N}} \times \mu)_\eta(\mathcal{F}) = 1$ such that

$$\{v^{\mathbf{N}} \times \rho : \rho \in \mathcal{M}'_e\} = \{(v^{\mathbf{N}} \times \mu)_C : C \in \mathcal{F}\}.$$

It is easy to see that $P^{-1}\eta$ is a measurable partition of $\Omega^{\mathbf{Z}} \times M$ fixed under G and

$$\{\rho^* : \rho \in \mathcal{M}_e\} = \{\mu^*_D : D \in \Omega^{\mathbf{Z}} \times M/P^{-1}\eta\} \mod 0.$$

(2.8) is then derived from Theorem 0.5.5. □

§3 Lyapunov Exponents of Random Diffeomorphisms

The linear theory of Lyapunov exponents and the associated nonlinear theory of stable manifolds play a fundamental role in smooth ergodic theory of deterministic dynamical systems. In this section we introduce the concept of Lyapunov exponents of random diffeomorphisms and give some useful properties of these exponents. The theory of stable manifolds of random diffeomorphisms associated with these exponents will be discussed in Chapter III.

Lyapunov exponents of random diffeomorphisms describe the exponential growth rates of the norms of vectors under successive actions of the derivatives of the random diffeomorphisms. The definition of them is based on the following proposition which can be easily derived from Theorem 3.2.

Proposition 3.1. *Let $\mathcal{X}^+(M, v, \mu)$ be given. Then for μ-a.e. $x \in M$ there exist numbers*

$$\lambda^{(1)}(x) < \lambda^{(2)}(x) < \cdots < \lambda^{(r(x))}(x)$$

($\lambda^{(1)}(x)$ may be $\{-\infty\}$) such that for $v^{\mathbf{N}}$-a.e. $\omega \in \Omega^{\mathbf{N}}$ there exists a sequence of linear subspaces of T_xM

$$\{0\} = V_{(\omega,x)}^{(0)} \subset V_{(\omega,x)}^{(1)} \subset \cdots \subset V_{(\omega,x)}^{(r(x))} = T_xM$$

satisfying

$$\lim_{n\to+\infty} \frac{1}{n} \log |T_x f_\omega^n \xi| = \lambda^{(i)}(x),$$

for all $\xi \in V_{(\omega,x)}^{(i)} \backslash V_{(\omega,x)}^{(i-1)}, 1 \le i \le r(x)$. In addition, $\dim V_{(\omega,x)}^{(i)} - dim V_{(\omega,x)}^{(i-1)} \stackrel{def.}{=} m_i(x)$ *depends only on x, $1 \le i \le r(x)$.*

Definition 3.1. *The numbers $\lambda^{(i)}(x) : 1 \le i \le r(x)$ given in Proposition 3.1 are called the Lyapunov exponents of $\mathcal{X}^+(M, v, \mu)$ at x. The number $m_i(x)$ is called the multiplicity of $\lambda^{(i)}(x)$.*

Before proceeding to discuss the properties of such exponents, we first review the well-known subadditive ergodic theorem which was first given by Kingman in [Kin]. It will play a very important role in this and the following chapters.

Theorem 3.1. (Subadditive ergodic theorem) *Let $(X, \mathcal{B}, \mu)$ be a probability space and T a measure-preserving transformation on $(X, \mathcal{B}, \mu)$. Let $\{g_n\}_{n=1}^{+\infty}$ be a sequence of measurable functions $g_n : X \to \mathbf{R} \cup \{-\infty\}$ satisfying the conditions:*

1) Integrability: $g_1^+ \in L^1(X, \mathcal{B}, \mu)$;

2) Subadditivity: $g_{m+n} \le g_m + g_n \circ T^m$ μ-a.e. for all $m, n \ge 1$.

Then there exists a measurable function $g : X \to \mathbf{R} \cup \{-\infty\}$ such that

$$g^+ \in L^1(X, \mathcal{B}, \mu), g \circ T = g \ \mu - a.e., \lim_{n\to+\infty} \frac{1}{n} g_n = g \ \mu - a.e.$$

and

$$\lim_{n\to+\infty}\frac{1}{n}\int g_n d\mu = \inf_n \frac{1}{n}\int g_n d\mu = \int g d\mu.$$

Now we formulate a kind of multiplicative ergodic theorem for random diffeomorphisms (Theorem 3.2) which is a reformulation in the case of $\mathcal{X}^+(M,v,\mu)$ of the Oseledec theorem stated in Appendix 2 of [Kat]. Let us first explain how the Oseledec theorem applies to the case of $\mathcal{X}^+(M,v,\mu)$. Take a system of charts $\{(U_i,\varphi_i)\}_{i=1}^l$ which covers M such that for each $x \in U_i$ there exists an orthonormal basis $\{e_s^{(i)}(x)\}_{s=1}^{m_0}$ of T_xM which depends continuously on $x \in U_i, 1 \le i \le l$. Put $C_{ij} = \{(\omega,x) \in \Omega^{\mathbf{N}} \times M : x \in U_i, f_0(\omega)x \in U_j\}$ for $1 \le i,j \le l$, and by $C_1', C_2', \cdots, C_{l^2}'$ we denote $C_{11},\cdots,C_{1l},\cdots,C_{l1},\cdots,C_{ll}$ respectively. For every C_k', assuming that $C_k' = C_{ij}$, if $(\omega,x) \in C_k'$ then with respect to bases $\{e_s^{(i)}(x)\}_{s=1}^{m_0}$ and $\{e_s^{(j)}(f_0(\omega)x)\}_{s=1}^{m_0}$ the map $T_xf_0(\omega)$ can be expressed as an $m_0 \times m_0$ matrix, written $A_k(\omega,x)$. Then define $A : \Omega^{\mathbf{N}} \times M \to M(m_0,\mathbf{R})$ by

$$A(\omega,x) = \begin{cases} A_1(\omega,x) & \text{if } (\omega,x) \in C_1', \\ A_k(\omega,x) & \text{if } (\omega,x) \in C_k' \setminus \bigcup_{r=1}^{k-1} C_r', \end{cases}$$

where $M(m_0,\mathbf{R})$ denotes the set of all $m_0 \times m_0$ matrices with real entries. Obviously it is a measurable map and by condition (1.1) we have

$$\int \log^+ \|A(\omega,x)\|_0 dv^{\mathbf{N}} \times \mu < +\infty \tag{3.1}$$

where $\|\cdot\|_0$ denotes the usual Eucledean norm. Then we can apply the Oseledec theorem in Appendix 2 of [Kat] together with Proposition 1.1 and Corollary 1.1 to get the following multiplicative ergodic theorem for $\mathcal{X}^+(M,v,\mu)$. Some explainations about it will be given in Remark 3.1.

Theorem 3.2. *For the given system $\mathcal{X}^+(M,v,\mu)$ there exists a Borel set $\Lambda_0 \subset \Omega^{\mathbf{N}} \times M$ with $v^{\mathbf{N}} \times \mu(\Lambda_0) = 1, F\Lambda_0 \subset \Lambda_0$ such that:*

1) For every $(\omega,x) \in \Lambda_0$ there exist a sequence of linear subspaces of T_xM

$$\{0\} = V_{(\omega,x)}^{(0)} \subset V_{(\omega,x)}^{(1)} \subset \cdots \subset V_{(\omega,x)}^{(r(x))} = T_xM \tag{3.2}$$

and numbers

$$\lambda^{(1)}(x) < \lambda^{(2)}(x) < \cdots < \lambda^{(r(x))}(x) \tag{3.3}$$

($\lambda^{(1)}(x)$ may be $\{-\infty\}$), which depend only on x, such that

$$\lim_{n\to+\infty}\frac{1}{n}\log|T_xf_\omega^n\xi| = \lambda^{(i)}(x) \tag{3.4}$$

for all $\xi \in V^{(i)}_{(\omega,x)} \backslash V^{(i-1)}_{(\omega,x)}, 1 \le i \le r(x)$, *and in addition,*

$$\lim_{n\to+\infty} \frac{1}{n} \log |T_x f^n_\omega| = \lambda^{(r(x))}(x), \tag{3.5}$$

$$\lim_{n\to+\infty} \frac{1}{n} \log |\det(T_x f^n_\omega)| = \sum_i \lambda^{(i)}(x) m_i(x) \tag{3.6}$$

where $m_i(x) = \dim V^{(i)}_{(\omega,x)} - \dim V^{(i-1)}_{(\omega,x)}$, *which depends only on* x *as well. Moreover,* $r(x), \lambda^{(i)}(x)$ *and* $V^{(i)}_{(\omega,x)}$ *depend measurably on* $(\omega, x) \in \Lambda_0$ *and*

$$r(f_0(\omega)x) = r(x), \lambda^{(i)}(f_0(\omega)x) = \lambda^{(i)}(x), T_x f_0(\omega) V^{(i)}_{(\omega,x)} = V^{(i)}_{F(\omega,x)} \tag{3.7}$$

for each $(\omega, x) \in \Lambda_0, 1 \le i \le r(x)$.

2) For each $(\omega, x) \in \Lambda_0$, *we introduce*

$$\rho^{(1)}(x) \le \rho^{(2)}(x) \le \cdots \le \rho^{(m_0)}(x)$$

to denote $\lambda^{(1)}(x), \cdots, \lambda^{(1)}(x), \cdots, \lambda^{(i)}(x), \cdots, \lambda^{(i)}(x), \cdots, \lambda^{(r(x))}(x), \cdots, \lambda^{(r(x))}(x)$ *with* $\lambda^{(i)}(x)$ *being repeated* $m_i(x)$ *times. Now, for* $(\omega, x) \in \Lambda_0$, *if* $\{\xi_1, \cdots, \xi_{m_0}\}$ *is a basis of* $T_x M$ *which satisfies*

$$\lim_{n\to+\infty} \frac{1}{n} \log |T_x f^n_\omega \xi_i| = \rho^{(i)}(x)$$

for every $1 \le i \le m_0$, *then for every two non-empty disjoint subsets* $P, Q \subset \{1, \cdots, m_0\}$ *we have*

$$\lim_{n\to+\infty} \frac{1}{n} \log \gamma(T_x f^n_\omega E_P, T_x f^n_\omega E_Q) = 0$$

where E_P *and* E_Q *denote the subspaces of* $T_x M$ *spanned by the vectors* $\{\xi_i\}_{i\in P}$ *and* $\{\xi_j\}_{j\in Q}$ *respectively and* $\gamma(\cdot,\cdot)$ *denotes the angle between the two associated subspaces.*

Remark 3.1. 1) Given $\mathcal{X}^+(M, v, \mu)$, according to the Oseledec theorem in Appendix 2 of [Kat] we can first find a Borel set $\Lambda'_0 \subset \Omega^{\mathbf{N}} \times M$ with $v^{\mathbf{N}} \times \mu(\Lambda'_0) = 1, F\Lambda'_0 \subset \Lambda'_0$ and with the following properties: For every $(\omega, x) \in \Lambda'_0$ there exist measurable (in (ω, x)) linear subspaces of $T_x M$

$$\{0\} = V^{(0)}_{(\omega,x)} \subset V^{(1)}_{(\omega,x)} \subset \cdots \subset V^{(r(\omega,x))}_{(\omega,x)} = T_x M$$

and measurable (in (ω, x)) numbers

$$\lambda^{(1)}(\omega, x) < \lambda^{(2)}(\omega, x) < \cdots < \lambda^{(r(\omega,x))}(\omega, x)$$

such that (3.4)–(3.6) and 2) of Theorem 3.2 hold true with $\lambda^{(i)}(x), 1 \le i \le r(x)$ and $m_i(x)$ being replaced by $\lambda^{(i)}(\omega, x), 1 \le i \le r(\omega, x)$ and $m_i(\omega, x)$ respectively, and that for each $1 \le i \le r(\omega, x)$

$$r(F(\omega, x)) = r(\omega, x), \lambda^{(i)}(F(\omega, x)) = \lambda^{(i)}(\omega, x), T_x f_0(\omega) V^{(i)}_{(\omega,x)} = V^{(i)}_{F(\omega,x)}.$$

According to the subadditive ergodic theorem, from (3.1) and (3.5) it follows that $\lambda^{(r(\omega,x))}(\omega,x)^+ \in L^1(\Omega^{\mathbf{N}} \times M, v^{\mathbf{N}} \times \mu)$, then by Corollary 1.1 we can easily find $\Lambda_0 \subset \Lambda_0'$ satisfying the requirements of Theorem 3.2.

2) For each $1 \le i \le m_0$, the measurability of $\lambda^{(i)}(x)$ and $V_{(\omega,x)}^{(i)}$ on $(\omega,x) \in \Lambda_0$ means that the maps

$$\lambda^{(i)} : \{(\omega,x) \in \Lambda_0 : r(x) \ge i\} \to \mathbf{R}, \quad (\omega,x) \mapsto \lambda^{(i)}(x),$$

$$\mathcal{J}_{ij} : \{(\omega,x) \in \Lambda_0 : r(x) \ge i, \dim V_{(\omega,x)}^{(i)} = j\} \to Gr(M,j), (\omega,x) \mapsto V_{(\omega,x)}^{(i)},$$

$i \le j \le m_0$ are measurable, where $Gr(M,j)$ denotes the Grassman bundle of M with the fibre at $x \in M$ being the Grassman manifold of j-dimensional subspaces of T_xM.

3) Let $x \in M$ and E, E' be two subspaces of T_xM. The *angle* between E and E' is defined by

$$\gamma(E,E') = \inf\{\cos^{-1} | < \xi, \xi' > | : \xi \in E, \xi' \in E', |\xi| = |\xi'| = 1\}$$

where $< \cdot, \cdot >$ is the Riemannian metric on M.

For $x \in M$ and $1 \le p \le m_0$, let $(T_xM)^{\wedge p}$ be the *p-th exterior power space* of T_xM, namely, $(T_xM)^{\wedge p}$ is the linear space of all linear combinations of elements in $\{\xi_1 \wedge \cdots \wedge \xi_p : \xi_i \in T_xM, 1 \le i \le p\}$ in which the following relations hold:

$$(1) \quad \xi_1 \wedge \cdots \wedge (\alpha\xi_i + \beta\xi_i') \wedge \cdots \wedge \xi_p = \alpha\xi_1 \wedge \cdots \wedge \xi_i \wedge \cdots \wedge \xi_p + \beta\xi_1 \wedge \cdots \wedge \xi_i' \wedge \cdots \wedge \xi_p$$

for all $\alpha, \beta \in \mathbf{R}$ and $1 \le i \le p$;

$$(2) \quad \xi_1 \wedge \cdots \wedge \xi_i \wedge \cdots \wedge \xi_j \wedge \cdots \wedge \xi_p = -\xi_1 \wedge \cdots \wedge \xi_j \wedge \cdots \wedge \xi_i \wedge \cdots \wedge \xi_p$$

for all $1 \le i,j \le p$.
Obviously, if $\{\xi_i : 1 \le i \le m_0\}$ is a basis of T_xM then $\{\xi_{i_1} \wedge \cdots \wedge \xi_{i_p} : 1 \le i_1 < \cdots < i_p \le m_0\}$ is a basis of $(T_xM)^{\wedge p}$. Now, if $\{e_i : 1 \le i \le m_0\}$ is an orthonormal basis of T_xM, then by letting

$$< e_{i_1} \wedge \cdots \wedge e_{i_p}, e_{j_1} \wedge \cdots \wedge e_{j_p} > = \begin{cases} 1 & \text{if } (i_1, \cdots, i_p) = (j_1, \cdots, j_p) \\ 0 & \text{otherwise} \end{cases}$$

we can define an inner product $\langle \cdot, \cdot \rangle$ on $(T_xM)^{\wedge p}$, and it is clearly independent of the choice of the orthonormal basis $\{e_i : 1 \le i \le m_0\}$. We shall denote also by $| \cdot |$ the norm on $(T_xM)^{\wedge p}$ induced by this inner product.

If $f : M \to M$ is a C^1 map, we define for $x \in M$ and $1 \le p \le m_0$

$$(T_xf)^{\wedge p} : \ (T_xM)^{\wedge p} \to (T_{f(x)}M)^{\wedge p},$$

$$\xi_1 \wedge \cdots \wedge \xi_p \mapsto (T_xf\xi_1) \wedge \cdots \wedge (T_xf\xi_p)$$

and define

$$|(T_x f)^\wedge| = 1 + \sum_{p=1}^{m_0} |(T_x f)^{\wedge_p}|.$$

Then a reformulation in the case of $\mathcal{X}^+(M, v, \mu)$ of an important conclusion from [Rue]$_1$ gives

Proposition 3.2. *Let $\mathcal{X}^+(M, v, \mu)$ be given. Then we have*

$$\lim_{n\to+\infty} \frac{1}{n} \log |(T_x f_\omega^n)^\wedge| = \sum_i \lambda^{(i)}(x)^+ m_i(x) \quad v^{\mathbf{N}} \times \mu - a.e.$$

and

$$\lim_{n\to+\infty} \frac{1}{n} \int \log |(T_x f_\omega^n)^\wedge| dv^{\mathbf{N}} \times \mu = \int \sum_i \lambda^{(i)}(x)^+ m_i(x) d\mu.$$

Before going further, we first give a lemma which comes from Part III of [Kat] with a slight modification.

Lemma 3.1. *Suppose that $(X, \mathcal{B}, \mu)$ is a probability space and T is a measure-preserving transformation on $(X, \mathcal{B}, \mu)$. Let g be a μ-a.e. positive finite measurable function defined on X such that*

$$\log^+ \frac{g \circ T}{g} \in L^1(X, \mathcal{B}, \mu). \tag{3.8}$$

Then

$$\lim_{n\to+\infty} \frac{1}{n} \log g(T^n x) = 0 \qquad \mu - a.e. \tag{3.9}$$

and

$$\int \log \frac{g \circ T}{g} d\mu = 0. \tag{3.10}$$

If (3.8) is replaced by $\log^- \frac{g \circ T}{g} \in L^1(X, \mathcal{B}, \mu)$ where $\log^- a = \min\{\log a, 0\}$, then (3.9) and (3.10) still hold true.

Proof. Since $\log^+ \frac{g\circ T}{g} \in L^1(X, \mathcal{B}, \mu)$, by the subadditive ergodic theorem the following limit exists μ-a.e.

$$\lim_{n\to+\infty} \frac{1}{n} \sum_{k=0}^{n-1} \log \frac{g \circ T^{k+1}}{g \circ T^k} = \lim_{n\to+\infty} \frac{1}{n} \log \frac{g \circ T^n}{g} \stackrel{def.}{=} h$$

and moreover

$$\int h d\mu = \int \log \frac{g \circ T}{g} d\mu$$

where both sides may be $-\infty$. Since

$$\lim_{n\to+\infty}\frac{1}{n}\log g=0 \quad \mu-a.e.,$$

we have

$$\lim_{n\to+\infty}\frac{1}{n}\log(g\circ T^n)=h \quad \mu-a.e..$$

On the other hand, from $0<g<+\infty$ $\mu-a.e.$ it follows that as $n\to+\infty$

$$\frac{1}{n}\log(g\circ T^n)$$

converges to 0 in measure because T is measure-preserving. Thus there is a sequence $\{n_i : i\in\mathbf{N}\}\subset\mathbf{N}$ such that

$$\lim_{i\to+\infty}\frac{1}{n_i}\log(g\circ T^{n_i})=0 \quad \mu-a.e..$$

This implies $h=0$ $\mu-a.e.$ and proves (3.9) and (3.10). When (3.8) is replaced by $\log^-\frac{g\circ T}{g}\in L^1(X,\mathcal{B},\mu)$, the discussions above remain true with $\log\frac{g\circ T}{g}$ being replaced by $-\log\frac{g\circ T}{g}$ and prove that (3.9) and (3.10) still hold true. □

The following result for ergodic case was given before in [Bax]₁ and Chapter V of [Kif]₁. Now we prove it for the general case. We denote by Leb. the normalized Lebesgue measure on M induced by the Riemannian metric.

Proposition 3.3. *Let $\mathcal{X}^+(M,v,\mu)$ be given. If $\mu<<$ Leb. , then*
1) $\sum_i\lambda^{(i)}(x)m_i(x)\le 0$ $\mu-a.e.$

2) $\sum_i\lambda^{(i)}(x)m_i(x)=0$ $\mu-a.e.$ *if and only if* $f\mu=\mu$ *for* v*-a.e.* $f\in\Omega$.

Proof. 1) In view of (3.7), in order to prove 1) it suffices to show that for any Borel set $\Lambda\subset\Lambda_0$ with $F\Lambda\subset\Lambda$ and $v^{\mathbf{N}}\times\mu(\Lambda)>0$ we have

$$\int_\Lambda\sum_i\lambda^{(i)}(x)m_i(x)dv^{\mathbf{N}}\times\mu\le 0. \tag{3.11}$$

Since for each $(\omega,x)\in\Omega^{\mathbf{N}}\times M$

$$|\det(T_xf_\omega^1)|\le|f_0(\omega)|_{C^1}^{m_0},$$

where $|f|_{C^1}=\max\{|T_xf|: x\in M\}$ for $f\in\Omega$, from condition (1.1) it follows that $\log^+|\det(T_xf_\omega^1)|\in L^1(\Omega^{\mathbf{N}}\times M, v^{\mathbf{N}}\times\mu)$. Since $F|_\Lambda:\Lambda\to\Lambda$ preserves $(v^{\mathbf{N}}\times\mu)|_\Lambda$, by the subadditive ergodic theorem

$$\begin{aligned}\int_\Lambda\log|\det(T_xf_\omega^1)|dv^{\mathbf{N}}\times\mu &= \lim_{n\to+\infty}\frac{1}{n}\int_\Lambda\log|\det(T_xf_\omega^n)|dv^{\mathbf{N}}\times\mu\\ &= \int_\Lambda\sum_i\lambda^{(i)}(x)m_i(x)dv^{\mathbf{N}}\times\mu.\end{aligned}\tag{3.12}$$

If $\int_\Lambda \log^- |\det(T_x f_\omega^1)| dv^{\mathbf{N}} \times \mu = -\infty$, then (3.11) holds true obviously. Now we assume that $\log^- |\det(T_x f_\omega^1)| \in L^1(\Lambda, (v^{\mathbf{N}} \times \mu)|_\Lambda)$.

Denote Leb. by λ and let $q = d\mu/d\lambda$, then

$$\begin{aligned}
1 &= \frac{1}{v^{\mathbf{N}} \times \mu(\Lambda)} \int_\Lambda dv^{\mathbf{N}} \times \mu = \frac{1}{v^{\mathbf{N}} \times \mu(\Lambda)} \int_\Lambda q(x) dv^{\mathbf{N}} \times \lambda \\
&= \frac{1}{v^{\mathbf{N}} \times \mu(\Lambda)} \int_{\Omega^{\mathbf{N}}} \int_{\Lambda_\omega} q(x) d\lambda dv^{\mathbf{N}} \\
&= \frac{1}{v^{\mathbf{N}} \times \mu(\Lambda)} \int_{\Omega^{\mathbf{N}}} \int_{(F^{-1}\Lambda)_\omega} q(f_\omega^1 x) |\det(T_x f_\omega^1)| d\lambda dv^{\mathbf{N}} \\
&= \frac{1}{v^{\mathbf{N}} \times \mu(\Lambda)} \int_{\Omega^{\mathbf{N}}} \int_{(F^{-1}\Lambda)_\omega} \frac{q(f_\omega^1 x)}{q(x)} |\det(T_x f_\omega^1)| d\mu dv^{\mathbf{N}}
\end{aligned}$$

where $\Lambda_\omega = \{x : (\omega, x) \in \Lambda\}$ and $(F^{-1}\Lambda)_\omega$ has the similar meaning. From the concavity of the function $\log x$ it follows that

$$\begin{aligned}
0 &= \log \frac{1}{v^{\mathbf{N}} \times \mu(\Lambda)} \int_{\Omega^{\mathbf{N}}} \int_{(F^{-1}\Lambda)_\omega} \frac{q(f_\omega^1 x)}{q(x)} |\det(T_x f_\omega^1)| d\mu dv^{\mathbf{N}} \\
&\geq \frac{1}{v^{\mathbf{N}} \times \mu(\Lambda)} \int_{\Omega^{\mathbf{N}}} \int_{(F^{-1}\Lambda)_\omega} \left[\log \frac{q(f_\omega^1 x)}{q(x)} + \log |\det(T_x f_\omega^1)|\right] d\mu dv^{\mathbf{N}} \\
&= \frac{1}{v^{\mathbf{N}} \times \mu(\Lambda)} \int_\Lambda \left[\log \frac{q(f_\omega^1 x)}{q(x)} + \log |\det(T_x f_\omega^1)|\right] dv^{\mathbf{N}} \times \mu.
\end{aligned} \tag{3.13}$$

Since

$$\log^+ \frac{q(f_\omega^1 x)}{q(x)} \leq \log^+ \left(\frac{q(f_\omega^1 x)}{q(x)} |\det(T_x f_\omega^1)|\right) + (-\log^- |\det(T_x f_\omega^1)|),$$

we have

$$\log^+ \frac{q(f_\omega^1 x)}{q(x)} \in L^1(\Lambda, (v^{\mathbf{N}} \times \mu)|_\Lambda).$$

Then by Lemma 3.1

$$\int_\Lambda \log \frac{q(f_\omega^1 x)}{q(x)} dv^{\mathbf{N}} \times \mu = 0$$

which together with (3.12) and (3.13) yields (3.11).

2) Since 1) holds true, $\sum_i \lambda^{(i)}(x) m_i(x) = 0$ $\mu-a.e.$ if and only if $\int \sum_i \lambda^{(i)}(x)$ $m_i(x)\, d\mu = 0$. From the proof of 1) we know that this is equivalent to

$$\frac{q(f_\omega^1 x)}{q(x)} |\det(T_x f_\omega^1)| = 1, \quad v^{\mathbf{N}} \times \mu - a.e.$$

which means $f\mu = \mu$ for v–a.e. $f \in \Omega$. □

Chapter II Estimation of Entropy from Above Through Lyapunov Exponents

The relationship between entropy and Lyapunov exponents has been well studied in smooth ergodic theory of deterministic dynamical systems. A well-known theorem in [Rue]$_1$ asserts that, if $f : M \to M$ is a C^1 map and μ is an f-invariant Borel probability measure on M, then

$$h_\mu(f) \leq \int \sum_i \lambda^{(i)}(x)^+ m_i(x) d\mu \tag{0.1}$$

where $\lambda^{(1)}(x) < \lambda^{(2)}(x) < \cdots < \lambda^{(r(x))}(x)$ denote the Lyapunov exponents of f at $x, m_i(x)$, $1 \leq i \leq r(x)$ their multiplicities respectively and $h_\mu(f)$ the usual entropy of the system $f : (M, \mu) \hookleftarrow$. (0.1) is sometimes called *Ruelle's inequality*.

Our purpose in this chapter is to prove that Ruelle's inequality remains true for $\mathcal{X}^+(M, v, \mu)$ generated by random diffeomorphisms as defined in Chapter I, that is, the following result holds true:

Theorem 0.1. *For any system* $\mathcal{X}^+(M, v, \mu)$ *it holds that*

$$h_\mu(\mathcal{X}^+(M, v)) \leq \int \sum_i \lambda^{(i)}(x)^+ m_i(x) d\mu.$$

The Lyapunov exponents $\lambda^{(1)}(x) < \cdots < \lambda^{(r(x))}(x)$ of $\mathcal{X}^+(M, v, \mu)$ at x, their multiplicities $m_i(x), 1 \leq i \leq r(x)$ and the entropy $h_\mu(\mathcal{X}^+(M, v))$ are as introduced in Chapter I.

Now we begin to prove Theorem 0.1 and we shall divide the proof into two sections.

§1 Preliminaries

By the compactness of M there exists $\rho_0 > 0$ such that the exponential map $\exp_x : \{\xi \in T_x M : |\xi| < \rho_0\} \to B(x, \rho_0)$ is a C^∞ diffeomorphism for every $x \in M$, where $B(x, \rho_0) = \{y \in M : d(y, x) < \rho_0\}$. The following lemma is a basic fact of differential geometry.

Lemma 1.1. *For every* $0 < r_0 < \rho_0$ *there exists* $b = b(r_0) > 0$ *such that for every* $x \in M$ *and any* $y, z \in \overline{B(x, r_0)}$

$$b^{-1} d(y, z) \leq |\exp_x^{-1} y - \exp_x^{-1} z| \leq b d(y, z).$$

In order to prove the main lemma (Lemma 1.3) of this section we need to introduce a new and equivalent definition of the C^2-norm for C^2 diffeomorphisms on M.

Let TTM be the tangent space of TM. On M and TM we can introduce respectively the Riemannian metrics Φ and Ψ to be defined in the following two paragraphs.

Take a system of charts $\{(U_i, \varphi_i)\}_{i=1}^{l}$ on M with $\varphi_i(U_i) = B_3^{m_0}(0), 1 \leq i \leq l$ such that $\{V_i = \varphi_i^{-1} B_1^{m_0}(0)\}_{i=1}^{l}$ cover M, where $B_r^{m_0}(0) = \{\xi \in \mathbf{R}^{m_0} : \|\xi\|_0 < r\}$. Let $\{p_i\}_{i=1}^{l}$ be a C^∞ partition of unity on M subordinate to $\{U_i, V_i, \varphi_i\}_{i=1}^{l}$, i.e. all p_i are C^∞ functions defined on M and satisfying $p_i \geq 0$ on $M, p_i > 0$ on V_i , $p_i = 0$ outside $\varphi_i^{-1} B_2^{m_0}(0)$ and $\Sigma_i p_i(x) = 1$ for every $x \in M$. For each $1 \leq i \leq l$ the bilinear form

$$\Phi_i : (\xi, \eta) \mapsto \ll T_x\varphi_i\xi, T_x\varphi_i\eta \gg_0 \quad \text{for} \quad \xi, \eta \in T_xM \quad \text{and} \quad x \in U_i$$

defines a Riemannian metric on U_i, and clearly $p_i\Phi_i$ can be extended to a C^∞ symmetric bilinear form on all of M which vanishes outside $\varphi_i^{-1} B_2^{m_0}(0)$ but is positive definite at every point of V_i. Then it is easy to check that $\Phi = \Sigma_{i=1}^{l} p_i\Phi_i$, defined precisely by

$$\Phi(\xi, \eta) = \sum_{i=1}^{l} p_i(x)\Phi_i(\xi, \eta) \quad \text{if} \quad \xi, \eta \in T_xM, x \in M,$$

is a C^∞ Riemannian metric on M.

Note that $\{(T_{U_i}M, T\varphi_i)\}_{i=1}^{l}$ is a system of charts on TM, and for each $1 \leq i \leq l$ the bilinear form

$$\Psi_i : (\zeta, \rho) \mapsto \ll T_\xi T\varphi_i\zeta, T_\xi T\varphi_i\rho \gg_0 \quad \text{for} \quad \zeta, \rho \in T_\xi TM \quad \text{and} \quad \xi \in T_{U_i}M$$

defines a Riemannian metric on $T_{U_i}M$. Now we define $P_i : TM \to \mathbf{R}, \xi \mapsto p_i(x)$ if $\xi \in T_xM$. Then $P_i\Psi_i$ can be extended to a C^∞ symmetric bilinear form on all of TM which vanishes outside $T_{\varphi_i^{-1} B_2^{m_0}(0)}M$ and is positive definite at every point of $T_{V_i}M$. Define $\Psi = \Sigma_{i=1}^{l} P_i\Psi_i$ by

$$\Psi(\zeta, \rho) = \sum_{i=1}^{l} P_i(\xi)\Psi_i(\zeta, \rho) \quad \text{if} \quad \zeta, \rho \in T_\xi TM, \xi \in TM,$$

then Ψ can be easily verified to be a C^∞ Riemannian metric on TM.

Note that our discussions about Lyapunov exponents and entropy in this book will not depend on the choice of a Riemannian metric on M since any two such metrics are equivalent. Hence we shall assume that the associated Riemannian metric on M is always Φ.

For simplicity of notation we denote by $|\cdot|$ both the norms on TM and TTM induced by Φ and Ψ respectively. Let $d_T(\ ,\ )$ denote the metric on

TM induced by Ψ. Then it is easy to see that for every $x \in M$ and any $\xi \in T_xM$ we have $d_T(\xi, 0) = |\xi|$ where $0 \in T_xM$ is the zero vector.

For $f \in \Omega$, define

$$\|f\|_{C^2} = \sup\{|T_\xi Tf| : \xi \in TM, |\xi| \leq 1\} \tag{1.1}$$

where $T_\xi Tf$ is the tangent map of Tf at $\xi \in TM$. Let $|f|_{C^2}$ be the C^2-norm defined in Section I.1.

Proposition 1.1. *There exists a number $C_0 > 0$ such that:*

1) For every $f \in \Omega$

$$C_0^{-1}|f|_{C^2} \leq \|f\|_{C^2} \leq C_0|f|_{C^2}; \tag{1.2}$$

2) For every $f \in \Omega$

$$\sup\{|T_\xi Tf| : \xi \in TM, |\xi| \leq r\} \leq C_0 r\|f\|_{C^2}, \forall r \in [1, +\infty); \tag{1.3}$$

3) For any $f, g \in \Omega$

$$\|f \circ g\|_{C^2} \leq C_0 \max\{|g|_{C^1}, 1\}\|f\|_{C^2}\|g\|_{C^2} \tag{1.4}$$

where $|g|_{C^1} = \sup_{x \in M} |T_x g|$.

Proof. 1) Let $\{U_i, V_i, \varphi_i\}_{i=1}^l$ be as taken above in the definitions of Φ and Ψ. For each $f \in \Omega$ define

$$|f|'_{C^2} = \sup_{x \in M} \max_{i,j} \left\{ \max_{1 \leq s,k,r \leq m_0} \left\{ \left| \frac{\partial f_{ij}^s}{\partial u_k}(\varphi_i(x)) \right|, \left| \frac{\partial^2 f_{ij}^s}{\partial u_k \partial u_r}(\varphi_i(x)) \right| \right\} \right.$$

$$\left. : x \in V_i \cap f^{-1}V_j \right\}$$

where $(f_{ij}^1(u), \cdots, f_{ij}^{m_0}(u)) = \varphi_j \circ f \circ \varphi_i^{-1}(u), u = (u_1, \cdots, u_{m_0})^\top \in \mathbf{R}^{m_0}$. Then it is easy to verify that $|\cdot|'_{C^2}$ is equivalent to $|\cdot|_{C^2}$.

Since for every $f \in \Omega$, if $x \in V_i \cap f^{-1}V_j$ and $\xi \in T_xM$, the local expression of $T_\xi Tf$ with respect to the charts $(T_{U_i}M, T\varphi_i)$ and $(T_{U_j}M, T\varphi_j)$ is given by the value at the point $T\varphi_i\xi = (u, v) \in \varphi_i(U_i) \times \mathbf{R}^{m_0}$ of the following $2m_0 \times 2m_0$ matrix function:

$$\begin{bmatrix} \frac{\partial f_{ij}^1}{\partial u_1} & \cdots & \frac{\partial f_{ij}^1}{\partial u_{m_0}} & 0 & \cdots & 0 \\ & \cdots & & & \cdots & \\ \frac{\partial f_{ij}^{m_0}}{\partial u_1} & \cdots & \frac{\partial f_{ij}^{m_0}}{\partial u_{m_0}} & 0 & \cdots & 0 \\ \sum_{k=1}^{m_0} \frac{\partial^2 f_{ij}^1}{\partial u_k \partial u_1} v_k & \cdots & \sum_{k=1}^{m_0} \frac{\partial^2 f_{ij}^1}{\partial u_k \partial u_{m_0}} v_k & \frac{\partial f_{ij}^1}{\partial u_1} & \cdots & \frac{\partial f_{ij}^1}{\partial u_{m_0}} \\ & \cdots & & & \cdots & \\ \sum_{k=1}^{m_0} \frac{\partial^2 f_{ij}^{m_0}}{\partial u_k \partial u_1} v_k & \cdots & \sum_{k=1}^{m_0} \frac{\partial^2 f_{ij}^{m_0}}{\partial u_k \partial u_{m_0}} v_k & \frac{\partial f_{ij}^{m_0}}{\partial u_1} & \cdots & \frac{\partial f_{ij}^{m_0}}{\partial u_{m_0}} \end{bmatrix},$$

and since on every $TT\varphi_i(T_{T_{V_i}M}TM) = T\varphi_i(T_{V_i}M) \times \mathbf{R}^{2m_0}$ the norm $|\cdot|$ induced by Ψ is equivalent to the usual Euclidean norm $\|\cdot\|_0$, i.e., there exists $a_i > 0$ such that

$$a_i^{-1}|\eta| \leq \|\eta\|_0 \leq a_i|\eta|, \forall \eta \in TT\varphi_i(T_{T_{V_i}M}TM),$$

we know that $\|\cdot\|_{C^2}$ is equivalent to $|\cdot|'_{C^2}$ and hence to $|\cdot|_{C^2}$. So there exists $C_1 > 0$ satisfying

$$C_1^{-1}|f|_{C^2} \leq \|f\|_{C^2} \leq C_1|f|_{C^2}, \quad \forall f \in \Omega.$$

2) It can also be easily seen from the above discussion that one can find a number $C_2 > 0$ such that for every $f \in \Omega$

$$\sup\{|T_\xi Tf| : \xi \in TM, |\xi| \leq r\} \leq C_2 r\|f\|_{C^2}, \forall r \in [1, +\infty). \tag{1.5}$$

3) For any $f, g \in \Omega$ we have

$$Tg\{\xi \in TM : |\xi| \leq 1\} \subset \{\eta \in TM : |\eta| \leq |g|_{C^1}\}$$

and

$$T_\xi T(f \circ g) = (T_{Tg\xi}Tf) \circ (T_\xi Tg), \forall \xi \in TM,$$

this together with (1.5) yields that

$$\|f \circ g\|_{C^2} \leq C_2 \max\{|g|_{C^1}, 1\}\|f\|_{C^2}\|g\|_{C^2}.$$

Finally, letting $C_0 = \max\{C_1, C_2\}$, we complete the proof. □

Lemma 1.2. *For any given $\mathcal{X}^+(M, v)$ we have*

$$\int \log^+ \|f_\omega^n\|_{C^2} dv^{\mathbf{N}}(\omega) < +\infty, \forall n \in \mathbf{N}.$$

Proof. Let $n \in \mathbf{N}$ be given arbitrarily. By 3) of Proposition 1.1, for each $\omega \in \Omega^{\mathbf{N}}$

$$\|f_\omega^n\|_{C^2} \leq C_0^{n-1} \prod_{k=0}^{n-2} \max\{|f_k(\omega)|_{C^1}, 1\} \prod_{k=0}^{n-1} \|f_k(\omega)\|_{C^2}$$

and therefore

$$\log^+ \|f_\omega^n\|_{C^2} \leq (n-1)\log^+ C_0 + \sum_{k=0}^{n-2} \log^+ |f_k(\omega)|_{C^1} + \sum_{k=0}^{n-1} \log^+ \|f_k(\omega)\|_{C^2}.$$

Then by condition (1.1) in Chapter I and 1) of Proposition 1.1 we have $\log^+ \|f_\omega^n\|_{C^2} \in L^1(\Omega^{\mathbf{N}}, v^{\mathbf{N}})$. The proof is completed. □

Let $a > 0$ be a number given arbitrarily . For each $f \in \Omega$ it can be easily verified that there exists $0 < r < \min\{\frac{\rho_0}{2}, 1\}$ such that for every $x \in M$ the map

$$H_{(f,x)} \stackrel{\text{def}}{=} \exp_{f(x)}^{-1} \circ f \circ \exp_x : \{\xi \in T_xM : |\xi| \le r\}$$
$$\to \{\eta \in T_{f(x)}M : |\eta| \le \tfrac{1}{2}\rho_0\}$$

is well-defined and

$$\sup_{\xi \in T_xM, |\xi| \le r} |T_\xi H_{(f,x)} - T_0 H_{(f,x)}| < a.$$

We denote by $r_a(f)$ the supremum of all possible r as explained above and call this the *relation number* of f with respect to $a > 0$. It is easy to show that $r_a(f)$ as a function of $f \in \Omega$ is lower-semicontinuous and so it is measurable with respect to $\mathcal{B}(\Omega)$.

Let $b_0 > 0$ be a number given by Lemma 1.1 associated with $r_0 = \frac{2}{3}\rho_0$ and let $a_0 = \min\{b_0^{-1}, 1\}$. One can easily check that for any $f \in \Omega$, if $x, y \in M$ satisfy $d(x,y) < r_{a_0}(f)$, then

$$d(f(y), \exp_{f(x)} \circ T_x f \circ \exp_x^{-1} y) \le d(x,y). \tag{1.6}$$

The following lemma will play an important role in the proof of Theorem A.

Lemma 1.3. *For any given* $\mathcal{X}^+(M,v)$ *we have*

$$-\int \log r_a(f_\omega^n) dv^{\mathbf{N}}(\omega) < +\infty \tag{1.7}$$

for all $a > 0$ *and all* $n \in \mathbf{N}$.

Proof. Let (U, φ) be a chart on M. In (U,φ) the system of geodesic equations is equivalent to the following

$$\begin{cases} \dfrac{du_k}{dt} = v_k \\ \dfrac{dv_k}{dt} + \displaystyle\sum_{i,j=1}^{m_0} \Gamma_{ij}^k(u) v_i v_j = 0, \end{cases} \qquad 1 \le k \le m_0, \tag{1.8}$$

where $(u,v) \in T\varphi T_U M = \varphi(U) \times \mathbf{R}^{m_0}$ and $\Gamma_{ij}^k(u), 1 \le i,j,k \le m_0$ are the Christoffel symbols (see, for instance, [Boo]). By $S(u,v,t) = (\alpha(u,v,t), \beta(u,v,t))$ we denote the solution of (1.8) with initial value (u,v) at $t = 0$, then for every $x \in U$ and any $0 < r < \rho_0$ with $B(x,r) \subset U$ the local expression of $\exp_x |_{T_xM(r)}$ in (U,φ) is

$$\exp_x : v \mapsto \alpha(\varphi(x), v, 1), \quad \forall v \in T_x\varphi T_x M(r)$$

where $T_xM(r) = \{\xi \in T_xM : |\xi| < r\}$. By this property of exponential maps and by the definitions of the Riemannian metrics Φ and Ψ one can find numbers $0 < r < \frac{1}{2}\rho_0$ and $A > 0$ such that:

(i) For every $x \in M$, if $\xi_1, \xi_2 \in T_x M(r)$ and $\xi \in T_x M$ with $|\xi| \geq 1$, then

$$|(T_{\xi_1} \exp_x)\xi| \leq A|\xi|$$

and

$$d_T((T_{\xi_1} \exp_x)\xi, (T_{\xi_2} \exp_x)\xi) \leq A|\xi_1 - \xi_2||\xi|,$$

where $T_\zeta \exp_x : T_x M \to T_{\exp_x \zeta} M$ is the derivative of $\exp_x$ at $\zeta \in T_x M(r)$.

(ii) For every $x \in M$, if $y, z \in \overline{B(x,r)}$, then for any $\eta \in T_y M$ and $\zeta \in T_z M$

$$|(T_y \exp_x^{-1})\eta - (T_z \exp_x^{-1})\zeta| \leq (1 + |\eta|)d_T(\eta, \zeta).$$

(iii) For any $t > 0$, if $\eta \in T_y M$ and $\zeta \in T_z M$ with $|\eta| \leq t, |\zeta| \leq t$ and $d(y,z) \leq 2r$, then any piecewise C^1 curve in TM which is from η to ζ and whose length is less than $2d_T(\eta, \zeta)$ lies in $\{\rho \in TM : |\rho| \leq At\}$.

Now for every $f \in \Omega$ the following hold true for any $x \in M$:

(a)

$$H_{(f,x)} = \exp_{f(x)}^{-1} of \circ \exp_x : \quad \left\{\eta \in T_x M : |\eta| \leq \frac{r}{\max\{|f|_{C^1}, 1\}}\right\}$$
$$\to \{\zeta \in T_{f(x)} M : |\zeta| \leq r\}$$

is well-defined.

(b) If $\xi_1, \xi_2 \in \left\{\eta \in T_x M : |\eta| \leq \frac{r}{\max\{|f|_{C^1}, 1\}}\right\}$ and $\xi \in T_x M$ with $1 \leq |\xi| \leq 2$, letting $y = \exp_x \xi_1$ and $z = \exp_x \xi_2$, we have

$$\begin{aligned}
& |T_{\xi_1} H_{(f,x)}\xi - T_{\xi_2} H_{(f,x)}\xi| \\
= \ & |(T_{f(y)} \exp_{f(x)}^{-1}) \circ (T_y f) \circ (T_{\xi_1} \exp_x)\xi - (T_{f(z)} \exp_{f(x)}^{-1}) \circ (T_z f) \circ (T_{\xi_2} \exp_x)\xi| \\
\leq \ & A[1 + |(T_y f) \circ (T_{\xi_1} \exp_x)\xi|]d_T((T_y f) \circ (T_{\xi_1} \exp_x)\xi, (T_z f) \circ (T_{\xi_2} \exp_x)\xi) \\
\leq \ & A[1 + |f|_{C^1} A|\xi|] \sup\{|T_\zeta Tf| : \zeta \in TM, |\zeta| \leq 2A^2\} \\
& \cdot 2d_T((T_{\xi_1} \exp_x)\xi, (T_{\xi_2} \exp_x)\xi) \\
\leq \ & 8C_0 A^5[1 + |f|_{C^1}]||f||_{C^2}|\xi_1 - \xi_2||\xi|.
\end{aligned}$$

From this it follows that

$$|T_{\xi_1} H_{(f,x)} - T_{\xi_2} H_{(f,x)}| \leq 8C_0 A^5[1 + |f|_{C^1}]||f||_{C^2}|\xi_1 - \xi_2|. \tag{1.9}$$

For any given $a > 0$ we have now the estimate

$$r_a(f) \geq \min\left\{\frac{a}{8C_0 A^5(1 + |f|_{C^1})||f||_{C^2}}, \frac{r}{\max\{|f|_{C^1}, 1\}}\right\}$$

which implies

$$-\log r_a(f) \le B + \log^+ |f|_{C^1} + \log^+ \|f\|_{C^2}$$

where B is a number independent of $f \in \Omega$. Lemma 1.3 follows then immediately from Lemma 1.2. □

§2 Proof of Theorem 0.1

The standard knowledge of linear algebra tells us that every $m_0 \times m_0$ invertible matrix A with real entries can be decomposed as $A = Q_1 \Delta Q_2$ where Δ is a diagonal matrix with positive diagonal elements and Q_1, Q_2 are unitary matrics. We denote by $0 < \delta_1(A) \le \delta_2(A) \le \cdots \le \delta_{m_0}(A)$ the diagonal elements of Δ. If $f \in \Omega$ and $x \in M, \delta_i(T_x f), 1 \le i \le m_0$ can be well defined by expressing $T_x f$ as a $m_0 \times m_0$ matrix with respect to an orthonormal basis of $T_x M$ and $T_{f(x)} M$ respectively.

Proof of Theorem 0.1. Fix $n \in \mathbf{N}$ arbitrarily. For any given $\varepsilon > 0$ small enough, define

$$\Gamma_1 = \{\omega \in \Omega^{\mathbf{N}} : r_{a_0}(f_\omega^n) > \varepsilon\},$$

$$\Gamma_k = \{\omega \in \Omega^{\mathbf{N}} : \frac{\varepsilon}{k} < r_{a_0}(f_\omega^n) \le \frac{\varepsilon}{k-1}\}, k = 2, 3, \cdots.$$

For every $k \in \mathbf{N}$, take a maximal $\frac{\varepsilon}{k}$-seperated set E_ε^k of M, i.e., a subset E_ε^k of M such that if $x, y \in E_\varepsilon^k$ with $x \ne y$ then $d(x,y) > \frac{\varepsilon}{k}$ and for any $z \in M$ there exists an element $x \in E_\varepsilon^k$ satisfying $d(x,z) \le \frac{\varepsilon}{k}$. We then define a measurable partition $\alpha_\varepsilon^k = \{\alpha_\varepsilon^k(x) : x \in E_\varepsilon^k\}$ of M such that $\alpha_\varepsilon^k(x) \subset \overline{int(\alpha_\varepsilon^k(x))}$ and $\mathrm{int}(\alpha_\varepsilon^k(x)) = \{y \in M : d(y,x) < d(y,x_i)$ if $x \ne x_i \in E_\varepsilon^k\}$ for every $x \in E_\varepsilon^k$. Furthermore, we take a countable measurable partition $\{\Gamma_{k1}, \Gamma_{k2}, \cdots\}$ of Γ_k satisfying $\mathrm{diam}(\Gamma_{ki}) < \varepsilon$ for all $i \in \mathbf{N}$. This is possible since $\Omega^{\mathbf{N}}$ is a separable metric space.

Put

$$\alpha_\varepsilon = \{\Gamma_{ki} \times \alpha_\varepsilon^k(x) : x \in E_\varepsilon^k, k, i \in \mathbf{N}\}.$$

Obviously it is a countable measurable partition of $\Omega^{\mathbf{N}} \times M$ and $\mathrm{diam}(\alpha_\varepsilon) < \varepsilon$. Now we show that $H_{{}_v\mathbf{N} \times \mu}(\alpha_\varepsilon | \sigma_0) < +\infty$.

In fact,

$$H_{{}_v\mathbf{N} \times \mu}(\alpha_\varepsilon | \sigma_0)$$

$$= \sum_{k=1}^{+\infty} \int_{\Gamma_k} H_\mu(\alpha_\varepsilon^k) dv^{\mathbf{N}}(\omega) \le \sum_{k=1}^{+\infty} v^{\mathbf{N}}(\Gamma_k) \log |\alpha_\varepsilon^k|$$

where $|\alpha_\varepsilon^k|$ denotes the number of elements in α_ε^k. It is easy to show that there exist $C > 0$ and $t_0 > 0$ such that for any $0 < t \leq t_0$ M contains at most $[C(\frac{1}{t})^{m_0}]$ disjoint balls with diameters t where [a] is the integral part of $a > 0$. Hence when ε is small enough we have $|\alpha_\varepsilon^k| < C(\frac{k}{\varepsilon})^{m_0}$. Therefore

$$\begin{aligned} & H_{{}_v\mathbf{N}\times\mu}(\alpha_\varepsilon|\sigma_0) \\ \leq & \sum_{k=1}^{+\infty} {}_v\mathbf{N}(\Gamma_k)\log C(\frac{k}{\varepsilon})^{m_0} = \log C\varepsilon^{-m_0} + \sum_{k=1}^{+\infty} {}_v\mathbf{N}(\Gamma_k)\log k. \end{aligned}$$

Since

$$\sum_{k=1}^{+\infty} {}_v\mathbf{N}(\Gamma_k)\log k \leq \log 2 - \int_{\{\omega: r_{a_0}(f_\omega^n)\leq\varepsilon\}} \log r_{a_0}(f_\omega^n)d{}_v\mathbf{N}, \tag{2.1}$$

by Lemma 1.3 we have $H_{{}_v\mathbf{N}\times\mu}(\alpha_\varepsilon|\sigma_0) < +\infty$.

Then by Theorem 0.4.3, Theorem I.2.4 and 2) of Theorem I.2.2 one has

$$nh_\mu(\mathcal{X}^+(M,v)) = \lim_{\varepsilon\to 0} h^{\sigma_0}_{{}_v\mathbf{N}\times\mu}(F^n,\alpha_\varepsilon). \tag{2.2}$$

From the properties of conditional entropy (see Section 0.3) it follows that

$$\begin{aligned} & H_{{}_v\mathbf{N}\times\mu}\left(\bigvee_{k=0}^{l-1} F^{-kn}\alpha_\varepsilon|\sigma_0\right) \\ = & H_{{}_v\mathbf{N}\times\mu}(\alpha_\varepsilon|\sigma_0) + H_{{}_v\mathbf{N}\times\mu}(F^{-n}\alpha_\varepsilon|\alpha_\varepsilon\vee\sigma_0) + \cdots \\ & + H_{{}_v\mathbf{N}\times\mu}\left(F^{-(l-1)n}\alpha_\varepsilon|\bigvee_{k=0}^{l-2} F^{-kn}\alpha_\varepsilon\vee\sigma_0\right) \\ \leq & H_{{}_v\mathbf{N}\times\mu}(\alpha_\varepsilon|\sigma_0) + (l-1)H_{{}_v\mathbf{N}\times\mu}(F^{-n}\alpha_\varepsilon|\alpha_\varepsilon\vee\sigma_0). \end{aligned} \tag{2.3}$$

Define $g(x) = \begin{cases} 0 & \text{if } x = 0 \\ -x\log x & \text{if } x > 0 \end{cases}$, by (2.3) we have

$$\begin{aligned} & h^{\sigma_0}_{{}_v\mathbf{N}\times\mu}(F^n,\alpha_\varepsilon) \\ \leq & H_{{}_v\mathbf{N}\times\mu}(F^{-n}\alpha_\varepsilon|\alpha_\varepsilon\vee\sigma_0) \\ = & \sum_{k,i}\sum_{l,j}\int_{\Gamma_{ki}\cap\tau^{-n}\Gamma_{lj}} \sum_{x\in E_\varepsilon^k}\mu(\alpha_\varepsilon^k(x))\sum_{y\in E_\varepsilon^l} g\left(\frac{\mu((f_\omega^n)^{-1}\alpha_\varepsilon^l(y)\cap\alpha_\varepsilon^k(x))}{\mu(\alpha_\varepsilon^k(x))}\right)d{}_v\mathbf{N}(\omega) \\ \leq & \sum_k\sum_l\sum_{x\in E_\varepsilon^k}\int_{\Gamma_k\cap\tau^{-n}\Gamma_l} \mu(\alpha_\varepsilon^k(x))\log N_x^{n,k,l}(\omega)d{}_v\mathbf{N}(\omega) \end{aligned}$$

where $N_x^{n,k,l}(\omega)$ is the number of elements of α_ε^l which intersect $f_\omega^n\alpha_\varepsilon^k(x)$. By (1.6)

$$f_\omega^n\alpha_\varepsilon^k(x) \subset f_\omega^n \exp_x \overline{B(0,\frac{\varepsilon}{k})} \subset \exp_{f_\omega^n x} B(T_x f_\omega^n B(0,\frac{\varepsilon}{k}), b_0\frac{\varepsilon}{k})$$

where $B(Q,\delta)$ is the δ–neighbourhood of $Q \subset T_{f_\omega^n x}M$ in $T_{f_\omega^n x}M$. If

$$\alpha_\varepsilon^l(y) \cap f_\omega^n \alpha_\varepsilon^k(x) \neq \phi,$$

we have

$$B(y, \frac{\varepsilon}{2l}) \cap \exp_{f_\omega^n x} B(T_x f_\omega^n B(0, \frac{\varepsilon}{k}), b_0 \frac{\varepsilon}{k} + b_0 \frac{\varepsilon}{l}) \neq \phi.$$

Then

$$B(\exp_{f_\omega^n x}^{-1} y, b_0^{-1} \frac{\varepsilon}{2l}) \cap B(T_x f_\omega^n B(0, \frac{\varepsilon}{k}), b_0 \frac{\varepsilon}{k} + b_0 \frac{\varepsilon}{l}) \neq \phi.$$

Since unitary operators preserve distances, it is easy to verify that the number of disjoint balls which intersect $B(T_x f_\omega^n B(0, \frac{\varepsilon}{k}), b_0 \frac{\varepsilon}{k} + b_0 \frac{\varepsilon}{l})$ and whose diameters are $b_0^{-1} \frac{\varepsilon}{l}$ does not exceed $C_1 \prod_{i=1}^{m_0} \max\{\delta_i(T_x f_\omega^n), 1\} \max\{\frac{l}{k}, 1\}$ where C_1 is a constant depending only on b_0 and m_0. Therefore

$$\begin{aligned}
& h^{\sigma_0}_{{}_v\mathbf{N} \times \mu}(F^n, \alpha_\varepsilon) \\
\leq\; & \sum_k \sum_l \sum_{x \in E_\varepsilon^k} \int_{\Gamma_k \cap \tau^{-n}\Gamma_l} \mu(\alpha_\varepsilon^k(x)) \log \Bigg[C_1 \prod_{i=1}^{m_0} \max\{\delta_i(T_x f_\omega^n), 1\} \\
& \qquad \max\{\frac{l}{k}, 1\} \Bigg] dv^{\mathbf{N}}(\omega) \\
\leq\; & \log C_1 + \sum_k \sum_l v^{\mathbf{N}}(\Gamma_k \cap \tau^{-n}\Gamma_l) \log \max\{\frac{l}{k}, 1\} \\
& + \sum_{i=1}^{m_0} \int H_\varepsilon^{(i)}(\omega, y) dv^{\mathbf{N}} \times \mu \\
\stackrel{def.}{=}\; & \log C_1 + \Delta_1 + \Delta_2
\end{aligned}$$

where $H_\varepsilon^{(i)}(\omega, y) = \log^+ \delta_i(T_x f_\omega^n)$ if $(\omega, y) \in \Gamma_k \times \alpha_\varepsilon^k(x)$. From the definition of $\Gamma_k, k \in \mathbf{N}$ one can deduce that for any $k, l \in \mathbf{N}$

$$v^{\mathbf{N}}(\Gamma_k \cap \tau^{-n}\Gamma_l) = v^{\mathbf{N}}(\Gamma_k) v^{\mathbf{N}}(\Gamma_l),$$

this together with (2.1) yields

$$\begin{aligned}
\Delta_1 \;& \leq \sum_k \sum_l v^{\mathbf{N}}(\Gamma_k) v^{\mathbf{N}}(\Gamma_l)(\log k + \log l) \\
& = 2 \sum_k v^{\mathbf{N}}(\Gamma_k) \log k \\
& \leq 2 \log 2 - 2 \int_{\{\omega : r_{a_0}(f_\omega^n) \leq \varepsilon\}} \log r_{a_0}(f_\omega^n) dv^{\mathbf{N}}.
\end{aligned}$$

By Lemma 1.3 one has

$$\limsup_{n \to +\infty} \Delta_1 \leq 2 \log 2. \tag{2.4}$$

As for Δ_2, since for every $1 \le i \le m_0$

$$|H_\varepsilon^{(i)}(\omega, y)| \le \sup_{x \in M} \log^+ \delta_1(T_x f_\omega^n) = \log^+ |f_\omega^n|_{C^1}$$

and

$$\lim_{\varepsilon \to 0} H_\varepsilon^{(i)}(\omega, y) = \log^+ \delta_i(T_y f_\omega^n)$$

for any $(\omega, y) \in \Omega^{\mathbf{N}} \times M$, by the dominated convergence theorem

$$\lim_{\varepsilon \to 0} \Delta_2 = \sum_{i=1}^{m_0} \int \log^+ \delta_i(T_y f_\omega^n) dv^{\mathbf{N}} \times \mu. \tag{2.5}$$

Since for every $(\omega, y) \in \Omega^{\mathbf{N}} \times M$

$$|(T_y f_\omega^n)^{\wedge p}| = \prod_{i=m_0-p+1}^{m_0} \delta_i(T_y f_\omega^n), \quad 1 \le p \le m_0,$$

one has

$$|(T_y f_\omega^n)^\wedge| \ge \prod_{i=1}^{m_0} \max\{\delta_i(T_y f_\omega^n), 1\}. \tag{2.6}$$

Letting $\varepsilon \to 0$, by (2.2)–(2.6) we get

$$nh_\mu(\mathcal{X}^+(M, v)) \le \log C_1 + 2 \log 2 + \int \log |(T_y f_\omega^n)^\wedge| dv^{\mathbf{N}} \times \mu,$$

then by Proposition I.3.2

$$h_\mu(\mathcal{X}^+(M, v)) \le \int \sum_i \lambda^{(i)}(x)^+ m_i(x) d\mu. \qquad \square$$

Remark 2.1. Ruelle's inequality for random diffeomorphisms was first considered by Y.Kifer for ergodic case (Theorem V.1.4 of [Kif]$_1$), but his proof of that theorem seems to be unacceptable. For example, the key estimate (1.23) on Page 164 of [Kif]$_1$ seems to be a nontrivial mistake. This led the authors of the present book to an essentially different treatment of this problem in the paper [Liu]$_1$ and this chapter comes from that paper. In our treatment here we make the assumption $\int \log^+ |f|_{C^2} dv(f) < +\infty$ rather than $\int \log^+ |f|_{C^1} dv(f) < +\infty$, which was set down in Kifer's theorem. A recent preprint [Bah] shows that our this assumption is extraneous and Theorem 0.1 also holds true under Kifer's assumption (see the Appendix for the argument). However, the treatment in this chapter (especially the introduction of the relation number $r_a(f)$ and the related estimates) will be, besides its own right, very useful in the later chapters.

Chapter III Stable Invariant Manifolds of Random Diffeomorphisms

In the development of smooth ergodic theory of deterministic dynamical systems, one of the remarkable landmarks is Ya. Pesin's work $[\mathrm{Pes}]_1$ which translated the linear theory of Lyapunov exponents into the non-linear theory of stable and unstable invariant manifolds. Pesin developed there a general theory concerning the existence and the so-called "absolute continuity" of invariant families of stable and unstable manifolds of a smooth dynamical system, corresponding to its non-zero Lyapunov exponents, and thus paved the way to deep results in ergodic theory of arbitrary diffeomorphisms preserving a smooth measure ($[\mathrm{Pes}]_2$). A theorem concerning the existence of such families was proved later by D. Ruelle ($[\mathrm{Rue}]_2$) for dynamical systems preserving only a Borel measure, through a rather different and in some sense more profitable approach. Pesin's above work has also been generalized to a broad class of dynamical systems with singularities ([Kat]) along his original scheme but in a technically much more detailed way.

The purpose of this chapter is to present an extension of Pesin's work $[\mathrm{Pes}]_1$ to the random case, i.e. to carry out along Pesin's scheme some results concerning the existence and absolute continuity of invariant families of stable manifolds for random diffeomorphisms. Let us emphasize that Ruelle's theorem mentioned above can also be adopted to our present situation by a trivialization argument to obtain the existence of the stable manifolds. However, in this chapter we will prove some more subtle results, which besides their own rights are necessary for the treatments of Chapters IV, VI and VII, and deal with the absolute continuity problem. Although we will not provide here the detailed proof of our absolute continuity theorem (Theorem 5.1) because it involves too much work and is completely parallel to the treatment in Part II of [Kat] of the deterministic case, we will present in Section 4 a detailed discussion of the Hölder continuity of tangent spaces of the random stable manifolds and this will naturally lead the reader to the absolute continuity theorem. Results of this chapter will play a fundamental role in Chapters IV, VI and VII.

§1 Some Preliminary Lemmas

Let $\mathcal{X}^+(M,v,\mu)$ be given as in Chapter I, and let $[a,b], a < b \leq 0$, be a closed interval of $\mathbf{R}$. Denote by $\Lambda_{a,b}$ the subset of Λ_0 (see Theorem I.3.2) which consists of points (ω,x) such that $\lambda^{(i)}(x) \notin [a,b], i = 1,\cdots,r(x)$. It is clear that $F\Lambda_{a,b} \subset \Lambda_{a,b}$. For $(\omega,x) \in \Lambda_{a,b}$ and $n,l \in \mathbf{Z}^+$ we sometimes use the following notations:

$$E_0(\omega,x) = \cup_{\lambda^{(i)}(x)<a} V^{(i)}_{(\omega,x)}, \qquad H_0(\omega,x) = E_0(\omega,x)^{\perp};$$

$$E_n(\omega,x) = T_x f_\omega^n E_0(\omega,x), \qquad H_n(\omega,x) = T_x f_\omega^n H_0(\omega,x), \quad n>0;$$

$$f_n^0(\omega) = id, \qquad f_n^l(\omega) = f_{n+l-1}(\omega)\circ\cdots\circ f_n(\omega), \quad l>0;$$

$$T_n^l(\omega,x) = T_{f_\omega^n x} f_n^l(\omega), \quad S_n^l(\omega,x) = T_n^l(\omega,x)|_{E_n(\omega,x)},$$

$$U_n^l(\omega,x) = T_n^l(\omega,x)|_{H_n(\omega,x)}.$$

We now fix arbitrarily $k \in \{1,\cdots,m_0\}$ and $0<\varepsilon\le\min\{1,(b-a)/200\}$ and assume that the set $\Lambda_{a,b,k} \stackrel{\text{def}}{=} \{(\omega,x)\in\Lambda_{a,b} : \dim E_0(\omega,x)=k\} \neq \phi$.

Lemma 1.1. *There exists a measurable function $l : \Lambda_{a,b,k}\times \mathbf{Z}^+ \to (0,+\infty)$ such that for each $(\omega,x)\in\Lambda_{a,b,k}$ and $n,l\in\mathbf{Z}^+$ we have*

1) $|S_n^l(\omega,x)\xi| \le l(\omega,x,n)e^{(a+\varepsilon)l}|\xi|, \xi\in E_n(\omega,x)$;

2) $|U_n^l(\omega,x)\eta| \ge l(\omega,x,n)^{-1}e^{(b-\varepsilon)l}|\eta|, \eta\in H_n(\omega,x)$;

3) $\gamma(E_{n+l}(\omega,x), H_{n+l}(\omega,x)) \ge l(\omega,x,n)^{-1}e^{-\varepsilon l}$;

4) $l(\omega,x,n+l)\le l(\omega,x,n)e^{\varepsilon l}$.

Proof. Let (ω,x) be a point in $\Lambda_{a,b,k}$ and let $n,l\in\mathbf{Z}^+$. We choose a basis $\{\xi_i\}_{i=1}^{m_0}$ of T_xM such that $\{\xi_i\}_{i=1}^k \subset E_0(\omega,x)$, $\{\xi_j\}_{j=k+1}^{m_0}\subset H_0(\omega,x)$ and for each ξ_i

$$\lim_{m\to+\infty}\frac{1}{m}\log|T_xf_\omega^m\xi_i| = \rho^{(i)}(x) \tag{1.1}$$

where $\rho^{(i)}(x), 1\le i\le m_0$ are as introduced in Theorem I.3.2. According to Theorem I.3.2, one has for every two non-empty disjoint subsets $P,Q\subset\{1,\cdots,m_0\}$

$$\lim_{m\to+\infty}\frac{1}{m}\log\gamma(T_xf_\omega^mE_P, T_xf_\omega^mE_Q) = 0$$

where E_P and E_Q denote respectively the subspaces of T_xM spanned by $\{\xi_i\}_{i\in P}$ and $\{\xi_j\}_{j\in Q}$. From this it follows that

$$A(\omega,x,n) \stackrel{\text{def}}{=} \inf_{P,Q}\inf_{r\in\mathbf{Z}^+}\gamma(T_xf_\omega^{n+r}E_P, T_xf_\omega^{n+r}E_Q)e^{\frac{\varepsilon}{2m_0}r} > 0, \tag{1.2}$$

and

$$A(\omega,x,n+l) \ge A(\omega,x,n)e^{-\frac{\varepsilon}{2m_0}l}. \tag{1.3}$$

Particularly, we define

$$l_1(\omega, x, n) = \inf_{r \in \mathbf{Z}^+} \gamma(E_{n+r}(\omega, x), H_{n+r}(\omega, x)) e^{\frac{\varepsilon}{2m_0} r},$$

then it is an everywhere positive measurable function on $\Lambda_{a,b,k} \times \mathbf{Z}^+$.

Let $m \in \mathbf{Z}^+$. For each $\xi = \sum_i \alpha_i T_x f_\omega^m \xi_i \in T_{f_\omega^m x} M$, from (1.2) one has

$$\begin{aligned}
|\xi| = & \left| \sum_i \alpha_i T_x f_\omega^m \xi_i \right| \leq \sum_i |\alpha_i| |T_x f_\omega^m \xi_i| \\
\leq & \ [4A(\omega, x, m)^{-1}]^{m_0} \left| \sum_i \alpha_i T_x f_\omega^m \xi_i \right| \\
= & \ B(\omega, x, m) |\xi|
\end{aligned} \tag{1.4}$$

where $B(\omega, x, m) = [4A(\omega, x, m)^{-1}]^{m_0}$. Indeed, let E be a vector space with an inner product $\langle \ , \ \rangle$, and let $\|\cdot\|$ be the norm deduced from $\langle \ , \ \rangle$. If $\eta, \zeta \in E$ satisfy $\gamma(\eta, \pm\zeta) \geq q^{-1}$, an simple geometrical consideration shows that $\|\eta\| + \|\xi\| \leq 4q\|\eta + \xi\|$. Then an easy induction yields (1.4).

It follows from (1.1) that there exists $C(\omega, x, n) > 0$ such that for each ξ_i and $r \in \mathbf{Z}^+$

$$C(\omega, x, n)^{-1} e^{(\rho^{(i)}(x) - \frac{\varepsilon}{4})r} \leq |T_x f_\omega^{n+r} \xi_i| \leq C(\omega, x, n) e^{(\rho^{(i)}(x) + \frac{\varepsilon}{4})r}. \tag{1.5}$$

By these inequalities a simple computation yields that for any $r, s \in \mathbf{Z}^+$ we have

$$|T_x f^{n+r+s} \xi_i| \leq C(\omega, x, n)^2 |T_x f_\omega^{n+r} \xi_i| e^{(a + \frac{\varepsilon}{4})s + \frac{\varepsilon}{2} r} \tag{1.6}$$

for each $\xi_i \in E_0(\omega, x)$ and

$$|T_x f_\omega^{n+r} \xi_j| \leq C(\omega, x, n)^2 |T_x f_\omega^{n+r+s} \xi_j| e^{-(b - \frac{\varepsilon}{4})s + \frac{\varepsilon}{2} r} \tag{1.7}$$

for each $\xi_j \in H_0(\omega, x)$. From (1.4) and (1.6) it follows that for each $\xi = \sum_i \alpha_i T_x f_\omega^n \xi_i \in E_n(\omega, x)$ and any $r, s \in \mathbf{Z}^+$

$$\begin{aligned}
|T_n^{r+s}(\omega, x)\xi| = & \left| \sum_i \alpha_i T_x f_\omega^{n+r+s} \xi_i \right| \leq \sum_i |\alpha_i| |T_x f_\omega^{n+r+s} \xi_i| \\
\leq & \left(\sum_i |\alpha_i| |T_x f_\omega^{n+r} \xi_i| \right) C(\omega, x, n)^2 e^{(a + \frac{\varepsilon}{4})s + \frac{\varepsilon}{2} r} \\
\leq & \ |T_n^r(\omega, x)\xi| [4^{m_0} A(\omega, x, n)^{-m_0} C(\omega, x, n)^2] e^{(a + \frac{\varepsilon}{4})s + \varepsilon r}.
\end{aligned}$$

Thus the function

$$l_2(\omega, x, n) = \sup \left\{ \frac{|T_n^{r+s}(\omega, x)\xi|}{|T_n^r(\omega, x)\xi|} e^{-(a + \frac{\varepsilon}{4})s - \varepsilon r} : r, s \in \mathbf{Z}^+, 0 \neq \xi \in E_n(\omega, x) \right\}$$

is finite at each point of $\Lambda_{a,b,k}$. By the same way one can show that the function

$$l_3(\omega,x,n)=\sup\left\{\frac{|T_n^r(\omega,x)\eta|}{|T_n^{r+s}(\omega,x)\eta|}e^{(b-\frac{3\varepsilon}{4})s-\varepsilon r}: r,s\in\mathbf{Z}^+, 0\neq\eta\in H_n(\omega,x)\right\}$$

is also finite at each point of $\Lambda_{a,b,k}$. Finally, we define

$$l(\omega,x,n)=\max\{l_1(\omega,x,n)^{-1}, l_2(\omega,x,n), l_3(\omega,x,n)\}.$$

Then it is easy to verify that it is a measurable function on $\Lambda_{a,b,k}\times\mathbf{Z}^+$ satisfying 1) –4). □

Let $l'\geq 1$ be a number such that the set $\Lambda^{l'}_{a,b,k,\varepsilon}\stackrel{\text{def}}{=}\{(\omega,x)\in\Lambda_{a,b,k}: l(\omega,x,0)\leq l'\}\neq\phi$.

Lemma 1.2. *$E_0(\omega,x)$ and $H_0(\omega,x)$ depend continuously on $(\omega,x)\in\Lambda^{l'}_{a,b,k,\varepsilon}$.*

Proof. If $\{(\omega_n,x_n)\}_{n=1}^{+\infty}$ is a sequence of points in $\Lambda^{l'}_{a,b,k,\varepsilon}$ such that (ω_n,x_n) converges to $(\omega,x)\in\Lambda^{l'}_{a,b,k,\varepsilon}$ and $E_0(\omega_n,x_n)$ converges to a subspace of T_xM as $n\to+\infty$, by 1) and 2) of Lemma 1.1 one can easily verify that this subspace coincides with $E_0(\omega,x)$. From this the continuity of $E_0(\omega,x)$, and also of $H_0(\omega,x)$, with respect to $(\omega,x)\in\Lambda^{l'}_{a,b,k,\varepsilon}$ follows obviously. □

Let $(\omega,x)\in\Lambda^{l'}_{a,b,k,\varepsilon}$ and $n\in\mathbf{Z}^+$. Lemma 1.1 allows us to introduce an inner product $\langle\ ,\ \rangle_{(\omega,x),n}$ on $T_{f^n_\omega x}M$ such that

$$\langle\xi,\xi'\rangle_{(\omega,x),n}=\sum_{l=0}^{+\infty}e^{-2(a+2\varepsilon)l}\langle S_n^l(\omega,x)\xi, S_n^l(\omega,x)\xi'\rangle,\quad \xi,\xi'\in E_n(\omega,x),\tag{1.8}$$

$$\langle\eta,\eta'\rangle_{(\omega,x),n}=\sum_{l=0}^{n}e^{2(b-2\varepsilon)l}\langle[U^l_{n-l}(\omega,x)]^{-1}\eta,[U^l_{n-l}(\omega,x)]^{-1}\eta'\rangle,$$
$$\eta,\eta'\in H_n(\omega,x),\tag{1.9}$$

and $E_n(\omega,x)$ and $H_n(\omega,x)$ are orthogonal with respect to $\langle\ ,\ \rangle_{(\omega,x),n}$. Then we define a norm $\|\cdot\|_{(\omega,x),n}$ on $T_{f^n_\omega x}M$ such that

$$\|\xi\|_{(\omega,x),n}=[\langle\xi,\xi\rangle_{(\omega,x),n}]^{\frac{1}{2}},\xi\in E_n(\omega,x),\tag{1.10}$$

$$\|\eta\|_{(\omega,x),n}=[\langle\eta,\eta\rangle_{(\omega,x),n}]^{\frac{1}{2}},\eta\in H_n(\omega,x),\tag{1.11}$$

$$\|\zeta\|_{(\omega,x),n}=\max\{\|\xi\|_{(\omega,x),n},\|\eta\|_{(\omega,x),n}\},$$
$$\zeta=\xi+\eta\in E_n(\omega,x)\oplus H_n(\omega,x).\tag{1.12}$$

The sequence of norms $\{\|\cdot\|_{(\omega,x),n}\}_{n=0}^{+\infty}$ is usually called a *Lyapunov metric* at the point (ω, x). It follows from Lemma 1.2 and (1.8), (1.9) that for each fixed $n \in \mathbf{Z}^+$ the inner product $\langle\ ,\ \rangle_{(\omega,x),n}$ depends continuously on $(\omega, x) \in \Lambda_{a,b,k,\varepsilon}^{l'}$.

Lemma 1.3. *Let $(\omega, x) \in \Lambda_{a,b,k,\varepsilon}^{l'}$. Then the sequence of norms $\{\|\cdot\|_{(\omega,x),n}\}_{n=0}^{+\infty}$ satisfies for each $n \in \mathbf{Z}^+$*

1) $\|S_n^1(\omega, x)\xi\|_{(\omega,x),n+1} \le e^{a+2\varepsilon}\|\xi\|_{(\omega,x),n}$, $\quad \xi \in E_n(\omega, x)$;

2) $\|U_n^1(\omega, x)\eta\|_{(\omega,x),n+1} \ge e^{b-2\varepsilon}\|\eta\|_{(\omega,x),n}$, $\quad \eta \in H_n(\omega, x)$;

3) $\frac{1}{2}|\zeta| \le \|\zeta\|_{(\omega,x),n} \le Ae^{2\varepsilon n}|\zeta|, \zeta \in T_{f_\omega^n x}M$, where $A = 4(l')^2(1-e^{-2\varepsilon})^{-\frac{1}{2}}$.

Proof. For each $\xi \in E_n(\omega, x)$ we have

$$\begin{aligned}
\|S_n^1(\omega, x)\xi\|_{(\omega,x),n+1} &= \left[\sum_{l=0}^{+\infty} e^{-2(a+2\varepsilon)l}|S_{n+1}^l(\omega, x)S_n^1(\omega, x)\xi|^2\right]^{\frac{1}{2}} \\
&= e^{a+2\varepsilon}\left[\sum_{l=0}^{+\infty} e^{-2(a+2\varepsilon)(l+1)}|S_n^{l+1}(\omega, x)\xi|^2\right]^{\frac{1}{2}} \\
&\le e^{a+2\varepsilon}\|\xi\|_{(\omega,x),n}
\end{aligned}$$

and hence 1) holds. Similarly for each $\eta \in H_n(\omega, x)$

$$\begin{aligned}
&\|U_n^1(\omega, x)\eta\|_{(\omega,x),n+1} \\
&= \left[\sum_{l=0}^{n+1} e^{2(b-2\varepsilon)l}|[U_{n+1-l}^l(\omega, x)]^{-1}U_n^1(\omega, x)\eta|^2\right]^{\frac{1}{2}} \\
&= \left[|U_n^1(\omega, x)\eta|^2 + \sum_{l=1}^{n+1} e^{2(b-2\varepsilon)l}|[U_{n-(l-1)}^{l-1}(\omega, x)]^{-1}\eta|^2\right]^{\frac{1}{2}} \\
&= \left[|U_n^1(\omega, x)\eta|^2 + e^{2(b-2\varepsilon)}\sum_{l=1}^{n+1} e^{2(b-2\varepsilon)(l-1)}|[U_{n-(l-1)}^{l-1}(\omega, x)]^{-1}\eta|^2\right]^{\frac{1}{2}} \\
&\ge e^{b-2\varepsilon}\|\eta\|_{(\omega,x),n}
\end{aligned}$$

which proves 2). Now let $\zeta \in T_{f_\omega^n x}M$ and write $\zeta = \xi + \eta$ with $\xi \in E_n(\omega, x)$ and $\eta \in H_n(\omega, x)$. From (1.8)– (1.12) it can be easily seen that

$$|\zeta| \le |\xi| + |\eta| \le \|\xi\|_{(\omega,x),n} + \|\eta\|_{(\omega,x),n} \le 2\|\zeta\|_{(\omega,x),n}$$

which implies the first inequality in 3). We now prove the second one. By 1) and 4) of Lemma 1.1 and (1.10)

$$\|\xi\|_{(\omega,x),n} \le \left[\sum_{l=0}^{+\infty} e^{-2(a+2\varepsilon)l}l(\omega, x, n)^2 e^{2(a+\varepsilon)l}|\xi|^2\right]^{\frac{1}{2}} \le [l'(1-e^{-2\varepsilon})^{-\frac{1}{2}}]e^{\varepsilon n}|\xi|,$$

and similarly, by 2) and 4) of Lemma 1.1 and (1.11),

$$\|\eta\|_{(\omega,x),n} \le \left[\sum_{l=0}^{n} e^{2(b-2\varepsilon)l} l(\omega,x,n-l)^2 e^{-2(b-\varepsilon)l}|\eta|^2\right]^{\frac{1}{2}} \le [l'(1-e^{-2\varepsilon})^{-\frac{1}{2}}]e^{\varepsilon n}|\eta|.$$

Since

$$\gamma(E_n(\omega,x), H_n(\omega,x)) \ge (l'e^{\varepsilon n})^{-1}$$

we have

$$\begin{aligned}\|\zeta\|_{(\omega,x),n} \le\ & \|\xi\|_{(\omega,x),n} + \|\eta\|_{(\omega,x),n} \le [l'(1-e^{-2\varepsilon})^{-\frac{1}{2}}]e^{\varepsilon n}(|\xi|+|\eta|)\\ \le\ & [l'(1-e^{-2\varepsilon})^{-\frac{1}{2}}]e^{\varepsilon n}4l'e^{\varepsilon n}|\xi+\eta| = Ae^{2\varepsilon n}|\zeta|.\end{aligned}$$

The proof is completed. □

Finally we prove the following important lemma. We use Lip $(\cdot)$ as usual to denote the Lipschitz constant of a Lipschitz map, and the norm we use is $|\cdot|$ except mentioned otherwise.

Lemma 1.4. *There exist a Borel set $\Gamma_0 \subset \Omega^{\mathbf{N}}$ (independent of ε) and a measurable function $r : \Gamma_0 \to (0,+\infty)$ such that $v^{\mathbf{N}}(\Gamma_0) = 1, \tau\Gamma_0 \subset \Gamma_0$ and the following hold true:*

1) For each $\omega \in \Gamma_0$ and $x \in M$, the map

$$F_{(\omega,x),0} \overset{\text{def}}{=} \exp^{-1}_{f_0(\omega)x} \circ f_0(\omega) \circ \exp_x : T_xM(r(\omega)^{-1}) \to T_{f_0(\omega)x}M$$

is well defined and

$$Lip(T.F_{(\omega,x),0}) \le r(\omega),$$

where $T_xM(r(\omega)^{-1}) = \{\xi \in T_xM : |\xi| < r(\omega)^{-1}\}$ and $T.F_{(\omega,x),0} : \xi \longmapsto T_\xi F_{(\omega,x),0}, \xi \in T_xM(r(\omega)^{-1})$;

2) $r(\tau^n\omega) \le r(\omega)e^{\varepsilon n}, n \in \mathbf{Z}^+, \omega \in \Gamma_0$.

Proof. From the proof of Lemma II. 1.3 we see that there exist numbers $r > 0$ and $C > 0$ such that for each $f \in \Omega$ and $x \in M$ the map

$$H_{(f,x)} = \exp^{-1}_{f(x)} \circ f \circ \exp_x : \left\{\xi \in T_xM : |\xi| \le \frac{r}{\max\{|f|_{C^1},1\}}\right\} \to T_{f(x)}M$$

makes sense and

$$\text{Lip}(T.H_{(f,x)}) \le C\max\{|f|_{C^1},1\}\max\{\|f\|_{C^2},1\}.$$

Define now $r' : \Omega^{\mathbf{N}} \to (0,+\infty)$ by the formula

$$r'(\omega) = \max\left\{r^{-1}\max\{|f_0(\omega)|_{C^1},1\}, C\max\{|f_0(\omega)|_{C^1},1\}\max\{\|f_0(\omega)\|_{C^2},1\}\right\}.$$

Then it follows that for each $\omega \in \Omega^{\mathbf{N}}$ and $x \in M$ the map

$$F'_{(\omega,x),0} \stackrel{\text{def}}{=} \exp^{-1}_{f_0(\omega)x} \circ f_0(\omega) \circ \exp_x : T_x M(r'(\omega)^{-1}) \to T_{f_0(\omega)x} M$$

is well defined and

$$\mathrm{Lip}(T.F'_{(\omega,x),0}) \leq r'(\omega),$$

and moreover, by condition (1.1) in Section I.1, $\log r' \in L^1(\Omega^{\mathbf{N}}, \mathcal{B}(\Omega^{\mathbf{N}}), v^{\mathbf{N}})$. According to Birkhoff ergodic theorem,

$$\lim_{n \to +\infty} \frac{1}{n} \log r'(\tau^n \omega) = 0 \quad v^{\mathbf{N}} - \text{a.e.}$$

Then there exists a Borel set $\Gamma_0 \subset \Omega^{\mathbf{N}}$ such that $v^{\mathbf{N}}(\Gamma_0) = 1, \tau\Gamma_0 \subset \Gamma_0$ and for each $\omega \in \Gamma_0$

$$\lim_{n \to +\infty} \frac{1}{n} \log r'(\tau^n \omega) = 0. \tag{1.13}$$

From (1.13) it follows that

$$r(\omega) \stackrel{\text{def}}{=} \sup\{r'(\tau^n \omega) e^{-n\varepsilon} : n \geq 0\}$$

is finite at each point $\omega \in \Gamma_0$. Then it can be easily verified that the Borel set Γ_0 and the function $r : \Gamma_0 \to (0, +\infty)$ satisfy the requirements of this lemma. □

§2 Some Technical Facts About Contracting Maps

In this section we introduce some additional preliminaries which consist of some technical facts about contracting maps. We first have the following simple result.

Lemma 2.1. *Let X and Y be complete metric spaces and let $X \times Y$ have the product metric, i.e.*

$$d((x,y),(x',y')) = \max\{d(x,x'), d(y,y')\}, \quad (x,y),(x',y') \in X \times Y.$$

Let $\theta : X \times Y \to Y$ be a continuous map. Suppose that θ is a uniform λ-contraction ($\lambda < 1$) on the second factor, i.e.

$$d(\theta(x,y), \theta(x,y')) \leq \lambda d(y,y'), \quad \forall x \in X, \forall y, y' \in Y.$$

For each $x \in X$, denote by θ_x the map: $Y \to Y, y \longmapsto \theta(x,y)$ and let $\varphi(x)$ be the unique fixed point of θ_x. Then the following hold true:

1) The map $\varphi : X \to Y, x \longmapsto \varphi(x)$ is continuous;

2) When θ is Lipschitz, φ is also Lipschitz with Lip $(\varphi) \leq \frac{1}{1-\lambda} Lip(\theta)$. Moreover, if $Lip(\theta) \leq \lambda$, then $Lip(\varphi) \leq \lambda$.

Proof. Let x and x' be two points of X. Then

$$\begin{aligned} d(\varphi(x),\varphi(x')) = &\ d(\theta(x,\varphi(x)),\theta(x',\varphi(x'))) \\ \leq &\ d(\theta(x,\varphi(x)),\theta(x',\varphi(x))) + d(\theta(x',\varphi(x)),\theta(x',\varphi(x'))) \\ \leq &\ d(\theta(x,\varphi(x)),\theta(x',\varphi(x))) + \lambda d(\varphi(x),\varphi(x')). \end{aligned} \tag{2.1}$$

Thus

$$d(\varphi(x),\varphi(x')) \leq \frac{1}{1-\lambda} d(\theta(x,\varphi(x)),\theta(x',\varphi(x))) \tag{2.2}$$

from which 1) follows. When θ is Lipschitz, from (2.2) we obtain

$$d(\varphi(x),\varphi(x')) \leq \frac{1}{1-\lambda} \mathrm{Lip}(\theta) d(x,x').$$

Hence φ is Lipschitz with Lip $(\varphi) \leq \frac{1}{1-\lambda}\mathrm{Lip}(\theta)$. Moreover, if $\mathrm{Lip}(\theta) \leq \lambda$, we have

$$\begin{aligned} d(\varphi(x),\varphi(x')) = &\ d(\theta(x,\varphi(x)),\theta(x',\varphi(x'))) \\ \leq &\ \lambda \max\{d(x,x'), d(\varphi(x),\varphi(x'))\}. \end{aligned}$$

Since $\lambda < 1$, we obtain

$$d(\varphi(x),\varphi(x')) \leq \lambda d(x,x').$$

Thus 2) is proved. □

Lemma 2.2. *Under the circumstances of Lemma 2.1, we suppose moreover that X and Y are closed subsets of two respective Banach spaces $(E, \|\cdot\|)$ and $(F, \|\cdot\|)$ and that θ is Lipschitz. We restrict below θ and φ respectively to int$(X \times Y)$ and int $(X) \cap \varphi^{-1}$int(X), where int$(\cdot)$ denotes the interior of a set. If θ is of class C^1, then φ is of class C^1. In addition, if $T.\theta$ is Lipschitz, then $T.\varphi$ is also Lipschitz with $Lip(T.\varphi) \leq C Lip(T.\theta)$, where $C = (1-\lambda)^{-3}[1 + Lip(\theta)]^2$.*

Proof. Let $(x,y) \in \mathrm{int}(X \times Y)$. Denote by $T_1\theta(x,y)$ the partial derivative of θ with respect to the first factor at the point (x,y).$T_2\theta(x,y)$ is defined analogously. We first remark that $\|T_2\theta(x,y)\| \leq \lambda$ since θ is a uniform λ-contraction on the second factor. Consequently id$-T_2\theta(x,y)$ is invertible and its inversion is given by the formula

$$[id - T_2\theta(x,y)]^{-1} = \sum_{n=0}^{+\infty} [T_2\theta(x,y)]^n. \tag{2.3}$$

We now prove that φ is of class C^1 with

$$T_x\varphi = [id - T_2\theta(x,\varphi(x))]^{-1} T_1\theta(x,\varphi(x)). \tag{2.4}$$

In fact, $(x, \varphi(x)) \in \text{int}(X \times Y)$ since $x \in \text{int}(X) \cap \varphi^{-1}\text{int}(Y)$. Let $\Delta x \in E$ such that $x + \Delta x \in \text{int}(X)$, then we have

$$\begin{aligned}
& \|\varphi(x+\Delta x) - \varphi(x) - [id - T_2\theta(x,\varphi(x))]^{-1}T_1\theta(x,\varphi(x))\Delta x\| \\
\leq\ & \|[id - T_2\theta(x,\varphi(x))]^{-1}\|\|\varphi(x+\Delta x) - \varphi(x) - T_2\theta(x,\varphi(x)) \\
& [\varphi(x+\Delta x) - \varphi(x)] - T_1\theta(x,\varphi(x))\Delta x\| \\
\leq\ & \|[id - T_2\theta(x,\varphi(x))]^{-1}\|\|\theta(x+\Delta x, \varphi(x+\Delta x)) - \theta(x,\varphi(x)) \\
& -T_{(x,\varphi(x))}\theta(\Delta x, \varphi(x+\Delta x) - \varphi(x))\|.
\end{aligned}$$

This last expression is $o(\|(\Delta x, \varphi(x+\Delta x) - \varphi(x))\|)$ and is hence $o(\|\Delta x\|)$ since φ is Lipschitz. Therefore, the derivative of φ at x exists and is given by formula (2.4). From (2.3) and (2.4) it follows that $T_x\varphi$ depends continuously on $x \in \text{int}(X) \cap \varphi^{-1}\text{int}(Y)$. Thus φ is of class C^1.

Now suppose that $T.\theta$ is Lipschitz. Let $x, x' \in \text{int}(X) \cap \varphi^{-1}\text{int}(Y)$, then by (2.3) and (2.4) we have

$$\begin{aligned}
& \|T_x\varphi - T_{x'}\varphi\| \\
\leq\ & \left\|\sum_{n=0}^{+\infty}[T_2\theta(x,\varphi(x))]^n - \sum_{n=0}^{+\infty}[T_2\theta(x',\varphi(x'))]^n\right\| \|T_1\theta(x,\varphi(x))\| \\
& + \left\|\sum_{n=0}^{+\infty}[T_2\theta(x',\varphi(x'))]^n\right\| \|T_1\theta(x,\varphi(x)) - T_1\theta(x',\varphi(x'))\| \\
\leq\ & \left[\sum_{n=0}^{+\infty}(n+1)\lambda^n\right] \text{Lip}(T.\theta)\text{Lip}(\theta)d((x,\varphi(x)),(x',\varphi(x'))) \\
& + \left[\sum_{n=0}^{+\infty}\lambda^n\right] \text{Lip}(T.\theta)d((x,\varphi(x)),(x',\varphi(x'))) \\
\leq\ & (1-\lambda)^{-3}[1+\text{Lip}(\theta)]^2\text{Lip}(T.\theta)d(x,x').
\end{aligned}$$

Thus $T.\varphi$ is Lipschitz with $\text{Lip}(T.\varphi) \leq (1-\lambda)^{-3}[1+\text{Lip}(\theta)]^2\text{Lip}(T.\theta)$. □

§3 Local and Global Stable Manifolds

Let $r', r' \geq \max\{1, 2\rho_0^{-1}\}$, be a number such that the Borel set

$$\Lambda^{l',r'}_{a,b,k,\varepsilon} \stackrel{\text{def}}{=} \{(\omega, x) \in \Lambda^{l'}_{a,b,k,\varepsilon} : \omega \in \Gamma_0 \quad \text{and} \quad r(\omega) < r'\}$$

is not empty. In this section we fix such a set $\Lambda_{a,b,k,\varepsilon}^{l',r'}$ and confine ourselves to it. We shall write

$$\Lambda' = \Lambda_{a,b,k,\varepsilon}^{l',r'}$$

for simplicity of notations.

Before formulating the main results of this section, we first recall the definition of a continuous family of C^1 embedded k-dimensional discs.

Definition 3.1. *Let X be a metric space and let $\{D_x\}_{x\in X}$ be a collection of subsets of M. We call $\{D_x\}_{x\in X}$ a continuous family of C^1 embedded k-dimensional discs in M if there is a finite open cover $\{U_i\}_{i=1}^{l}$ of X such that for each U_i there exists a continuous map $\theta_i : U_i \to Emb^1(D^k, M)$ such that $\theta_i(x)D^k = D_x, x \in U_i$, where $D^k = \{\xi \in \mathbf{R}^k : \|\xi\|_0 < 1\}$.*

About local stable manifolds of random diffeomorphisms we have the following

Theorem 3.1 . *For each $n \in \mathbf{Z}^+$, there exists a continuous family of C^1 embedded k-dimensional discs $\{W_n(\omega, x)\}_{(\omega,x)\in\Lambda'}$ in M and there exist numbers α_n, β_n and γ_n which depend only on a, b, k, ε, l' and r' such that the following hold true for every $(\omega, x) \in \Lambda'$:*

1) There exists a $C^{1,1}$ map

$$h_{(\omega,x),n} : O_n(\omega, x) \to H_n(\omega, x),$$

where $O_n(\omega, x)$ is an open subset of $E_n(\omega, x)$ which contains $\{\xi \in E_n(\omega, x) : |\xi| < \alpha_n\}$, such that

i) $h_{(\omega,x),n}(0) = 0$;

ii) $Lip(h_{(\omega,x),n}) \le \beta_n$, $Lip(T.h_{(\omega,x),n}) \le \beta_n$;

iii) $W_n(\omega, x) = \exp_x$ Graph $(h_{(\omega,x),n})$ and $W_n(\omega, x)$ is tangent to $E_n(\omega, x)$ at the point $f_\omega^n x$;

2) $f_n(\omega)W_n(\omega, x) \subset W_{n+1}(\omega, x)$;

3) $d^s(f_n^l(\omega)y, f_n^l(\omega)z) \le \gamma_n e^{(a+4\varepsilon)l} d^s(y, z), y, z \in W_n(\omega, x), l \in \mathbf{Z}^+$, where $d^s(\ ,\ )$ is the distance along $W_m(\omega, x)$ for $m \in \mathbf{Z}^+$;

4) $\alpha_{n+1} = \alpha_n e^{-5\varepsilon}, \beta_{n+1} = \beta_n e^{7\varepsilon}, \gamma_{n+1} = \gamma_n e^{2\varepsilon}$.

Proof. We complete the proof by several steps.

Step 1. Let

$$\varepsilon_0 = e^{a+4\varepsilon} - e^{a+2\varepsilon}, \quad c_0 = 4Ar'e^{2\varepsilon}, \quad r_0 = c_0^{-1}\varepsilon_0.$$

Then for every $(\omega, x) \in \Lambda'$ and $l \in \mathbf{Z}^+$, by 3) of Lemma 1.3 and Lemma 1.4 we have that

$$F_{(\omega,x),l} \stackrel{\text{def}}{=} \exp_{f_\omega^{l+1}x}^{-1} \circ f_l(\omega) \circ \exp_{f_\omega^l x} :$$

$$\{\xi \in T_{f_\omega^l x}M : \|\xi\|_{(\omega,x),l} \le r_0 e^{-3\varepsilon l}\} \to T_{f_\omega^{l+1}x}M$$

is well defined and

$$\mathrm{Lip}_{\|\cdot\|}(T.F_{(\omega,x),l}) \le c_0 e^{3\varepsilon l},$$

where $\mathrm{Lip}_{\|\cdot\|}(\cdot)$ is defined with respect to $\|\cdot\|_{(\omega,x),l}$ and $\|\cdot\|_{(\omega,x),l+1}$. Then from this it follows that

$$\mathrm{Lip}_{\|\cdot\|}(F_{(\omega,x),l} - T_0F_{(\omega,x),l}) \le \varepsilon_0.$$

Step 2. Let $(\omega,x) \in \Lambda'$ and $n \in \mathbf{Z}^+$ be fixed once and for all from this step to Step 5. We sometimes denote $\|\cdot\|_{(\omega,x),l}$ simply by $\|\cdot\|_l$.

Let $t = e^{a+6\varepsilon}$ and define

$$\begin{aligned}\Gamma(E) = \ & \{\sigma = \{\sigma(l)\}_{l=1}^{+\infty} : \sigma(l) \in E_{n+l}(\omega,x), l \ge 1,\\ & \text{and } \|\sigma\| \overset{\text{def}}{=} \sup_{l\ge 1} \|t^{-l}\sigma(l)\|_{n+l} < +\infty\},\\ \Gamma(H) = \ & \{\tau = \{\tau(l)\}_{l=0}^{+\infty} : \tau(l) \in H_{n+l}(\omega,x), l \ge 0,\\ & \text{and } \|\tau\| \overset{\text{def}}{=} \sup_{l\ge 0} \|t^{-l}\tau(l)\|_{n+l} < +\infty\}.\end{aligned}$$

It can be easily verified that $(\Gamma(E),\|\cdot\|)$ and $(\Gamma(H),\|\cdot\|)$ are both Banach spaces with respect to the natural operations of addition and scalar multiplication. Define

$$X = \{\sigma \in \Gamma(E) : \|\sigma\| \le r_0 e^{-3\varepsilon n}\},$$

$$Y = \{\tau \in \Gamma(H) : \|\tau\| \le r_0 e^{-3\varepsilon n}\},$$

$$Z = \{\xi \in E_n(\omega,x) : \|\xi\|_n \le r_0 e^{-3\varepsilon n}\}.$$

They are obviously closed subsets of $\Gamma(E), \Gamma(H)$ and $E_n(\omega,x)$ respectively.

Step 3. For the point (ω,x), we denote $F_{(\omega,x),m}$ simply by F_m and denote $U_m^l(\omega,x)$ by U_m^l for $m,l \in \mathbf{Z}^+$. We now define

$$\theta : Z \times X \times Y \to X \times Y, \quad (\xi,\sigma,\tau) \longmapsto (\sigma',\tau')$$

where

$$\begin{cases}\sigma'(1) = \pi_1 F_n(\xi,\tau(0)),\\ \sigma'(l) = \pi_1 F_{n+l-1}(\sigma(l-1),\tau(l-1)), l \ge 2,\end{cases}$$

$$\begin{cases}\tau'(0) = [U_n^1]^{-1}[\tau(1) + U_n^1\tau(0) - \pi_2 F_n(\xi,\tau(0))],\\ \tau'(l) = [U_{n+l}^1]^{-1}[\tau(l+1) + U_{n+l}^1\tau(l) - \pi_2 F_{n+l}(\sigma(l),\tau(l))], l \ge 1,\end{cases}$$

where π_1 and π_2 are respectively the projections of $E_m(\omega,x) \oplus H_m(\omega,x)$ to $E_m(\omega,x)$ and $H_m(\omega,x), m \in \mathbf{Z}^+$.

It is easy to verify that θ is well defined and is a Lipschitz map satisfying

$$\mathrm{Lip}(\theta) \le e^{-2\varepsilon}$$

with respect to the product metrics on $Z \times X \times Y$ and $X \times Y$. We now show that θ is of class C^1 in $\mathrm{int}(Z \times X \times Y)$ and $T.\theta$ is Lipschitz. Let $(\xi_0, \sigma_0, \tau_0)$ be a point in $\mathrm{int}(Z \times X \times Y)$. Define a map

$$A(\xi_0, \sigma_0, \tau_0) : E_n(\omega, x) \times \Gamma(E) \times \Gamma(H) \to \Gamma(E) \times \Gamma(H), (\xi, \sigma, \tau) \longmapsto (\sigma'', \tau'')$$

by letting

$$\begin{cases} \sigma''(1) = \pi_1 TF_n(\xi_0, \tau_0(0))(\xi, \tau(0)), \\ \sigma''(l) = \pi_1 TF_{n+l-1}(\sigma_0(l-1), \tau_0(l-1))(\sigma(l-1), \tau(l-1)), l \geq 2, \end{cases}$$

where $T_\eta F_m$ is denoted by $TF_m(\eta), \eta \in T_{f_\omega^m x}M, m \geq 0$, and

$$\begin{cases} \tau''(0) = [U_n^1]^{-1}[\tau(1) + U_n^1 \tau(0) - \pi_2 TF_n(\xi_0, \tau_0(0))(\xi, \tau(0))], \\ \tau''(l) = [U_{n+l}^1]^{-1}[\tau(l+1) + U_{n+l}^1 \tau(l) - \pi_2 TF_{n+l}(\sigma_0(l), \tau_0(l))(\sigma(l), \tau(l))], l \geq 1. \end{cases}$$

From Step 1 it follows that $A(\xi_0, \sigma_0, \tau_0)$ is a well-defined bounded linear operator and is the derivative of θ at the point $(\xi_0, \sigma_0, \tau_0)$ and obviously $A(0,0,0) = 0$. By a simple computation we obtain that $T.\theta$, which is defined on $\mathrm{int}(Z \times X \times Y)$, is a Lipschitz map such that

$$\mathrm{Lip}(T.\theta) \leq c_0 e^{-a} e^{3\varepsilon n}.$$

Step 4. Since $\mathrm{Lip}(\theta) \leq e^{-2\varepsilon}$, by Lemma 2.1 there exists a Lipschitz map

$$\varphi : Z \to X \times Y$$

with

$$\mathrm{Lip}(\varphi) \leq e^{-2\varepsilon}$$

such that for each $\xi \in Z$, $\varphi(\xi)$ is the unique fixed point of the map $\theta_\xi : X \times Y \to X \times Y, (\sigma, \tau) \longmapsto \theta(\xi, \sigma, \tau)$. Since $\varphi(0) = (0,0)$ and $\mathrm{Lip}(\varphi) \leq e^{-2\varepsilon}$, we have $\varphi(\mathrm{int}(Z)) \subset \mathrm{int}(X \times Y)$. Thus by Lemma 2.2 the map $T.\varphi$ is well defined on $\mathrm{int}(Z)$ and is a Lipschitz map such that

$$\mathrm{Lip}(T.\varphi) \leq (1 - e^{-2\varepsilon})^{-3}(1 + e^{-2\varepsilon})^2 \mathrm{Lip}(T.\theta) \leq De^{3\varepsilon n}$$

where $D = (1 - e^{-2\varepsilon})^{-3}(1 + e^{-2\varepsilon})^2 c_0 e^{-a}$.

Step 5. Let $\varphi(\xi) = (\sigma(\xi), \tau(\xi))$ for $\xi \in \mathrm{int}(Z)$. Define

$$h_{(\omega,x),n} : \{\xi \in E_n(\omega, x) : \|\xi\|_{(\omega,x),n} < r_0 e^{-3\varepsilon n}\} \to H_n(\omega, x), \quad \xi \longmapsto \tau(\xi)(0).$$

From Step 4 it follows that $h_{(\omega,x),n}$ is a $C^{1,1}$ map satisfying $h_{(\omega,x),n}(0) = 0, T_0 h_{(\omega,x),n} = 0$ and

$$\mathrm{Lip}_{\|\cdot\|}(h_{(\omega,x),n}) \leq e^{-2\varepsilon}, \tag{3.1}$$

$$\mathrm{Lip}_{\|\cdot\|}(T.h_{(\omega,x),n}) \leq De^{3\varepsilon n}. \tag{3.2}$$

For $l \in \mathbf{Z}^+$, let

$$F_n^0(\omega, x) = id, \quad F_n^l(\omega, x) = F_{(\omega,x),n+l-1} \circ \cdots \circ F_{(\omega,x),n}, \quad l > 0,$$

defined wherever they make sense. Since for every $\xi \in \text{int}(Z)$ there is an only point $\eta \in H_n(\omega, x)$ such that

$$\|F_n^l(\omega, x)(\xi, \eta)\|_{(\omega,x),n+l} < r_0 e^{-3\varepsilon n} e^{(a+6\varepsilon)l}, \quad l \in \mathbf{Z}^+,$$

we have

$$\begin{aligned}\text{Graph}(h_{(\omega,x),n}) = \quad & \{(\xi, \eta) \in E_n(\omega, x) \oplus H_n(\omega, x) : \\ & \|F_n^l(\omega, x)(\xi, \eta)\|_{(\omega,x),n+l} < r_0 e^{-3\varepsilon n} e^{(a+6\varepsilon)l}, l \in \mathbf{Z}^+\}.\end{aligned} \tag{3.3}$$

Step 6. First let us notice that what we have done in Steps 1–5 holds for every $(\omega, x) \in \Lambda'$ and $n \in \mathbf{Z}^+$. Now let $(\omega, x) \in \Lambda'$ and $n \in \mathbf{Z}^+$. Since for each $(\xi, \eta) \in \text{Graph}(h_{(\omega,x),n})$

$$\begin{aligned}\|F_{n+1}^l(\omega, x)F_n^1(\omega, x)(\xi, \eta)\|_{(\omega,x),n+l+1} &= \|F_n^{l+1}(\omega, x)(\xi, \eta)\|_{(\omega,x),n+l+1} \\ &\leq r_0 e^{-3\varepsilon n} e^{(a+6\varepsilon)(l+1)} \\ &\leq r_0 e^{-3\varepsilon(n+1)} e^{(a+6\varepsilon)l}, \quad l \in \mathbf{Z}^+,\end{aligned}$$

then by (3.3) we have

$$F_n^1(\omega, x)\text{Graph}(h_{(\omega,x),n}) \subset \text{Graph}(h_{(\omega,x),n+1}). \tag{3.4}$$

We now prove that for every $\zeta \in \text{Graph}(h_{(\omega,x),n})$

$$\|T_\zeta F_n^1(\omega, x)|_{T_\zeta \text{Graph}(h_{(\omega,x),n})}\| \leq e^{a+4\varepsilon}, \tag{3.5}$$

where $\|\cdot\|$ is defined with respect to $\|\cdot\|_{(\omega,x),n}$ and $\|\cdot\|_{(\omega,x),n+1}$ and T_ζ Graph$(h_{(\omega,x),n})$ is the tangent space of Graph$(h_{(\omega,x),n})$ at the point ζ. In fact, let $(\xi_n, \eta_n) \in T_\zeta\text{Graph}(h_{(\omega,x),n})$ and let $(\xi_{n+1}, \eta_{n+1}) = T_\zeta F_n^1(\omega, x)(\xi_n, \eta_n)$. It follows from Step 1 and (3.1) and (3.4) that

$$\begin{aligned}& \|(\xi_{n+1}, \eta_{n+1})\|_{(\omega,x),n+1} \\ = \quad & \|\xi_{n+1}\|_{(\omega,x),n+1} \\ = \quad & \|S_n^1(\omega, x)\xi_n + \pi_1(T_\zeta F_n^1(\omega, x) - T_0 F_n^1(\omega, x))(\xi_n, \eta_n)\|_{(\omega,x),n+1} \\ \leq \quad & e^{a+2\varepsilon}\|\xi_n\|_{(\omega,x),n} + c_0 e^{3\varepsilon n}\|\zeta\|_{(\omega,x),n}\|(\xi_n, \eta_n)\|_{(\omega,x),n} \\ \leq \quad & (e^{a+2\varepsilon} + \varepsilon_0)\|(\xi_n, \eta_n)\|_{(\omega,x),n} \\ = \quad & e^{a+4\varepsilon}\|(\xi_n, \eta_n)\|_{(\omega,x),n}\end{aligned}$$

which implies (3.5).

Step 7. In this step we present the counterpart of the above results in terms of the norm $|\cdot|$.

Let $(\omega, x) \in \Lambda'$ and $n \in \mathbf{Z}^+$. Define

$$O_n(\omega, x) = \{\xi \in E_n(\omega, x) : \|\xi\|_{(\omega,x),n} < r_0 e^{-3\varepsilon n}\},$$

$$W_n(\omega, x) = \exp_{f_\omega^n x} \mathrm{Graph}(h_{(\omega,x),n})$$

and let

$$\alpha_n = A^{-1} r_0 e^{-5\varepsilon n}.$$

Then by 3) of Lemma 1.3 we have

$$\{\xi \in E_n(\omega, x) : |\xi| < \alpha_n\} \subset O_n(\omega, x).$$

And since $T_0 h_{(\omega,x),n} = 0, W_n(\omega, x)$ is tangent to $E_n(\omega, x)$ at the point $f_\omega^n x$. Moreover, from (3.4) we obtain immediately

$$f_n(\omega) W_n(\omega, x) \subset W_{n+1}(\omega, x).$$

Let

$$\beta_n = 2DA^2 e^{7\varepsilon n}.$$

Then from (3.1), (3.2) and 3) of Lemma 1.3 it follows that

$$\mathrm{Lip}(h_{(\omega,x),n}) \le \beta_n, \mathrm{Lip}(T.h_{(\omega,x),n}) \le \beta_n.$$

By (3.5), (3.4) and 3) of Lemma 1.3 we obtain that for each $\zeta \in \mathrm{Graph}(h_{(\omega,x),n})$ and $l \in \mathbf{Z}^+$

$$|T_\zeta F_n^l(\omega, x)|_{T_\zeta \mathrm{Graph}(h_{(\omega,x),n})}| \le 2Ae^{2\varepsilon n} e^{(a+4\varepsilon)l}$$

which implies that

$$d^s(f_n^l(\omega, x)y, f_n^l(\omega, x)z) \le \gamma_n e^{(a+4\varepsilon)l} d^s(y, z), \quad y, z \in W_n(\omega, x), \quad l \in \mathbf{Z}^+,$$

where

$$\gamma_n = 2[b(\rho_0/2)]^2 A e^{2\varepsilon n}$$

and $b(\rho_0/2)$ is as introduced in Lemma II.1.1. Hence, for every $(\omega, x) \in \Lambda'$ and $n \in \mathbf{Z}^+, W_n(\omega, x)$ and the numbers α_n, β_n and γ_n satisfy 1) – 4) of the theorem.

Step 8. In this step we complete the proof by showing that $\{W_n(\omega, x)\}_{(\omega,x)\in\Lambda'}$ is a continuous family of C^1 embedded k-dimensional discs in M for each $n \in \mathbf{Z}^+$.

Let $n \in \mathbf{Z}^+$. By Lemma 1.2 we know that $E_n(\omega, x)$ and $H_n(\omega, x)$ depend continuously on $(\omega, x) \in \Lambda'$. Then there exists a finite open cover $\{\Lambda'_l\}_{l=1}^r$ of Λ' such that for each Λ'_l we can find a basis of $E_n(\omega, x)$ and a basis of $H_n(\omega, x), (\omega, x) \in \Lambda'_l$ which are continuous with respect to $(\omega, x) \in \Lambda'_l$. Let Λ'_p be a set in $\{\Lambda'_l\}_{l=1}^r$. Since $\langle\ ,\ \rangle_{(\omega,x),n}$ depends continuously on $(\omega, x) \in \Lambda'$, then for each $(\omega, x) \in \Lambda'_p$ there exist, with respect to $\langle\ ,\ \rangle_{(\omega,x),n}$, an orthonormal basis $\{\xi_i(\omega, x)\}_{i=1}^k$ of $E_n(\omega, x)$ and an orthonormal basis $\{\xi_j(\omega, x)\}_{j=k+1}^{m_0}$ of

$H_n(\omega, x)$ such that they are continuous with respect to $(\omega, x) \in \Lambda'_p$. For each $(\omega, x) \in \Lambda'_p$, let

$$A(\omega, x) : \mathbf{R}^k \oplus \mathbf{R}^{m_0-k} \to E_n(\omega, x) \oplus H_n(\omega, x)$$

be a linear map satisfying $A(\omega, x)e_s = \xi_s(\omega, x), 1 \leq s \leq m_0$, where $\{e_s\}_{s=1}^{m_0}$ is the natural basis of $\mathbf{R}^k \oplus \mathbf{R}^{m_0-k}$. Define a map

$$\theta_p : \Lambda'_p \to Emb^1(D^k, M)$$

by the formula

$$\theta_p(\omega, x) = \exp_{f_\omega^n x} \circ (id, h_{(\omega,x),n}) \circ A(\omega, x)|_{D^k}.$$

Then it is clear that for each $(\omega, x) \in \Lambda'_p, \theta_p(\omega, x)$ is a $C^{1,1}$ embedding with

$$\theta_p(\omega, x)D^k = W_n(\omega, x).$$

Now we show that θ_p is continuous. For $(\omega, x) \in \Lambda'_p$, let

$$h'_{(\omega,x),n} = A(\omega, x)^{-1} \circ h_{(\omega,x),n} \circ A(\omega, x)|_{D^k}.$$

Suppose that $\{(\omega_m, x_m)\}_{m=1}^{+\infty}$ is a sequence of points in Λ'_p such that $(\omega_m, x_m) \to (\omega_0, x_0) \in \Lambda'_p$ as $m \to +\infty$. According to Arzela-Ascoli Theorem, by (3.1)-(3.3) we know that $h'_{(\omega_m,x_m),n}$ and $T.h'_{(\omega_m,x_m),n}$ converge uniformly to $h'_{(\omega_0,x_0),n}$ and $T.h'_{(\omega_0,x_0),n}$ respectively. This together with the continuity of $A(\omega, x)$ with respect to $(\omega, x) \in \Lambda'_p$ implies that θ_p is continuous. □

Let us remark that the set $\Lambda_{a,b,k,\varepsilon}^{l',r'}$ defined above depends only on a, b, k, ε, l' and r'. Denote

$$\widehat{\Lambda}_0 = \Lambda_0 \cap (\Gamma_0 \times M), \quad \widehat{\Lambda}_{a,b,k} = \Lambda_{a,b,k} \cap \widehat{\Lambda}_0 \tag{3.6}$$

and let $\{l'_m\}_{m=1}^{+\infty}$ and $\{r'_m\}_{m=1}^{+\infty}$ be sequences of positive numbers such that $l'_m \nearrow +\infty$ and $r'_m \nearrow +\infty$ as $m \to +\infty$. Then we have

$$\Lambda_{a,b,k,\varepsilon}^{l'_m,r'_m} \subset \Lambda_{a,b,k,\varepsilon}^{l'_{m+1},r'_{m+1}}, \qquad m \in \mathbf{N}$$

and

$$\widehat{\Lambda}_{a,b,k} = \bigcup_{m=1}^{+\infty} \Lambda_{a,b,k,\varepsilon}^{l'_m,r'_m}. \tag{3.7}$$

If we write

$$\{(a_n, b_n)\}_{n=1}^{+\infty} = \{(a, b) : a < b \leq 0, a \text{ and } b \text{ are rational}\}$$

and let

$$\varepsilon_n = \frac{1}{2}\min\{1, \frac{1}{200}(b_n - a_n)\},$$

then

$$\widehat{\Lambda}_0 = \left\{ \bigcup_{n=1}^{+\infty} \bigcup_{k=1}^{m_0} \bigcup_{m=1}^{+\infty} \Lambda_{a_n,b_n,k,\varepsilon_n}^{l'_m,r'_m} \right\} \cup \{(\omega, x) \in \widehat{\Lambda}_0 : \lambda^{(i)}(x) \geq 0, 1 \leq i \leq r(x)\}. \tag{3.8}$$

Needless to say, Theorem 3.1 holds for each $\Lambda_{a_n,b_n,k,\varepsilon_n}^{l'_m,r'_m}$.

The following is a theorem about global stable manifolds of random diffeomorphisms.

Theorem 3.2. *Let $(\omega, x) \in \widehat{\Lambda}_0 \backslash \{(\omega, x) \in \widehat{\Lambda}_0 : \lambda^{(i)}(x) \geq 0, 1 \leq i \leq r(x)\}$ and let $\lambda^{(1)}(x) < \cdots < \lambda^{(p)}(x)$ be the strictly negative Lyapunov exponents at (ω, x). Define $W^{s,1}(\omega, x) \subset \cdots \subset W^{s,p}(\omega, x)$ by*

$$W^{s,i}(\omega, x) = \left\{ y \in M : \limsup_{n \to +\infty} \frac{1}{n} \log d(f_\omega^n x, f_\omega^n y) \leq \lambda^{(i)}(x) \right\} \tag{3.9}$$

for $1 \leq i \leq p$. Then $W^{s,i}(\omega, x)$ is the image of $V_{(\omega,x)}^{(i)}$ under an injective immersion of class $C^{1,1}$ and is tangent to $V_{(\omega,x)}^{(i)}$ at x. In addition, if $y \in W^{s,i}(\omega, x)$, then

$$\limsup_{n \to +\infty} \frac{1}{n} \log d^s(f_\omega^n x, f_\omega^n y) \leq \lambda^{(i)}(x) \tag{3.10}$$

where $d^s(\ ,\ )$ is the distance along the submanifold $f_\omega^n W^{s,i}(\omega, x)$.

Proof. We carry out the proof by several steps. Now let $i \in \{1, \cdots, p\}$ be fixed arbitrarily and let $k = \dim V_{(\omega,x)}^{(i)}$.

Step 1. Let a, b, ε, l' and r' be numbers with the following properties: $\lambda^{(i)}(x) < a < b < \min\{0, \lambda^{(i+1)}(x)\}$ (we admit here $\lambda^{(r(x)+1)}(x) = +\infty$), $0 < \varepsilon < \min\{1, (b-a)/200\}$ and $(\omega, x) \in \Lambda_{a,b,k,\varepsilon}^{l',r'}$. Corresponding to these numbers, we have the sequence of norms $\{\|\cdot\|_{(\omega,x),n}\}_{n=0}^{+\infty}$ and numbers A and r_0 as defined in the previous sections. Let $\{W_n(\omega, x)\}_{n=0}^{+\infty}$ be the sequence of embedded k-dimensional discs obtained by applying Theorem 3.1 to the set $\Lambda_{a,b,k,\varepsilon}^{l',r'}$, and finally we put

$$W = \bigcup_{n=0}^{+\infty} [f_\omega^n]^{-1} W_n(\omega, x).$$

Since for each $n \in \mathbf{Z}^+$

$$f_n(\omega) W_n(\omega, x) \subset W_{n+1}(\omega, x),$$

we know W is the union of an increasing sequence of $C^{1,1}$ embedded discs $\{[f_\omega^n]^{-1} W_n(\omega, x)\}_{n=0}^{+\infty}$ and hence is the image of $V_{(\omega,x)}^{(i)}$ under an injective immersion of class $C^{1,1}$ (see Chapter VIII of [Hir]$_1$).

From 3) of Theorem 3.1 it follows clearly that for every $y \in W$

$$\limsup_{n \to +\infty} \frac{1}{n} \log d^s(f_\omega^n x, f_\omega^n y) \leq a + 4\varepsilon. \tag{3.11}$$

We now show that

$$W^{s,i}(\omega, x) \subset W. \tag{3.12}$$

In fact, if $y \in W^{s,i}(\omega, x)$, it is clear that there exists $n_0 \in \mathbf{N}$ such that

$$d(f_\omega^n x, f_\omega^n y) \le e^{(a+\varepsilon)n}, \quad \forall n \ge n_0$$

and

$$e^{(a+\varepsilon)n_0} \le A^{-1} r_0 e^{-5\varepsilon n_0}.$$

Then we have for each $l \in \mathbf{Z}^+$

$$d(f_\omega^{n_0+l} x, f_\omega^{n_0+l} y) \le e^{(a+\varepsilon)n_0} e^{(a+\varepsilon)l} \le A^{-1} r_0 e^{-5\varepsilon n_0} e^{(a+6\varepsilon)l}$$

which together with (3.3) and 3) of Lemma 1.3 proves that $f_\omega^{n_0} y \in W_{n_0}(\omega, x)$ and hence $y \in W$. Therefore, (3.12) holds true.

Step 2. Let a', ε', l'' and r'' be numbers such that $\lambda^{(i)}(x) < a' \le a, \varepsilon' \le \varepsilon, l'' \ge l', r'' \ge r'$, and $(\omega, x) \in \Lambda_{a',b,k,\varepsilon'}^{l'',r''}$. Corresponding to $\Lambda_{a',b,k,\varepsilon'}^{l'',r''}$, we obtain analogously the sequence of norms $\{\|\cdot\|'_{(\omega,x),n}\}_{n=0}^{+\infty}$, the numbers A' and r'_0 and the sequence of embedded discs $\{W'_n(\omega, x)\}_{n=0}^{+\infty}$. Define

$$W' = \bigcup_{n=0}^{+\infty} [f_\omega^n]^{-1} W'_n(\omega, x).$$

We now claim that

$$W' = W. \tag{3.13}$$

Indeed, it is clear that for every $n \in \mathbf{Z}^+$

$$\|\zeta\|_{(\omega,x),n} \le \|\zeta\|'_{(\omega,x),n} \le 2A' e^{2\varepsilon' n} \|\zeta\|_{(\omega,x),n}, \qquad \zeta \in T_{f_\omega^n x} M.$$

For each $n \in \mathbf{Z}^+$, write

$$W_n(\omega, x) = \exp_{f_\omega^n x} \mathrm{Graph}(h_{(\omega,x),n}), W'_n(\omega, x) = \exp_{f_\omega^n x} \mathrm{Graph}(h'_{(\omega,x),n})$$

and define

$$U_n(\omega, x) = \{\xi \in E_n(\omega, x) : \|\xi\|'_{(\omega,x),n} \le \min\{r_0 e^{-3\varepsilon n}, r'_0 e^{-3\varepsilon' n}\}\}.$$

Since for each $\zeta \in \mathrm{Graph}(h'_{(\omega,x),n}|_{U_n(\omega,x)})$

$$\begin{aligned} \|F_n^l(\omega, x)\zeta\|_{(\omega,x),n+l} \le & \ \|F_n^l(\omega, x)\zeta\|'_{(\omega,x),n+l} \\ \le & \ \|\zeta\|'_{(\omega,x),n} e^{(a'+4\varepsilon')l} \le r_0 e^{-3\varepsilon n} e^{(a+6\varepsilon)l} \end{aligned}$$

for all $l \in \mathbf{Z}^+$, we have

$$\mathrm{Graph}(h'_{(\omega,x),n} |_{U_n(\omega,x)}) \subset \mathrm{Graph}(h_{(\omega,x),n})$$

and hence

$$\mathrm{Graph}(h'_{(\omega,x),n}|_{U_n(\omega,x)}) = \mathrm{Graph}(h_{(\omega,x),n}|_{U_n(\omega,x)}). \tag{3.14}$$

For each $n \in \mathbf{Z}^+$, when l is sufficiently large we have for all $\zeta \in \mathrm{Graph}(h'_{(\omega,x),n})$

$$\begin{aligned} \|F_n^l(\omega,x)\zeta\|'_{(\omega,x),n+l} &\leq r'_0 e^{-3\varepsilon' n} e^{(a'+4\varepsilon')l} \\ &\leq \min\{r_0 e^{-3\varepsilon(n+l)}, r'_0 e^{-3\varepsilon'(n+l)}\} \end{aligned}$$

and hence, by (3.14),

$$f_n^l(\omega) W'_n(\omega,x) \subset W_{n+l}(\omega,x).$$

Therefore, we obtain

$$W' \subset W. \tag{3.15}$$

Similarly, for each $n \in \mathbf{Z}^+$, when l is large enough we have

$$\begin{aligned} \|F_n^l(\omega,x)\zeta\|'_{(\omega,x),n+l} &\leq 2A' e^{2\varepsilon' n} \|F_n^l(\omega,x)\zeta\|_{(\omega,x),n+l} \\ &\leq 2A' e^{2\varepsilon' n} r_0 e^{-3\varepsilon n} e^{(a+4\varepsilon)l} \\ &\leq \min\{r_0 e^{-3\varepsilon(n+l)}, r'_0 e^{-3\varepsilon'(n+l)}\} \end{aligned}$$

for all $\zeta \in \mathrm{Graph}(h_{(\omega,x),n})$, and hence, by (3.14),

$$f_n^l(\omega) W_n(\omega,x) \subset W'_{n+l}(\omega,x).$$

Thus we obtain

$$W \subset W'$$

which together with (3.15) yields (3.13).

Step 3. Considering that (3.11) holds for every $y \in W'$ with $a + 4\varepsilon$ being replaced by $a' + 4\varepsilon'$, by (3.13) we have for every $y \in W$

$$\limsup_{n \to +\infty} \frac{1}{n} \log d^s(f_\omega^n x, f_\omega^n y) \leq a' + 4\varepsilon'.$$

Since a' and ε' can be chosen such that $a' + 4\varepsilon'$ is arbitrarily close to $\lambda^{(i)}(x)$, then $y \in W$ implies

$$\limsup_{n \to +\infty} \frac{1}{n} \log d^s(f_\omega^n x, f_\omega^n y) \leq \lambda^{(i)}(x). \tag{3.16}$$

Thus we obtain

$$W \subset W^{s,i}(\omega,x). \tag{3.17}$$

Since W is tangent to $V^{(i)}_{(\omega,x)}$ at x, the theorem follows from (3.12), (3.17) and (3.16). □

Remark 3.1. Let $\Lambda' = \Lambda^{l',r'}_{a,b,k,\varepsilon}$ be a set as considered in Theorem 3.1. For $(\omega, x) \in \Lambda'$ let $\lambda^{(1)}(x) < \cdots < \lambda^{(i)}(x)$ be the Lyapunov exponents which are smaller than a. From the proof of Theorem 3.2 it follows obviously that

$$W^{s,i}(\omega, x) = \{y \in M : \limsup_{n\to+\infty} \frac{1}{n} \log d(f^n_\omega x, f^n_\omega y) \le a\}. \tag{3.18}$$

Remark 3.2. Given $\mathcal{X}^+(M, v, \mu)$ and a point $(\omega, x) \in \Omega^{\mathbf{N}} \times M$, the *global stable manifold* $W^s(\omega, x)$ is defined by

$$W^s(\omega, x) \stackrel{\text{def}}{=} \{y \in M : \limsup_{n\to+\infty} \frac{1}{n} \log d(f^n_\omega x, f^n_\omega y) < 0\}. \tag{3.19}$$

From (3.18) we know that, if $(\omega, x) \in \widehat{\Lambda}_0 \backslash \{(\omega, x) \in \widehat{\Lambda}_0 : \lambda^{(i)}(x) \ge 0, 1 \le i \le r(x)\}$ and $\lambda^{(1)}(x) < \cdots < \lambda^{(p)}(x)$ are the strictly negative exponents at (ω, x), then

$$W^s(\omega, x) = W^{s,p}(\omega, x)$$

and hence $W^s(\omega, x)$ is the image of $V^{(p)}_{(\omega,x)}$ under an injective immersion of class $C^{1,1}$ and is tangent to $V^{(p)}_{(\omega,x)}$ at the point x.

§4 Hölder Continuity of Subbundles

Here we consider the Hölder continuity of subbundles of TM constituted of tangent spaces of the submanifolds $W^{s,i}(\omega, x)$. This kind of continuity is necessary for dealing with the absolute continuity of foliations formed by invariant manifolds as introduced in Section 3. Some ideas of this section are adopted from [Bri]. We first introduce the notion of Hölder continuity for subbundles of a trivial vector bundle.

Definition 4.1. *Let Δ be a metric space, H a Hilbert space and $\{E_x\}_{x\in\Delta}$ a family of subspaces of H. The family $\{E_x\}_{x\in\Delta}$ is called Hölder continuous in x with exponent $\alpha, \alpha > 0$, and constant $L, L > 0$, if for any $x, y \in \Delta$*

$$d(E_x, E_y) \stackrel{\text{def}}{=} \max\{\Gamma(E_x, E_y), \Gamma(E_y, E_x)\} \le L[d(x, y)]^\alpha,$$

where we define for two subspaces E and F of H

$$\Gamma(E, F) = \sup_{\substack{\xi\in E \\ \|\xi\|=1}} \inf_{\eta\in F} \|\xi - \eta\|$$

and call it the aperture between E and F.

Let X be a metric space with $\mathrm{diam}(X) \le 1$, H a Hilbert space and $\{T_i(x)\}_{x\in X}$, $i = 1, 2, \cdots$ a sequence of families of bounded linear operators $T_i(x) : H \to H$. For each $x \in X$ and $n \in \mathbf{N}$, put

$$T^0(x) = id, \quad T^n(x) = T_n(x) \circ \cdots \circ T_1(x).$$

Proposition 4.1. *For numbers $\widehat{C} \ge 1$ and $\widehat{a} < \widehat{b}$ let $\Delta_{\widehat{C},\widehat{a},\widehat{b}} \subset X$ be the (maybe empty) set of points x for which there exist splittings*

$$H = E_x \oplus E_x^{\perp}$$

such that for any positive integer n

$$\|T^n(x)\xi\| \le \widehat{C}e^{\widehat{a}n}\|\xi\|, \quad \xi \in E_x,$$

$$\|T^n(x)\eta\| \ge \widehat{C}^{-1}e^{\widehat{b}n}\|\eta\|, \quad \eta \in E_x^{\perp}.$$

Suppose that there is a number $\widehat{c} > \widehat{a}$ and $\beta > 0$ such that

$$\|T^n(x) - T^n(y)\| \le e^{\widehat{c}n}[d(x,y)]^{\beta}$$

for any positive integer n and any $x, y \in \Delta_{\widehat{C},\widehat{a},\widehat{b}}$. Then the family $\{E_x\}_{x\in\Delta_{\widehat{C},\widehat{a},\widehat{b}}}$ is Hölder continuous in x on $\Delta_{\widehat{C},\widehat{a},\widehat{b}}$ with exponent $[(\widehat{a}-\widehat{b})/(\widehat{a}-\widehat{c})]\beta$ and constant $3\widehat{C}^2e^{\widehat{b}-\widehat{a}}$.

Proof. For $z \in \Delta_{\widehat{C},\widehat{a},\widehat{b}}$ and $n \in \mathbf{N}$, set

$$K_z^n = \{\zeta \in H : \|T^n(z)\zeta\| \le 2\widehat{C}e^{\widehat{a}n}\|\zeta\|\}.$$

Let $\zeta \in K_z^n$ and let $\zeta = \xi' + \eta'$ where $\xi' \in E_z$ and $\eta' \in E_z^{\perp}$. We have

$$\begin{aligned}\|T^n(z)\zeta\| = \|T^n(z)(\xi' + \eta')\| \ge\ & \|T^n(z)\eta'\| - \|T^n(z)\xi'\| \\ \ge\ & \widehat{C}^{-1}e^{\widehat{b}n}\|\eta'\| - \widehat{C}e^{\widehat{a}n}\|\zeta\|.\end{aligned}$$

Hence

$$\|\eta'\| \le \widehat{C}e^{-\widehat{b}n}(\|T^n(z)\zeta\| + \widehat{C}e^{\widehat{a}n}\|\zeta\|) \le 3\widehat{C}^2e^{(\widehat{a}-\widehat{b})n}\|\zeta\|$$

which implies

$$d(\zeta, E_z) \stackrel{\text{def}}{=} \inf_{\xi\in E_z} \|\zeta - \xi\| \le 3\widehat{C}^2e^{(\widehat{a}-\widehat{b})n}\|\zeta\|. \tag{4.1}$$

Let $x, y \in \Delta_{\widehat{C},\widehat{a},\widehat{b}}$. Define

$$\gamma = (\widehat{a} - \widehat{c})/\beta.$$

Since $d(x,y) \leq 1$ and $\gamma < 0$, there is a unique non-negative integer $m = m(x,y)$ such that

$$e^{\gamma(m+1)} < d(x,y) \leq e^{\gamma m}. \tag{4.2}$$

Then for any $\xi \in E_y$

$$\begin{aligned}||T^m(x)\xi|| \leq\ & ||T^m(y)\xi|| + ||T^m(x) - T^m(y)||||\xi|| \\ \leq\ & \widehat{C}e^{\widehat{a}m}||\xi|| + e^{\widehat{c}m}[d(x,y)]^{\beta}||\xi|| \\ \leq\ & (\widehat{C}e^{\widehat{a}m} + e^{\widehat{c}m}e^{\beta\gamma m})||\xi|| \leq 2\widehat{C}e^{\widehat{a}m}||\xi||.\end{aligned}$$

Thus $\xi \in K_x^m$ and $E_y \subset K_x^m$. By symmetry we also have $E_x \subset K_y^m$. It follows then from (4.1) and (4.2) that

$$d(E_x, E_y) \leq 3\widehat{C}^2 e^{(\widehat{a}-\widehat{b})m} \leq 3\widehat{C}^2 e^{\widehat{b}-\widehat{a}}[d(x,y)]^{[(\widehat{a}-\widehat{b})/(\widehat{a}-\widehat{c})]\beta}.$$

The proof is completed. □

Now we turn to the case of subbundles of TM. As before, let $\rho_0 > 0$ be as introduced in Section II.1. If $x, y \in M$ with $d(x,y) < \rho_0$, we denote by $P(x,y)$ the isometry from T_xM to T_yM defined by the parallel displacement along the unique shortest geodesic connecting x and y. Then for any $x, y \in M$, if E_x is a subspace of T_xM and E_y is a subspace of T_yM, define

$$d(E_x, E_y) = \begin{cases} 1 & \text{if } d(x,y) \geq \rho_0/4, \\ d(E_x, P(y,x)E_y) & \text{if } d(x,y) < \rho_0/4. \end{cases} \tag{4.3}$$

Analogously to Definition 4.1 we introduce

Definition 4.2. *Let $\Delta \subset M$ be a set. A family $\{E_x\}_{x\in\Delta}$ of subspaces $E_x \subset T_xM$ is called Hölder continuous in x on Δ with exponent $\alpha > 0$ and constant $L > 0$, if for any $x, y \in \Delta$*

$$d(E_x, E_y) \leq L[d(x,y)]^{\alpha}.$$

Let $x, x', y, y' \in M$ and let $A : T_xM \to T_{x'}M, B : T_yM \to T_{y'}M$ be linear maps. We introduce the following distance:

$$d(A,B) = \begin{cases} |A| + |B| & \text{if } \max\{d(x,y), d(x',y')\} \geq \rho_0/4, \\ |A - P(y',x') \circ B \circ P(x,y)| & \text{otherwise.} \end{cases} \tag{4.4}$$

And then for a differentiable map $f : M \to M$ and for a number $\sigma, 0 < \sigma \leq 1$, we define

$$|Tf|_{H^\sigma} = |f|_{C^1} + \sup_{x,y\in M} \frac{d(T_xf, T_yf)}{[d(x,y)]^{\sigma}} \tag{4.5}$$

where $|f|_{C^1}$ is defined to be $\sup\{|T_x f| : x \in M\}$ as usual and we admit $d(T_x f, T_y f)/[d(x,y)]^\sigma = 1$ if $x = y$.

As a consequence of Proposition 4.1 we have the following

Corollary 4.1. *Let $\{f_i\}_{i=1}^{+\infty}$ be a sequence of differentiable maps $f_i : M \to M$ such that the derivatives Tf_i satisfy*

$$\prod_{i=1}^{n} |Tf_i|_{H^\sigma} \le \widehat{C}_0 e^{\widehat{c}n}, \quad \forall n \in \mathbf{N} \tag{4.6}$$

where $\widehat{C}_0 \ge 1, \widehat{c} > 0, 0 < \sigma \le 1$. For each positive integer n, put

$$f^0 = id, f^n = f_n \circ \cdots \circ f_1.$$

Fix $\widehat{C} \ge 1$ and $\widehat{a} < \widehat{b}$ and let $\Delta_{\widehat{C},\widehat{a},\widehat{b}}$ be the (maybe empty) set of points x for which there exist splittings

$$T_x M = E_x \oplus E_x^\perp$$

such that for any $n \in \mathbf{N}$

$$|T_x f^n \xi| \le \widehat{C} e^{\widehat{a}n} |\xi|, \quad \xi \in E_x,$$

$$|T_x f^n \eta| \ge \widehat{C}^{-1} e^{\widehat{b}n} |\eta|, \quad \eta \in E_x^\perp.$$

Then the family $\{E_x\}$ is Hölder continuous in x on $\Delta_{\widehat{C},\widehat{a},\widehat{b}}$ with constant $3\widehat{C}^2 e^{\widehat{b}-\widehat{a}}$ and exponent $\alpha = [(\widehat{a} - \widehat{b})/(\widehat{a} - \widehat{d})]\sigma$, where $\widehat{d} = \ln(2\widehat{C}_0^2) + 2\widehat{c} + |\sigma \ln(\rho_0/4)| + |\widehat{a}|$.

In order to deduce this corollary from Proposition 4.1 we need the following lemma.

Lemma 4.1. *For any $x, y \in M$ and any $n \in \mathbf{N}$*

$$d(T_x f^n, T_y f^n) \le e^{\widehat{d}n} [d(x,y)]^\sigma. \tag{4.7}$$

Proof. Let $x, y \in M$. Assume that $d(f^m x, f^m y) \ge \rho_0/4$ for some $m \ge 0$. Then

$$\rho_0/4 \le d(f^m x, f^m y) \le |f^m|_{C^1} d(x,y)$$

$$\le \begin{cases} d(x,y) & \text{if } m = 0 \\ \prod_{i=1}^{m} |Tf_i|_{H^\sigma} d(x,y) & \text{if } m > 0, \end{cases}$$

and hence, by (4.6), we have

$$d(x,y) \ge \frac{\rho_0}{4} \widehat{C}_0^{-1} e^{-\widehat{c}m}.$$

Therefore, by (4.4)-(4.6) and by the choice of the number $\hat{d}$, we have for any $n \geq \max\{m, 1\}$

$$\begin{aligned} d(T_x f^n, T_y f^n) \leq & \ |T_x f^n| + |T_y f^n| \leq 2 \prod_{i=1}^{n} |Tf_i|_{H^\sigma} \\ \leq & \ 2\widehat{C}_0 e^{\widehat{c}n} \leq e^{\widehat{d}n} [d(x,y)]^\sigma . \end{aligned}$$

Hence it suffices to prove (4.7) for all positive integers n which satisfy $d(f^i x, f^i y) < \rho_0/4$ for every $0 \leq i \leq n$. It is easy to see that for such an integer n, (4.7) follows from (4.6) and the following inequality

$$d(T_x f^n, T_y f^n) \leq \left[\prod_{i=1}^{n} |Tf_i|_{H^\sigma} \right]^2 [d(x,y)]^\sigma \tag{4.8}$$

which we now prove by induction. For $n = 1$ (4.8) follows from (4.5). Suppose now that (4.8) holds true for $n = k$. Then, by (4.4), (4.5) and the inductive assumption, we have for $n = k + 1$

$$\begin{aligned}
& d(T_x f^{k+1}, T_y f^{k+1}) \\
= & \ |T_x f^{k+1} - P(f^{k+1}y, f^{k+1}x) \circ T_y f^{k+1} \circ P(x,y)| \\
= & \ |T_{f^k x} f_{k+1} \circ T_x f^k - P(f^{k+1}y, f^{k+1}x) \circ T_{f^k y} f_{k+1} \circ P(f^k x, f^k y) \\
& \circ P(f^k y, f^k x) \circ T_y f^k \circ P(x,y)| \\
\leq & \ |T_{f^k x} f_{k+1} \circ T_x f^k - P(f^{k+1}y, f^{k+1}x) \circ T_{f^k y} f_{k+1} \circ P(f^k x, f^k y) \circ T_x f^k| \\
& + |P(f^{k+1}y, f^{k+1}x) \circ T_{f^k y} f_{k+1} \circ P(f^k x, f^k y) \circ T_x f^k \\
& - P(f^{k+1}y, f^{k+1}x) \circ T_{f^k y} f_{k+1} \circ P(f^k x, f^k y) \circ P(f^k y, f^k x) \circ T_y f^k \circ P(x,y)| \\
\leq & \ d(T_{f^k x} f_{k+1}, T_{f^k y} f_{k+1}) |T_x f^k| + |T_{f^k y} f_{k+1}| d(T_x f^k, T_y f^k) \\
\leq & \ [|Tf_{k+1}|_{H^\sigma} - |f_{k+1}|_{C^1}][d(f^k x, f^k y)]^\sigma |T_x f^k| \\
& + |f_{k+1}|_{C^1} \left[\prod_{i=1}^{k} |Tf_i|_{H^\sigma} \right]^2 [d(x,y)]^\sigma
\end{aligned}$$

$$\leq \left[|Tf_{k+1}|_{H^\sigma} - |f_{k+1}|_{C^1}\right]\left[\prod_{i=1}^{k}|f_i|_{C^1}\right]^{1+\sigma}[d(x,y)]^\sigma$$

$$+|f_{k+1}|_{C^1}\left[\prod_{i=1}^{k}|Tf_i|_{H^\sigma}\right]^2[d(x,y)]^\sigma$$

$$\leq \left[\prod_{i=1}^{k+1}|Tf_i|_{H^\sigma}\right]^2[d(x,y)]^\sigma.$$

The proof is completed. □

Proof of Corollary 4.1. Let x and y be two points in $\Delta_{\widehat{C},\widehat{a},\widehat{b}}$. If $d(x,y) \geq \rho_0/4$ or $d(x,y) \geq 1$, it holds obviously that

$$d(E_x, E_y) \leq 3\widehat{C}^2 e^{\widehat{b}-\widehat{a}}[d(x,y)]^\alpha. \tag{4.9}$$

We now assume that $d(x,y) < \min\{\rho_0/4, 1\}$. By means of parallel displacement we can find for every $n \geq 0$ an isometry $P(f^n x, x) : T_{f^n x}M \to T_x M$ and an isometry $P(f^n y, f^n x) : T_{f^n y}M \to T_{f^n x}M$ such that $P(x,x) = id$ and, if $d(f^n y, f^n x) < \rho_0/4, P(f^n y, f^n x)$ is the isometry defined by the parallel displacement along the unique shortest geodesic connecting $f^n y$ and $f^n x$. Set $X = \{x,y\}, H = T_x M$ and for every $i \in \mathbf{N}$ define

$$T_i(x) = P(f^i x, x) \circ T_{f^{i-1}x} f_i \circ P(f^{i-1}x, x)^{-1},$$

$$T_i(y) = P(f^i x, x) \circ P(f^i y, f^i x) \circ T_{f^{i-1}y} f_i \circ P(f^{i-1}y, f^{i-1}x)^{-1} \circ P(f^{i-1}x, x)^{-1}.$$

For each $n \in \mathbf{N}$, by (4.4) and Lemma 4.1, we have

$$\begin{aligned}
&|T^n(x) - T^n(y)| \\
&= |P(f^n x, x) \circ T_x f^n - P(f^n x, x) \circ P(f^n y, f^n x) \circ T_y f^n \circ P(x,y)| \\
&= |T_x f^n - P(f^n y, f^n x) \circ T_y f^n \circ P(x,y)| \\
&\leq d(T_x f^n, T_y f^n) \leq e^{\widehat{d}n}[d(x,y)]^\sigma.
\end{aligned}$$

Then, by (4.3) and Proposition 4.1, we obtain (4.9) in this case. Thus, the corollary is proved. □

Lemma 4.2. *For system $\mathcal{X}^+(M,v)$ the following hold true:*

1) $\int \log|Tf_0(\omega)|_{H^1} dv^{\mathbf{N}} \stackrel{\text{def}}{=} \widehat{c}_0 < +\infty;$ *(4.10)*

2) There exists a Borel set $\Gamma'_0 \subset \Omega^{\mathbf{N}}$ with $v^{\mathbf{N}}(\Gamma'_0) = 1, \tau\Gamma'_0 \subset \Gamma'_0$ and there exists a Borel function $C : \Gamma'_0 \to (0, +\infty)$ such that for every $\omega \in \Gamma'_0$ and $n \in \mathbf{N}$

$$\prod_{i=0}^{n-1}|Tf_i(\omega)|_{H^1} \leq C(\omega)e^{2\widehat{c}_0 n}. \tag{4.11}$$

Proof. 1) Choose a finite number of points $\{x_i\}_{i=1}^l$ such that $\{B(x_i, \rho_0/4)\}_{i=1}^l$ cover M. For each $1 \le i \le l$, put $U_i = B(x_i, \rho_0)$, $V_i = B(x_i, 7\rho_0/12)$ and $W_i = B(x_i, \rho_0/2)$ and let (U_i, φ_i) be a normal coordinate neighborhood. Given $1 \le i \le l$, for $z \in V_i$ let $p(t), 0 \le t \le 1$ be a geodesic satisfying $p(0) = z$ and $(dp/dt)_{t=0} = \xi$ with $|\xi| < \rho_0/3$. Set $p(1) = z'$. We write $\varphi_i(z) = q, T\varphi_i\xi = u$ and $\varphi_i(z') = q'$ and, with respect to the natural basis $\{e_j\}_{j=1}^{m_0}$ of $\mathbf{R}^{m_0}$, we express the linear map $T_{z'}\varphi_i \circ P(z, z') \circ (T_z\varphi_i)^{-1}$ as an $m_0 \times m_0$ matrix $A_i(q, u)$, where $P(z, z')$ is the isometry from T_zM to $T_{z'}M$ defined by the parallel displacement along $p(t), 0 \le t \le 1$. From the characterizations in local charts of geodesics and parallel displacement (see Chapter VII of [Boo]) it follows that $A_i(q, u)$ is a C^∞ matrix function of $(q, u) \in T\varphi_i\{\eta \in T_{V_i}M : |\eta| < \rho_0/3\}$. Therefore, using the notation $A_i(q, q')$ to denote $A_i(q, u)$, we have

$$||A_i(q, q') - id||_0 \le C_i||q - q'||_0 \tag{4.12}$$

for any $(q, u) \in T\varphi_i\{\eta \in T_{W_i}M : |\eta| < \rho_0/4\}$, where C_i is a constant depending only on the chart (U_i, φ_i).

Now let $x, y \in M$ and $f \in \Omega$. If $\max\{d(x, y), d(fx, fy)\} \ge \rho_0/4$, then

$$\frac{d(T_xf, T_yf)}{d(x, y)} = \frac{|T_xf| + |T_yf|}{d(x, y)} \le 8(\max\{1, |f|_{C^1}\})^2\rho_0^{-1}. \tag{4.13}$$

We assume now that $\max\{d(x, y), d(fx, fy)\} < \rho_0/4$ and $x \ne y$. Suppose that $x \in B(x_i, \rho_0/4)$ and $fx \in B(x_j, \rho_0/4)$ and let F_{ij} denote the restriction of $\varphi_j \circ f \circ \varphi_i^{-1}$ to $\varphi_i(W_i \cap f^{-1}W_j)$. Put $p = \varphi_i(x)$ and $q = \varphi_i(y)$. Then

$$\begin{aligned}
&||T_pF_{ij} - A_j(F_{ij}(q), F_{ij}(p)) \circ T_qF_{ij} \circ A_i(p, q)||_0 \\
\le\ & ||T_pF_{ij} - T_qF_{ij}||_0 + ||T_qF_{ij} - T_qF_{ij} \circ A_i(p, q)||_0 \\
&+ ||T_qF_{ij} \circ A_i(p, q) - A_j(F_{ij}(q), F_{ij}(p)) \circ T_qF_{ij} \circ A_i(p, q)||_0 \\
\le\ & ||T_pF_{ij} - T_qF_{ij}||_0 + ||T_qF_{ij}||_0||id - A_i(p, q)||_0 \\
&+ ||T_qF_{ij} \circ A_i(p, q)||_0||id - A_j(F_{ij}(q), F_{ij}(p))||_0
\end{aligned}$$

which together with (4.12) yields

$$|T_xf - P(fy, fx) \circ T_yf \circ P(x, y)| \le C_{ij}(\max\{1, |f|_{C^2}\})^2 d(x, y) \tag{4.14}$$

where C_{ij} is a constant depending only on the charts $(U_i, \varphi_i), (U_j, \varphi_j)$ and the definition of $|f|_{C^2}$ (see Section I. 1). Therefore, by (4.5), (4.13) and (4.14), there exists a number $C_0' > 0$ such that for every $f \in \Omega$

$$|Tf|_{H^1} \le C_0'(\max\{1, |f|_{C^2}\})^2$$

which together with the condition $\int \log^+ |f|_{C^2} dv(f) < +\infty$ yields (4.10).

2) Since $\tau : (\Omega^{\mathbf{N}}, v^{\mathbf{N}}) \hookleftarrow$ is ergodic, by Birkhoff ergodic theorem and (4.10) we have as $n \to +\infty$

$$\frac{1}{n} \log \prod_{k=0}^{n-1} |f_0(\tau^k \omega)|_{H^1} \to \int \log |f_0(\omega)|_{H^1} dv^{\mathbf{N}} \qquad v^{\mathbf{N}} - \text{a.e.} \tag{4.15}$$

Then it follows that there exists a Borel set $\Gamma'_0 \subset \Omega^{\mathbf{N}}$ such that $v^{\mathbf{N}}(\Gamma'_0) = 1, \tau\Gamma'_0 \subset \Gamma'_0$ and (4.15) holds for each $\omega \in \Gamma'_0$. For each $\omega \in \Gamma'_0$ define

$$C(\omega) = \sup \left\{ \left(\prod_{k=0}^{n-1} |f_k(\omega)|_{H^1} \right) e^{-2\hat{c}_0 n} : \quad n = 1, 2, \cdots \right\}.$$

It is obvious that Γ'_0 and the function $C(\omega)$ satisfy our requirements. □

In order to prove the main result of this section (Theorem 4.1) we also need the following two auxiliary lemmas.

Lemma 4.3. *Let $(E_i, \|\cdot\|)$ and $(F_i, \|\cdot\|), i = 1, 2$ be Banach spaces, and let $E = E_1 \oplus E_2$ and $F = F_1 \oplus F_2$ with respectively the norms $\|(\xi, \eta)\| = \max\{\|\xi\|, \|\eta\|\}, (\xi, \eta) \in E_1 \oplus E_2$ and $\|(\xi', \eta')\| = \max\{\|\xi'\|, \|\eta'\|\}, (\xi', \eta') \in F_1 \oplus F_2$. Assume that*

$$A = \begin{bmatrix} A_{11} & A_{12} \\ A_{21} & A_{22} \end{bmatrix} : E_1 \oplus E_2 \to F_1 \oplus F_2, (\xi, \eta) \longmapsto (A_{11}\xi + A_{12}\eta, A_{21}\xi + A_{22}\eta)$$

is a linear map satisfying

$$\|A_{11}^{-1}\| \le b_0^{-1}, \quad \|A_{22}\| \le a_0, \|A_{12}\| \le \delta_0, \|A_{21}\| \le \delta_0$$

where $0 < a_0 < b_0, 0 < \delta_0 < b_0$ and $d_0 \stackrel{\text{def}}{=} (a_0 + \delta_0)/(b_0 - \delta_0) \le 1$. Then, if $P : E_1 \to E_2$ is a linear map with $\|P\| \le 1$, there exists a linear map $Q : F_1 \to F_2$ such that $\|Q\| \le 1$ and

$$A\text{Graph}(P) = \text{Graph}(Q).$$

Moreover, for each $(\xi, \eta) \in \text{Graph}(P)$ we have

$$\|A(\xi, \eta)\| \ge (b_0 - \delta_0)\|(\xi, \eta)\|.$$

Proof. Since

$$\|A_{12}P\| \le \|A_{12}\|\|P\| \le \delta_0 < \|A_{11}^{-1}\|^{-1},$$

we know that $A_{11} + A_{12}P : E_1 \to F_1$ is invertible and $\|(A_{11} + A_{12}P)^{-1}\| \le (b_0 - \delta_0)^{-1}$. Define

$$Q = (A_{21} + A_{22}P)(A_{11} + A_{12}P)^{-1} : F_1 \to F_2.$$

Then it is clear that $||Q|| \le d_0 \le 1$ and

$$A\text{Graph}(P) = \text{Graph}(Q).$$

In addition, since $||P|| \le 1$ and $||Q|| \le 1$, for each $(\xi, \eta) \in \text{Graph}(P)$ we have

$$\begin{aligned} ||A(\xi,\eta)|| = & \quad ||(A_{11}\xi + A_{12}\eta, A_{21}\xi + A_{22}\eta)|| \\ = & \quad ||A_{11}\xi + A_{12}\eta|| \ge b_0||\xi|| - \delta_0||\eta|| \\ = & \quad b_0||(\xi,\eta)|| - \delta_0||\eta|| \ge (b_0 - \delta_0)||(\xi,\eta)||. \end{aligned}$$

The proof is completed. □

Lemma 4.4. *Let $(E_1, \langle \ , \ \rangle), (E_2, \langle \ , \ \rangle)$ be inner product spaces of finite dimension and let $F_1 \subset E_1, F_2 \subset E_2$ be subspaces. Write $E_1 = F_1 \oplus F_1^{\perp}$ and $E_2 = F_2 \oplus F_2^{\perp}$. Given $0 < \alpha \le 1$, suppose that $A : E_1 \to E_2$ is an invertible linear map such that*

$$\max\{||A||, ||A^{-1}||\} \le (1 + \frac{\alpha^2}{4})^{\frac{1}{4}} \tag{4.16}$$

and

$$AF_1 = \text{Graph}(P)$$

where $P : F_2 \to F_2^{\perp}$ is a linear map with $||P|| \le \alpha/2$. Then there exists a linear map $Q : F_2^{\perp} \to F_2$ such that $||Q|| \le \alpha$ and

$$AF_1^{\perp} = \text{Graph}(Q).$$

Proof. Let $\eta \in AF_1^{\perp} \cap F_2$ and assume that $\eta \ne 0$. Since $A^{-1}\eta \in F_1^{\perp}$ and $A^{-1}(\eta + P\eta) \in F_1$, we have

$$\langle A^{-1}\eta, A^{-1}(\eta + P\eta)\rangle = \langle A^{-1}\eta, A^{-1}\eta\rangle + \langle A^{-1}\eta, A^{-1}P\eta\rangle = 0$$

which together with (4.16) and $||P|| \le \alpha/2$ yields that

$$1 \le \frac{4}{6}.$$

This is a contradiction. Hence

$$AF_1^{\perp} \cap F_2 = \{0\}. \tag{4.17}$$

(4.17) together with $\dim AF_1^{\perp} = \dim F_2^{\perp}$ yields immediately that there exists a linear map $Q : F_2^{\perp} \to F_2$ such that

$$AF_1^{\perp} = \text{Graph}(Q).$$

We now show that $||Q|| \le \alpha$. Let $\eta \in F_2^\perp$ with $||\eta|| = 1$. Since $A^{-1}(Q\eta+\eta) \in F_1^\perp$ and $A^{-1}(Q\eta + PQ\eta) \in F_1$, we have

$$||A^{-1}(Q\eta+\eta) + A^{-1}(Q\eta + PQ\eta)||^2 = ||A^{-1}(Q\eta+\eta)||^2 + ||A^{-1}(Q\eta + PQ\eta)||^2.$$

Using (4.16) and $||P|| \le \alpha/2$, by a simple calculation we obtain from the above equation

$$||Q\eta|| \le \alpha$$

which implies

$$||Q|| \le \alpha,$$

completing the proof. □

From now on we keep the notations of Sections 1 and 3. Fix $\rho'_0, 0 < \rho'_0 < \rho_0$, such that for every $x \in M$, if $\xi \in T_xM$ and $|\xi| \le \rho'_0$, then

$$\max\{|T_\xi \exp_x|, |(T_\xi \exp_x)^{-1}|\} \le [1 + \frac{1}{4}(\frac{1}{2A})^2]^{\frac{1}{4}}. \tag{4.18}$$

And define

$$r'_0 = \min\left\{ \left(\frac{1}{4A} \cdot \frac{1-e^{-2\varepsilon}}{1+e^{2\varepsilon}}\right)^2 r_0, \frac{1}{4}\rho'_0 \right\}. \tag{4.19}$$

Let $\Lambda^{l',r'}_{a,b,k,\varepsilon}$ be a set as considered in Theorem 3.1. For point (ω, x) in this set we put

$$\widehat{W}(\omega, x) = \exp_x \mathrm{Graph}(h_{(\omega,x),0}|_{\{\xi \in E_0(\omega,x): ||\xi||_{(\omega,x),0} < r'_0\}}) \tag{4.20}$$

which is obviously an open subset of $W_0(\omega, x)$.

Choose arbitrarily $C' \ge 1$ such that the Borel set

$$\Lambda^{l',r',C'}_{a,b,k,\varepsilon} \stackrel{\text{def}}{=} \{(\omega, x) \in \Lambda^{l',r'}_{a,b,k,\varepsilon} : \omega \in \Gamma'_0 \text{ and } C(\omega) \le C'\} \tag{4.21}$$

is not empty. From now up to the end of this chapter we fix such a set $\Lambda^{l',r',C'}_{a,b,k,\varepsilon}$ and write

$$\Delta = \Lambda^{l',r',C'}_{a,b,k,\varepsilon} \tag{4.22}$$

for simplicity of notations.

Given $\omega \in \Omega^{\mathbf{N}}$, we put

$$\Delta_\omega = \{x : (\omega, x) \in \Delta\}$$

and assume that $\Delta_\omega \ne \phi$. We then introduce the Borel set

$$\widehat{W}(\Delta_\omega) = \bigcup_{x \in \Delta_\omega} \widehat{W}(\omega, x). \tag{4.23}$$

Let $y \in \widehat{W}(\Delta_\omega)$ and assume that $y \in \widehat{W}(\omega, x)$. Denote by $E_\omega(y)$ the subspace of T_yM tangent to $\widehat{W}(\omega, x)$. Notice that if y also lies in $\widehat{W}(\omega, x')$ then the tangent spaces of $\widehat{W}(\omega, x')$ and $\widehat{W}(\omega, x)$ at point y coincide, hence $E_\omega(y)$ is defined independent of the choice of $\widehat{W}(\omega, x)$ which contains y.

Theorem 4.1. *Let $\{E_\omega(y)\}_{y \in \widehat{W}(\Delta_\omega)}$ be as defined above. Then this family is Hölder continuous in y on $\widehat{W}(\Delta_\omega)$ with constant $12A^2[b(\rho_0/2)]^2$ and exponent $\alpha = (a - b + 20\varepsilon)/(a + 10\varepsilon - d)$, where $d = \ln(2C')^2 + 2\widehat{c}_0 + |\ln(\rho_0/4)| + |a + 10\varepsilon|$.*

Proof. Let $y \in \widehat{W}(\Delta_\omega)$ and assume that $y \in \widehat{W}(\omega, x)$. Write

$$T_x M = E_\omega(x) \oplus E_\omega(x)^\perp, \quad T_y M = E_\omega(y) \oplus E_\omega(y)^\perp.$$

By 3) of Lemma 1.3, (3.2) and (4.19) we have $d(x, y) < \rho'_0$ and

$$(T_y \exp_x^{-1}) E_\omega(y) = \mathrm{Graph}(P)$$

where $P : E_\omega(x) \to E_\omega(x)^\perp$ is a linear map satisfying $|P| \leq (4A)^{-1}$. According to Lemma 4.4, from (4.18) it follows then that there exists a linear map $Q : E_\omega(x)^\perp \to E_\omega(x)$ such that $|Q| \leq (2A)^{-1}$ and

$$(T_y \exp_x^{-1}) E_\omega(y)^\perp = \mathrm{Graph}(Q).$$

By 3) of Lemma 1.3 we have $\|Q\|_{(\omega,x),0} \leq 1$. Denote $\zeta = \exp_x^{-1} y$ and write for $l \in \mathbf{Z}^+$

$$T_{F_0^l(\omega,x)\zeta} F_{(\omega,x),l} = \begin{bmatrix} A_{11}^{(l)} & A_{12}^{(l)} \\ A_{21}^{(l)} & A_{22}^{(l)} \end{bmatrix} : H_l(\omega, x) \oplus E_l(\omega, x) \to H_{l+1}(\omega, x) \oplus E_{l+1}(\omega, x).$$

From Step 1 of the proof of Theorem 3.1 and (3.5) and Lemma 1.3 it follows that

$$\|(A_{11}^{(l)})^{-1}\| \leq (e^{b-2\varepsilon} - \varepsilon_0)^{-1}, \quad \|A_{22}^{(l)}\| \leq e^{a+2\varepsilon} + \varepsilon_0, \quad \max\{\|A_{12}^{(l)}\|, \|A_{21}^{(l)}\|\} \leq \varepsilon_0,$$

where $\|\cdot\|$ is defined with respect to $\|\cdot\|_{(\omega,x),l}$ and $\|\cdot\|_{(\omega,x),l+1}$. Applying Lemma 4.3, we obtain for every $\eta \in \mathrm{Graph}(Q)$ and $n \in \mathbf{Z}^+$

$$\|T_\zeta F_0^n(\omega, x)\eta\|_{(\omega,x),n} \geq (e^{b-2\varepsilon} - \varepsilon_0)^n \|\eta\|_{(\omega,x),0}$$

which together with 3) of Lemma 1.3 yields

$$|T_y f_\omega^n \eta'| \geq C^{-1} e^{(b-10\varepsilon)n} |\eta'|, \quad \forall \eta' \in E_\omega(y)^\perp$$

where $C = 2A[b(\rho_0/2)]^2$. (3.5) together with 3) of Lemma 1.3 results in

$$|T_y f_\omega^n \xi'| \leq C e^{(a+10\varepsilon)n} |\xi'|, \quad \forall \xi' \in E_\omega(y).$$

The theorem follows then from Corollary 4.1. □

§5 Absolute Continuity of Families of Submanifolds

We keep here the previous notations. Let $\Delta = \Lambda_{a,b,k,\varepsilon}^{l',r',C'}$ be the Borel set introduced in (4.22). Theorem 0.1.4 allows us to choose a sequence of compact sets $\{\Delta^l\}_{l=1}^{+\infty}$ such that $\Delta^l \subset \Delta, \Delta^l \subset \Delta^{l+1}$ and $v^{\mathbf{N}} \times \mu(\Delta \backslash \Delta^l) \leq l^{-1}$ for every $l \geq 1$. We now fix arbitrarily such a set Δ^l.

For $(\omega, x) \in \Delta$ and sufficiently small $r > 0$ we put

$$U_{\Delta,\omega}(x, r) = \exp_x\{\zeta \in T_x M : \|\zeta\|_{(\omega,x),0} < r\}, \tag{5.1}$$

and if $(\omega, x) \in \Delta^l$ we put

$$V_{\Delta^l}((\omega, x), r) = \{(\omega', x') \in \Delta^l : d(\omega', \omega) < r, x' \in U_{\Delta,\omega}(x, r)\}. \tag{5.2}$$

From the formulation and proof of Theorem 3.1 together with the compactness of Δ^l it follows immediately that there exists a number $\delta_{\Delta^l} > 0$ such that for each $(\omega, x) \in \Delta^l$, if $(\omega', x') \in V_{\Delta^l}((\omega, x), q/2), 0 < q \leq \delta_{\Delta^l}$, there is a C^1 map $\varphi : \{\xi \in E_0(\omega, x) : \|\xi\|_{(\omega,x),0} < q\} \to H_0(\omega, x)$ satisfying

$$\exp_x^{-1}[\widehat{W}(\omega', x') \cap U_{\Delta,\omega}(x, q)] = \text{Graph}(\varphi) \tag{5.3}$$

and

$$\sup\{\|T_\xi \varphi\|_{(\omega,x),0} : \xi \in E_0(\omega, x), \|\xi\|_{(\omega,x),0} < q\} \leq \frac{1}{3}. \tag{5.4}$$

Let $(\omega, x) \in \Delta^l$ and $0 < q \leq \delta_{\Delta^l}$. We denote by $\mathcal{F}_{\Delta^l_\omega}(x, q)$ the collection of submanifolds $\widehat{W}(\omega, y)$ passing through $y \in \Delta^l_\omega \cap U_{\Delta,\omega}(x, q/2)$. Set

$$\widetilde{\Delta}^l_\omega(x, q) = \bigcup_{y \in \Delta^l_\omega \cap U_{\Delta,\omega}(x, q/2)} \widehat{W}(\omega, y) \cap U_{\Delta,\omega}(x, q). \tag{5.5}$$

Definition 5.1. *A submanifold W of M is called transversal to the family $\mathcal{F}_{\Delta^l_\omega}(x, q)$ if the following hold true: i) $W \subset U_{\Delta,\omega}(x, q)$ and $\exp_x^{-1} W$ is the graph of a C^1 map $\psi : \{\eta \in H_0(\omega, x) : \|\eta\|_{(\omega,x),0} < q\} \to E_0(\omega, x)$; ii) W intersects any $\widehat{W}(\omega, y), y \in \Delta^l_\omega \cap U_{\Delta,\omega}(x, q/2)$, at exactly one point and this intersection is transversal, i.e. $T_z W \oplus T_z \widehat{W}(\omega, y) = T_z M$ where $z = W \cap \widehat{W}(\omega, y)$.*

For a submanifold W transversal to $\mathcal{F}_{\Delta^l_\omega}(x, q)$ we define

$$\|W\| = \sup \|\psi(\eta)\|_{(\omega,x),0} + \sup \|T_\eta \psi\|_{(\omega,x),0} \tag{5.6}$$

where the supremums are taken over the set $\{\eta \in H_0(\omega, x) : \|\eta\|_{(\omega,x),0} < q\}$ and ψ is defined as above. And we shall denote by λ_W the Lebesgue measure on W induced by the Riemannian metric on W inherited from M.

Consider now two submanifolds W^1 and W^2 transversal to $\mathcal{F}_{\Delta^l_\omega}(x,q)$. Since, by Theorem 3.1 and (4.20), $\{\widehat{W}(\omega,y)\}_{y\in\Delta_\omega}$ is a continuous family of C^1 embedded discs, there exist two open submanifolds $\widehat{W}^1$ and $\widehat{W}^2$ respectively of W^1 and W^2 such that we can well define a so-called Poincaré map

$$P_{\widehat{W}^1,\widehat{W}^2}:\widehat{W}^1\cap\tilde{\Delta}^l_\omega(x,q)\to\widehat{W}^2\cap\tilde{\Delta}^l_\omega(x,q)$$

by letting

$$P_{\widehat{W}^1,\widehat{W}^2}: z\longmapsto\widehat{W}^2\cap\widehat{W}(\omega,y)$$

for $z=\widehat{W}^1\cap\widehat{W}(\omega,y), y\in\Delta^l_\omega\cap U_{\Delta,\omega}(x,q/2)$, and moreover, $P_{\widehat{W}^1,\widehat{W}^2}$ is a homeomorphism.

Definition 5.2. *The family $\mathcal{F}_{\Delta^l_\omega}(x,q)$ is said to be absolutely continuous if there exists a number $\varepsilon_{\Delta^l_\omega}(x,q)>0$ such that, for any two submanifolds W^1 and W^2 transversal to $\mathcal{F}_{\Delta^l_\omega}(x,q)$ and satisfying $\|W^i\|\le\varepsilon_{\Delta^l_\omega}(x,q), i=1,2$, every Poincaré map $P_{\widehat{W}^1,\widehat{W}^2}$ constructed as above is absolutely continuous with respect to λ_{W^1} and λ_{W^2}.*

Besides Theorem 3.1, which describes the existence of invariant families of local stable manifolds corresponding to Lyapunov exponents smaller than a fixed number $a<0$, we have now the following other main result of this chapter which deals with absolute continuity of such families. As before, let λ denote the Lebesgue measure on M.

Theorem 5.1. *Let Δ^l be given as above. There exist constants $0<q_{\Delta^l}\le\delta_{\Delta^l},\varepsilon_{\Delta^l}>0$ and $J_{\Delta^l}>0$ such that the following hold true for each $(\omega,x)\in\Delta^l$:*

1) The family $\mathcal{F}_{\Delta^l_\omega}(x,q_{\Delta^l})$ is absolutely continuous.

2) If $\lambda(\Delta^l_\omega)>0$ and x is a density point of Δ^l_ω with respect to λ, then for every two submanifolds W^1 and W^2 transversal to $\mathcal{F}_{\Delta^l_\omega}(x,q_{\Delta^l})$ and satisfying $\|W^i\|\le\varepsilon_{\Delta^l}, i=1,2$ any Poincaré map $P_{\widehat{W}^1,\widehat{W}^2}$ is absolutely continuous and the Jacobian $J(P_{\widehat{W}^1,\widehat{W}^2})$ satisfies the inequality

$$J^{-1}_{\Delta^l}\le J(P_{\widehat{W}^1,\widehat{W}^2})(y)\le J_{\Delta^l}\tag{5.7}$$

for λ_{W^1}-almost all points $y\in\widehat{W}^1\cap\tilde{\Delta}^l_\omega(x,q_{\Delta^l})$.

Because a detailed proof of the above absolute continuity theorem would involve too much work and it can be carried out by a completely parallel argument with that of Part II of [Kat] (for the deterministic case), we omit it here and we refer the interested reader to Part II of [Kat] for an analogous argument.

Roughly speaking, in smooth ergodic theory of deterministic dynamical systems, presence in a smooth system of an absolutely continuous family of local stable manifolds ensures many important ergodic properties of the system, for example, the positiveness of entropy and the Bernoulli property etc. (see [Pes]$_2$).

It would be seen that, in ergodic theory of systems generated by random diffeomorphisms as considered in this book, the presence in such a system of an absolutely continuous family of local stable manifolds also plays a very important role. For example, it will enable us to estimate the entropy of the system through its Lyapunov exponents (see Chapter IV).

§6 Absolute Continuity of Conditional Measures

The purpose of the present section is to prove an important theorem (Theorem 6.1) which is a consequence of the absolute continuity theorem (Theorem 5.1) and of Fubini Theorem. Roughly speaking, this theorem asserts that the conditional measures induced on local stable manifolds of random diffeomorphisms by an absolutely continuous measure on M are absolutely continuous on these submanifolds. The proof of this theorem presented here is adopted from that of a similar result for the deterministic case (Theorem II.11.1) in [Kat], whose idea goes back to Ya. G. Sinai ([Ano]).

We shall need the following basic proposition which is a straightforward corollary of the definition of conditional measures (see Section 0.2).

Proposition 6.1. Let $(X, \mathcal{B}, \nu)$ be a Lebesgue space and let α be a measurable partition of X. If $\widehat{\nu}$ is another probability measure on $\mathcal{B}$ which is absolutely continuous with respect to ν, then for $\widehat{\nu}$-almost all $x \in X$ the conditional measure $\widehat{\nu}_{\alpha(x)}$ is absolutely continuous with respect to $\nu_{\alpha(x)}$ and

$$\frac{d\widehat{\nu}_{\alpha(x)}}{d\nu_{\alpha(x)}} = \frac{g|_{\alpha(x)}}{\displaystyle\int_{\alpha(x)} g d\nu_{\alpha(x)}} \tag{6.1}$$

where $g = d\widehat{\nu}/d\nu$.

Let Δ^l be a set as introduced in Section 5. Beginning from now, we suppose that the numbers q_{Δ^l} and ε_{Δ^l} in Theorem 5.1 satisfy $q_{\Delta^l} = \varepsilon_{\Delta^l}$. Note that this last assumption does not present any restriction of generality.

Henceforth, we confine ourselves to an arbitrarily fixed point $(\omega, x) \in \Delta^l$ which is such that $\lambda(\Delta^l_\omega) > 0$ and x is a density point of Δ^l_ω with respect to λ. We introduce now the following notations.

$\widehat{U} : U_{\Delta,\omega}(x, q_{\Delta^l})$;
$\widehat{B}^1 : \{\xi \in E_0(\omega, x) : \|\xi\|_{(\omega,x),0} < q_{\Delta^l}\}$;
$\widehat{B}^2 : \{\eta \in H_0(\omega, x) : \|\eta\|_{(\omega,x),0} < q_{\Delta^l}\}$;
β: the measurable partition $\{\exp_x(\{\xi\} \times \widehat{B}^2)\}_{\xi \in \widehat{B}^1}$ of $\widehat{U}$;
$I : \beta(x) \cap \widetilde{\Delta}^l_\omega(x, q_{\Delta^l})$;
α : the partition $\{\widehat{W}(\omega, y) \cap \widehat{U}\}_{y \in \Delta^l_\omega \cap U_{\Delta,\omega}(x, q_{\Delta^l}/2)}$ of $\widetilde{\Delta}^l_\omega(x, q_{\Delta^l})$;
$[N] : \cup_{z \in N} \alpha(z)$ for $N \subset I$;

β_I : the restriction of β to $[I]$;

λ^X: the normalized Lebesgue measure $\lambda/\lambda(X)$ on a Borel subset X of M with $\lambda(X) > 0$;

λ_y^β: the normalized Lebesgue measure on $\beta(y), y \in \widehat{U}$ induced by the inherited Riemannian metric;

$\lambda_z^{\beta_I} : \lambda_z^\beta / \lambda_z^\beta(\beta_I(z))$ for $z \in [I]$;

λ_z^α: the normalized Lebesgue measure on $\alpha(z), z \in [I]$ induced by the inherited Riemannian metric.

Remark 6.1. 1) It is easy to see that $[I] = \widetilde{\Delta}_\omega^l(x, q_{\Delta^l})$.

2) Since x is a density point of Δ_ω^l with respect to λ, one has $\lambda(\Delta_\omega^l \cap U_{\Delta,\omega}(x, q_{\Delta^l}/2)) > 0$ and hence $\lambda([I]) > 0$. In addition, one easily sees that α is a measurable partition of $[I]$ since $\{\widehat{W}(\omega, y)\}_{y \in \Delta_\omega^l}$ is a continuous family of C^1 embedded discs.

3) From Proposition 6.1, Fubini Theorem, Theorem 5.1 and from the fact $\lambda([I]) > 0$ it follows clearly that $\lambda_z^\beta(\beta_I(z)) > 0$ for every $z \in [I]$.

Now we formulate our main result of this section as follows:

Theorem 6.1. *Let $\{\lambda_{\alpha(z)}^{[I]}\}_{z \in [I]}$ be a canonical system of conditional measures of $\lambda^{[I]}$ associated with the measurable partition α. Then for λ-almost every $z \in [I]$ the measure $\lambda_{\alpha(z)}^{[I]}$ is equivalent to λ_z^α, moreover, the following estimate*

$$R_{\Delta^l}^{-1} \leq \frac{d\lambda_{\alpha(z)}^{[I]}}{d\lambda_z^\alpha} \leq R_{\Delta^l} \tag{6.2}$$

holds λ_z^α-almost everywhere on $\alpha(z)$, where $R_{\Delta^l} > 0$ is a number depending only on the set Δ^l but not on individual $(\omega, x) \in \Delta^l$.

Proof. In the present proof let us admit that a number marked with the subscript Δ^l such as R_{Δ^l} means that it is a number which depends only on Δ^l but not on individual $(\omega, x) \in \Delta^l$. We complete the proof by two steps.

Step 1. Let us first notice that for every $z \in [I]$ there exist $\overline{x} \in \alpha(x)$ and $\overline{y} \in I$ such that $z = \beta(\overline{x}) \cap \alpha(\overline{y})$, and $\overline{x}$ and $\overline{y}$ are uniquely determined by z. Thus we may use $(\overline{x}, \overline{y})$ as coordinates of $z \in [I]$ and we shall sometimes write $(\overline{x}, \overline{y})$ instead of z.

Let $\{\lambda_{\beta(y)}^{\widehat{U}}\}_{y \in \widehat{U}}$ be a canonical system of conditional measures of $\lambda^{\widehat{U}}$ associated with the partition β. From Fubini Theorem, Proposition 6.1 and 3) of Lemma 1.3 it follows clearly that for λ-almost all $y \in \widehat{U}$ the measure $\lambda_{\beta(y)}^{\widehat{U}}$ is equivalent to λ_y^β and there exists a number $R_{\Delta^l}^{(1)} > 0$ such that

$$(R_{\Delta^l}^{(1)})^{-1} \leq \frac{d\lambda_{\beta(y)}^{\widehat{U}}}{d\lambda_y^\beta} \leq R_{\Delta^l}^{(1)} \tag{6.3}$$

holds λ_y^β-almost everywhere on $\beta(y)$.

Let $\{\lambda_{\beta_I(z)}^{[I]}\}_{z\in[I]}$ be a canonical system of conditional measures of $\lambda^{[I]}$ associated with the partition β_I. Then by Proposition 6.1 and (6.3) we obtain immediately that for λ-almost all $z \in [I]$ the measure $\lambda_{\beta_I(z)}^{[I]}$ is equivalent to $\lambda_z^{\beta_I}$ and there exists a number $R_{\Delta^l}^{(2)} > 0$ such that

$$(R_{\Delta^l}^{(2)})^{-1} \leq \frac{d\lambda_{\beta_I(z)}^{[I]}}{d\lambda_z^{\beta_I}} \stackrel{\text{def}}{=} h_z^{(1)} \leq R_{\Delta^l}^{(2)} \tag{6.4}$$

holds $\lambda_z^{\beta_I}$-almost everywhere on $\beta_I(z)$.

Now let $\overline{x} \in \alpha(x)$. Noticing that $I = \beta(x) \cap [I]$, we define a Poincaré map

$$P_{x\overline{x}}^\alpha : I \to \beta(\overline{x}) \cap [I], \quad \overline{y} \longmapsto \beta(\overline{x}) \cap \alpha(\overline{y}).$$

Since we admit $q_{\Delta^l} = \varepsilon_{\Delta^l}$, then by the absolute continuity theorem (Theorem 5.1) we know that $\lambda_{\overline{x}}^{\beta_I} \circ P_{x\overline{x}}^\alpha$ is equivalent to $\lambda_x^{\beta_I}$ and there exists a number $R_{\Delta^l}^{(3)} > 0$ such that

$$(R_{\Delta^l}^{(3)})^{-1} \leq \frac{d(\lambda_{\overline{x}}^{\beta_I} \circ P_{x\overline{x}}^\alpha)}{d\lambda_x^{\beta_I}} \stackrel{\text{def}}{=} h_{\overline{x}}^{(2)} \leq R_{\Delta^l}^{(3)} \tag{6.5}$$

holds $\lambda_x^{\beta_I}$-almost everywhere on $\beta_I(x)$.

For every $\overline{y} \in I$ we also define a Poincaré map

$$P_{x\overline{y}}^\beta : \alpha(x) \to \alpha(\overline{y}), \quad \overline{x} \longmapsto \beta(\overline{x}) \cap \alpha(\overline{y}).$$

From the definition of the partition β it follows clearly that $\lambda_{\overline{y}}^\alpha \circ P_{x\overline{y}}^\beta$ is equivalent to λ_x^α and there exists a number $R_{\Delta^l}^{(4)} > 0$ such that

$$(R_{\Delta^l}^{(4)})^{-1} \leq \frac{d\lambda_x^\alpha}{d(\lambda_{\overline{y}}^\alpha \circ P_{x\overline{y}}^\beta)} \stackrel{\text{def}}{=} h_{\overline{y}}^{(3)} \leq R_{\Delta^l}^{(4)} \tag{6.6}$$

holds λ_x^α-almost everywhere on $\alpha(x)$.

We introduce below two other measures on $\alpha(x)$ and I. Firstly, for a Borel set $K \subset \alpha(x)$ we put $K(\beta) = \cup_{\overline{x}\in K}\beta(\overline{x})$ and define

$$\nu_x(K) = \lambda^{[I]}(K(\beta) \cap [I]).$$

Secondly, for a Borel set $N \subset I$ we define

$$\nu_I(N) = \lambda^{[I]}([N]).$$

Then ν_x and ν_I are clearly Borel probability measures on $\alpha(x)$ and I respectively. The absolute continuity theorem easily implies that ν_x is equivalent to λ_x^α and

ν_I to $\lambda_x^{\beta_I}$. Moreover, (5.7) together with Fubini Theorem yields that there exists a number $R_{\Delta^I}^{(5)} > 0$ such that

$$(R_{\Delta^I}^{(5)})^{-1} \le \frac{d\nu_x}{d\lambda_x^\alpha} \stackrel{\text{def}}{=} h^{(4)} \le R_{\Delta^I}^{(5)} \tag{6.7}$$

holds λ_x^α-almost everywhere on $\alpha(x)$ and

$$(R_{\Delta^I}^{(5)})^{-1} \le \frac{d\lambda_x^{\beta_I}}{d\nu_I} \stackrel{\text{def}}{=} h^{(5)} \le R_{\Delta^I}^{(5)} \tag{6.8}$$

holds ν_I-almost everywhere on I.

Step 2. Let $Q \subset [I]$ be an arbitrary Borel subset. From the uniqueness of the canonical system of conditional measures we know that to prove Theorem 6.1 it is sufficient to show that

$$\lambda^{[I]}(Q) = \int_I \left\{ \int_{\alpha(\overline{y})} [\chi_{Q\cap\alpha(\overline{y})}(\widehat{z}) G_{\overline{y}}(\widehat{z})] d\lambda_{\overline{y}}^\alpha(\widehat{z}) \right\} d\nu_I(\overline{y}) \tag{6.9}$$

where $\{G_{\overline{y}} : \alpha(\overline{y}) \to [0, +\infty)\}_{\overline{y}\in I}$ is a family of functions which are such that the right-hand side of (6.9) is well defined and for ν_I-almost every $\overline{y} \in I$ the following estimate

$$R_{\Delta^I}^{-1} \le G_{\overline{y}}(\widehat{z}) \le R_{\Delta^I}$$

holds for $\lambda_{\overline{y}}^\alpha$-almost all $\widehat{z} \in \alpha(\overline{y})$, where $R_{\Delta^I} > 0$ is a number as described in the formulation of Theorem 6.1.

We now begin to prove (6.9). From the properties of conditional measures, from Fubini Theorem and from (6.4)–(6.8) it follows that

$$\begin{aligned}
&\lambda^{[I]}(Q) \\
&= \int_{[I]} \left\{ \int_{\beta_I(z)} [\chi_{Q\cap\beta_I(z)}(\widehat{z})] d\lambda_{\beta_I(z)}^{[I]}(\widehat{z}) \right\} d\lambda^{[I]}(z) \\
&= \int_{\alpha(x)} \left\{ \int_{\beta_I(\overline{x})} [\chi_{Q\cap\beta_I(\overline{x})}(\widehat{z})] d\lambda_{\beta_I(\overline{x})}^{[I]}(\widehat{z}) \right\} d\nu_x(\overline{x}) \\
&= \int_{\alpha(x)} \left\{ \int_{\beta_I(\overline{x})} [\chi_{Q\cap\beta_I(\overline{x})}(\widehat{z}) h_{\overline{x}}^{(1)}(\widehat{z})] d\lambda_{\overline{x}}^{\beta_I}(\widehat{z}) \right\} h^{(4)}(\overline{x}) d\lambda_x^\alpha(\overline{x}) \\
&= \int_{\alpha(x)} \left\{ \int_I [\chi_{Q\cap\beta_I(\overline{x})}(P_{x\overline{x}}^\alpha(\overline{y})) h_{\overline{x}}^{(1)}(P_{x\overline{x}}^\alpha(\overline{y})) h_{\overline{x}}^{(2)}(\overline{y}) h^{(4)}(\overline{x})] d\lambda_{\overline{x}}^{\beta_I}(\overline{y}) \right\} d\lambda_x^\alpha(\overline{x}) \\
&= \int_{\alpha(x)} \left\{ \int_I [\chi_{Q\cap\beta_I(\overline{x})}(P_{x\overline{x}}^\alpha(\overline{y})) H(\overline{x}, \overline{y})] d\nu_I(\overline{y}) \right\} d\lambda_x^\alpha(\overline{x})
\end{aligned}$$

(where $H(\overline{x},\overline{y}) = h^{(1)}_{\overline{x}}(P^{\alpha}_{x\overline{x}}(\overline{y}))h^{(2)}_{\overline{x}}(\overline{y})h^{(4)}(\overline{x})h^{(5)}(\overline{y})$)

$$= \int_I \left\{ \int_{\alpha(x)} [\chi_{Q\cap\beta_I(\overline{x})}(P^{\alpha}_{x\overline{x}}(\overline{y}))H(\overline{x},\overline{y})]d\lambda^{\alpha}_x(\overline{x}) \right\} d\nu_I(\overline{y})$$

$$= \int_I \left\{ \int_{\alpha(x)} [\chi_{Q\cap\alpha(\overline{y})}(P^{\beta}_{x\overline{y}}(\overline{x}))H(\overline{x},\overline{y})h^{(3)}_{\overline{y}}(\overline{x})]d(\lambda^{\alpha}_{\overline{y}} \circ P^{\beta}_{x\overline{y}})(\overline{x}) \right\} d\nu_I(\overline{y})$$

$$= \int_I \left\{ \int_{\alpha(\overline{y})} [\chi_{Q\cap\alpha(\overline{y})}(\widehat{z})H((P^{\beta}_{x\overline{y}})^{-1}\widehat{z},\overline{y})h^{(3)}_{\overline{y}}((P^{\beta}_{x\overline{y}})^{-1}\widehat{z})]d\lambda^{\alpha}_{\overline{y}}(\widehat{z}) \right\} d\nu_I(\overline{y}).$$

The fact that the last integral is equal to $\lambda^{[I]}(Q)$ implies (6.9) with the family $\{G_{\overline{y}} : \alpha(\overline{y}) \to [0,+\infty)\}_{\overline{y}\in I}$ being defined by

$$G_{\overline{y}} : \widehat{z} \longmapsto H((P^{\beta}_{x\overline{y}})^{-1}\widehat{z},\overline{y})h^{(3)}_{\overline{y}}((P^{\beta}_{x\overline{y}})^{-1}\widehat{z}), \quad \widehat{z} \in \alpha(\overline{y})$$

for each $\overline{y} \in I$ and with $R_{\Delta^I} = \prod_{i=1}^{5} R^{(i)}_{\Delta^I}$. The proof is completed. □

Chapter IV Estimation of Entropy from Below Through Lyapunov Exponents

In this chapter we carry out the estimation of the entropy $h_\mu(\mathcal{X}^+(M,v))$ from below through the Lyapunov exponents for a system $\mathcal{X}^+(M,v,\mu)$. This together with the estimation from above (see Theorem II.0.1) leads to a generalization of the well-known Pesin's entropy formula to the random case. A significant gain of our elaboration is when applied to the case of diffusion processes no conditions whatever like hyperbolicity are needed anymore (see Corollary 1.1 below and see also Chapter V).

§1 Introduction and Formulation of the Main Result

In smooth ergodic theory of deterministic dynamical systems, Pesin's entropy formula asserts that, if f is a C^2 (or Hölder C^1) diffeomorphism on a compact Riemannian manifold N and if it preserves an absolutely continuous (with respect to the Lebesgue measure on N induced by the Riemannian metric) Borel probability measure μ, then

$$h_\mu(f) = \int_N \sum_i \lambda^{(i)}(x)^+ m_i(x) d\mu$$

where $\lambda^{(1)}(x) < \lambda^{(2)}(x) < \cdots < \lambda^{(r(x))}(x)$ denote the Lyapunov exponents of f at point x, $\{m_i(x)\}_{i=1}^{r(x)}$ their multiplicities respectively, and $h_\mu(f)$ denotes the usual measure-theoretic entropy of the system (N, f, μ).

The purpose of this chapter is to prove the above entropy formula in the random case of $\mathcal{X}^+(M,v,\mu)$. The formula takes the same form as in the deterministic case, but the meaning of μ is quite different since it is no longer invariant for individual random diffeomorphisms. The main result is the following

Theorem 1.1. *Let $\mathcal{X}^+(M,v,\mu)$ be given and assume that v also satisfies the condition $\log|\det T_x f| \in L^1(\Omega \times M, v \times \mu)$. If μ is absolutely continuous with respect to the Lebesgue measure on M, then*

$$h_\mu(\mathcal{X}^+(M,v)) = \int_M \sum_i \lambda^{(i)}(x)^+ m_i(x) d\mu. \tag{1.1}$$

We also call (1.1) *Pesin's entropy formula.*

Remark 1.1. Given $\mathcal{X}^+(M, v, \mu)$, if v satisfies $\log^+ |f^{-1}|_{C^1} \in L^1(\Omega, v)$, then we have $\log|\det T_x f| \in L^1(\Omega \times M, v \times \mu)$, since $\log^+ |\det T_x f| \leq \log^+ |f|_{C^1}^{m_0}$ and $\log^- |\det T_x f| \geq -\log^+ |f^{-1}|_{C^1}^{m_0}$ for every $(f, x) \in \Omega \times M$.

The above theorem was actually first given by F.Ledrappier and L.-S. Young in article $[\text{Led}]_1$. But their result was formulated and proved in the setting of a system generated by two-sided (time set $T = \mathbf{Z}$ or $\mathbf{R}$) compositions of random diffeomorphisms, and the proof is somewhat sketchy. Considering that working on time set $T = \mathbf{Z}$ or $\mathbf{R}$ is unusual from the point of view of diffusion processes, we present here a detailed treatment of this result and carry out it mainly within the more general one-sided setting of $\mathcal{X}^+(M, v, \mu)$. We hope it will be accessible to a larger audience.

We first consider a case when the condition "$\mu << \text{Leb.}$" in Theorem 1.1 is met. Suppose that $\mathcal{X}^+(M, v)$ is a system as introduced in Section I.1. The *transition probabilities* $P(x, \cdot), x \in M$ of $\mathcal{X}^+(M, v)$ are defined by

$$P(x, A) = v(\{f \in \Omega : fx \in A\}) \tag{1.2}$$

for $x \in M$ and Borel subset A of M. For each $x \in M$ the formula (1.2) defines a probability measure on M, and for each $A \in \mathcal{B}(M)$ the function $P(x, A)$ is measurable in x. We say that the transition probabilities of $\mathcal{X}^+(M, v)$ have a *density* if there exists a Borel function $p : M \times M \to \mathbf{R}^+$ such that for every $x \in M$ one has

$$P(x, A) = \int_A p(x, y) d\lambda(y)$$

for each $A \in \mathcal{B}(M)$, where λ denotes as before the Lebesgue measure on M. In this case, any $\mathcal{X}^+(M, v)$-invariant measure is actually absolutely continuous with respect to λ. In fact, let μ be such an invariant measure. Then for any $A \in \mathcal{B}(M)$

$$\begin{aligned}
\mu(A) &= \int_\Omega f\mu(A) dv(f) = \int_\Omega \int_M \chi_A(fx) d\mu(x) dv(f) \\
&= \int_M \int_\Omega \chi_A(fx) dv(f) d\mu(x) = \int_M P(x, A) d\mu(x) \\
&= \int_M \int_A p(x, y) d\lambda(y) d\mu(x) = \int_A \int_M p(x, y) d\mu(x) d\lambda(y)
\end{aligned}$$

which implies $\mu << \lambda$. Thus a consequence of Theorem 1.1 is

Corollary 1.1. *Let $\mathcal{X}^+(M, v, \mu)$ be given with v satisfying* $\log|\det T_x f| \in L^1(\Omega \times M, v \times \mu)$. *If the transition probabilities of $\mathcal{X}^+(M, v)$ have a density, Pesin's entropy formula (1.1) holds true.*

Now let us show how the random case is different from the deterministic one. The assumption that $P(x, \cdot), x \in M$ have a density is natural from the probabilistic point of view. For example, it is satisfied when $\mathcal{X}^+(M, v)$ is derived from a diffusion process with elliptic generator. In contrast, for a deterministic map $P(x,\cdot)$ is always a δ-measure.

The remainning part of this chapter is devoted to the proof of Theorem 1.1. Since we have proved in Chapter II that Ruelle's inequality holds true for any system $\mathcal{X}^+(M,v,\mu)$, it remains to prove

$$h_\mu(\mathcal{X}^+(M,v)) \geq \int \sum_i \lambda^{(i)}(x)^+ m_i(x) d\mu \tag{1.3}$$

under the conditions stated in the formulation of Theorem 1.1.

§2 Construction of A Measurable Partition

In this section we construct a special measurable partition of $\Omega^{\mathbf{N}} \times M$, by means of which the estimate (1.3) will be achieved. Roughly speaking, this partition is invariant under the action of $F : \Omega^{\mathbf{N}} \times M \hookleftarrow$ and almost every element of it is an mod 0 open piece of a stable manifold of the random diffeomorphisms. The construction is accomplished by using local stable manifolds as introduced in Section III.3. Now we go into the precise treatment.

Let $\mathcal{X}^+(M,v,\mu)$ be given and let $\hat{\Lambda}_0$ be as defined by (3.6) in Chapter III. To repeat, from the definition of $\hat{\Lambda}_0$ it is easy to see that $v^{\mathbf{N}} \times \mu(\hat{\Lambda}_0) = 1$ and $F\hat{\Lambda}_0 \subset \hat{\Lambda}_0$. Put

$$\hat{\Lambda}_1 = \{(\omega,x) \in \hat{\Lambda}_0 : \lambda^{(1)}(x) < 0\}. \tag{2.1}$$

If $(\omega,x) \in \hat{\Lambda}_1$, Remark III.3.2 asserts that the (global) stable manifold $W^s(\omega,x)$ is the image of $\mathbf{R}^{\sum_{\lambda^{(i)}(x)<0} m_i(x)}$ under an injective immersion of class $C^{1,1}$. We put $W^s(\omega,x) = \{x\}$, if $(\omega,x) \notin \hat{\Lambda}_1$.

Definition 2.1. *A measurable partition η of $\Omega^{\mathbf{N}} \times M$ is said to be subordinate to W^s-manifolds of $\mathcal{X}^+(M,v,\mu)$, if for $v^{\mathbf{N}} \times \mu$-a.e. (ω,x), $\eta_\omega(x) \overset{\text{def.}}{=} \{y : (\omega,y) \in \eta(\omega,x)\} \subset W^s(\omega,x)$ and it contains an open neighbourhood of x in $W^s(\omega,x)$, this neighbourhood being taken in the submanifold topology of $W^s(\omega,x)$.*

Definition 2.2. *We say that the Borel probability measure μ has absolutely continuous conditional measures on W^s-manifolds of $\mathcal{X}^+(M,v,\mu)$, if for any measurable partition η subordinate to W^s-manifolds of $\mathcal{X}^+(M,v,\mu)$ one has for $v^{\mathbf{N}}$-a.e. $\omega \in \Omega^{\mathbf{N}}$*

$$\mu_x^{\eta_\omega} << \lambda^s_{(\omega,x)}, \qquad \mu - a.e. x \in M \tag{2.2}$$

where $\{\mu_x^{\eta_\omega}\}_{x\in M}$ is a (essentially unique) canonical system of conditional measures of μ associated with the partition $\{\eta_\omega(x)\}_{x\in M}$ of M, and $\lambda^s_{(\omega,x)}$ is the Lebesgue measure on $W^s(\omega,x)$ induced by its inherited Riemannian metric as a submanifold of M ($\lambda^s_{(\omega,x)} = \delta_x$ if $(\omega,x) \notin \hat{\Lambda}_1$).

Remark 2.1. It is a fairly straightforward fact that, if there is a measurable partition η subordinate to W^s-manifolds of $\mathcal{X}^+(M,v,\mu)$ such that μ satisfies

(2.2) for $v^{\mathbf{N}}$-a.e. $\omega \in \Omega^{\mathbf{N}}$, then μ has generally absolutely continuous conditional measures on W^s-manifolds of $\mathcal{X}^+(M, v, \mu)$. In fact, let η' be another partition subordinate to W^s-manifolds of $\mathcal{X}^+(M, v, \mu)$. Since for $v^{\mathbf{N}} \times \mu$-a.e. (ω, x) the partitions $(\eta \vee \eta')_\omega|_{\eta_\omega(x)}$ and $(\eta \vee \eta')_\omega|_{\eta'_\omega(x)}$ are countable, from (2.1) and from the transitivity of conditional measures it follows immediately that

$$\mu_x^{(\eta\vee\eta')_\omega} << \lambda^s_{(\omega,x)}, \qquad v^{\mathbf{N}} \times \mu - \text{a.e.} \quad (\omega, x)$$

and then

$$\mu_x^{\eta'_\omega} << \lambda^s_{(\omega,x)}, \qquad v^{\mathbf{N}} \times \mu - \text{a.e.} \quad (\omega, x).$$

This confirms the above assertion.

The main purpose of this section is to prove the following

Proposition 2.1. *Let $\mathcal{X}^+(M, v, \mu)$ be given. Then there exists a measurable partition η of $\Omega^{\mathbf{N}} \times M$ which has the following properties:*

1) $F^{-1}\eta \leq \eta, \{\{\omega\} \times M : \omega \in \Omega^{\mathbf{N}}\} \leq \eta$;

2) η is subordinate to W^s-manifolds of $\mathcal{X}^+(M, v, \mu)$;

3) For every $B \in \mathcal{B}(\Omega^{\mathbf{N}} \times M)$ the function

$$P_B(\omega, x) = \lambda^s_{(\omega,x)}(\eta_\omega(x) \cap B_\omega)$$

is measurable and $v^{\mathbf{N}} \times \mu$ almost everywhere finite, where B_ω is the section $\{y : (\omega, y) \in B\}$;

4) If $\mu <<$ Leb., then for $v^{\mathbf{N}} \times \mu$-a.e. (ω, x)

$$\mu_x^{\eta_\omega} << \lambda^s_{(\omega,x)}.$$

Before going to the proof, let us first remark that property 4) above together with Remark 2.1 implies that, if $\mu <<$ Leb., then μ has absolutely continuous conditional measures on W^s-manifolds of $\mathcal{X}^+(M, v, \mu)$. This is actually a consequence of the absolute continuity theorem (Theorem III. 5.1) (see the proof of Proposition 2.1).

In order to prove Proposition 2.1 we need some preliminaries. We first formulate a general lemma from measure theory (see also [Led]$_3$). A proof is included here for the sake of completeness.

Lemma 2.1. *Let $r_0 > 0$ be given and let ν be a finite Borel measure on $\mathbf{R}$ such that $\nu(\mathbf{R}\backslash[0, r_0]) = 0$. Then for any $\alpha, 0 < \alpha < 1$, the Lebesgue measure of the set*

$$L_\alpha = \left\{ r : r \in [0, r_0], \sum_{k=0}^{+\infty} \nu([r - \alpha^k, r + \alpha^k]) < +\infty \right\}$$

is equal to r_0.

Proof. Let $\alpha \in (0,1)$ be given. Define for $k \in \mathbf{N}$

$$N_{\alpha,k} = \{r : r \in [0, r_0], \nu([r-\alpha^k, r+\alpha^k]) > \frac{1}{k^2}\nu([0,r_0])\}.$$

It is easy to see that $N_{\alpha,k}$ can be covered by a finite number of intervals $[r_i - \alpha^k, r_i + \alpha^k], 1 \le i \le s(k)$ such that each r_i lies in $N_{\alpha,k}$ and any point of $\mathbf{R}$ meets at most two of these intervals. It follows then that

$$\frac{s(k)}{k^2}\nu([0,r_0]) \le \sum_{i=1}^{s(k)} \nu([r_i - \alpha^k, r_i + \alpha^k]) \le 2\nu([0,r_0])$$

which implies

$$s(k) \le 2k^2.$$

Thus

$$|N_{\alpha,k}| \le 2s(k)\alpha^k \le 4k^2\alpha^k$$

where $|K|$ denotes the Lebesgue measure for Borel subset K of $\mathbf{R}$. We then obtain

$$\sum_{k=1}^{+\infty} |N_{\alpha,k}| < +\infty.$$

According to Borel-Cantelli lemma, it follows that Lebesgue almost every $r \in [0, r_0]$ belongs only to a finite number of $N_{\alpha,k}$ and thus satisfies

$$\sum_{k=1}^{+\infty} \nu([r-\alpha^k, r+\alpha^k]) < +\infty.$$

The proof is completed. □

Let $\mathcal{X}^+(M, v, \mu)$ be given. Notice that the partition of $\Omega^{\mathbf{N}} \times M$ into global stable manifolds $\{\omega\} \times W^s(\omega, x), (\omega, x) \in \Omega^{\mathbf{N}} \times M$ is in general not measurable. But we may consider the σ-algebra consisting of measurable subsets of $\Omega^{\mathbf{N}} \times M$ which are unions of some global stable manifolds, i.e. the σ-algebra

$$\mathcal{B}^s = \left\{ B \in \mathcal{B}_{v^{\mathbf{N}} \times \mu}(\Omega^{\mathbf{N}} \times M) : B = \bigcup_{(\omega,x)\in B} \{\omega\} \times W^s(\omega, x) \right\}$$

where $\mathcal{B}_{v^{\mathbf{N}} \times \mu}(\Omega^{\mathbf{N}} \times M)$ is the completion of $\mathcal{B}(\Omega^{\mathbf{N}} \times M)$ with respect to $v^{\mathbf{N}} \times \mu$. Put now

$$\mathcal{B}^I = \{A \in \mathcal{B}_{v^{\mathbf{N}} \times \mu}(\Omega^{\mathbf{N}} \times M) : F^{-1}A = A\},$$

we have then the following useful fact.

Lemma 2.2. $\mathcal{B}^I \subset \mathcal{B}^s, v^{\mathbf{N}} \times \mu -$ mod 0.

Proof. Put $\Omega^{\mathbf{N}} \times \mathcal{B}_\mu(M) = \{\Omega^{\mathbf{N}} \times B : B \in \mathcal{B}_\mu(M)\}$ where $\mathcal{B}_\mu(M)$ is the completion of $\mathcal{B}(M)$ with respect to μ. Since M is a compact metric space, there exists a countable set

$$\mathcal{F} = \{g_i : g_i : \Omega^{\mathbf{N}} \times M \to \mathbf{R} \text{ is a continuous function and } g_i(\omega, x) \text{ depends only on } x \text{ for each } (\omega, x) \in \Omega^{\mathbf{N}} \times M, i \in \mathbf{N}\}$$

which is dense in $L^2(\Omega^{\mathbf{N}} \times M, \Omega^{\mathbf{N}} \times \mathcal{B}_\mu(M), v^{\mathbf{N}} \times \mu)$. For each $g_i \in \mathcal{F}$, according to Birkhoff ergodic theorem and the general properties of conditional expectations, one has

$$\lim_{n \to +\infty} \frac{1}{n} \sum_{k=0}^{n-1} g_i \circ F^k(\omega, x) = E(g_i|\mathcal{B}^I)(\omega, x) \tag{2.3}$$

for each point (ω, x) of a set $\Lambda_{g_i} \in \mathcal{B}^I$ with $v^{\mathbf{N}} \times \mu(\Lambda_{g_i}) = 1$. Denote $\Lambda_{\mathcal{F}} = \cap_i \Lambda_{g_i}$. If two points $(\omega, y), (\omega, z) \in \Lambda_{\mathcal{F}}$ belong to a same stable manifold, i.e. there exists (ω, x) such that $(\omega, y), (\omega, z) \in \{\omega\} \times W^s(\omega, x)$, by (2.3) we have for each g_i

$$E(g_i|\mathcal{B}^I)(\omega, y) = E(g_i|\mathcal{B}^I)(\omega, z)$$

since $\lim_{n \to +\infty} d(f_\omega^n y, f_\omega^n z) = 0$. Therefore, each $E(g_i|\mathcal{B}^I)|_{\Lambda_{\mathcal{F}}}$ (the restriction of $E(g_i|\mathcal{B}^I)$ to $\Lambda_{\mathcal{F}}$) is measurable with respect to $\mathcal{B}^s|_{\Lambda_{\mathcal{F}}}$ and hence

$$\{E(g_i|\mathcal{B}^I)|_{\Lambda_{\mathcal{F}}} : g_i \in \mathcal{F}\} \subset L^2(\Lambda_{\mathcal{F}}, \mathcal{B}^s|_{\Lambda_{\mathcal{F}}}, v^{\mathbf{N}} \times \mu). \tag{2.4}$$

On the other hand, according to Corollary I.1.1, we have

$$L^2(\Omega^{\mathbf{N}} \times M, \mathcal{B}^I, v^{\mathbf{N}} \times \mu) \subset L^2(\Omega^{\mathbf{N}} \times M, \Omega^{\mathbf{N}} \times \mathcal{B}_\mu(M), v^{\mathbf{N}} \times \mu). \tag{2.5}$$

Since $\mathcal{F}$ is a dense subset of the right-hand space in (2.5), by Theorem 0.1.6, $\{E(g_i|\mathcal{B}^I) : g_i \in \mathcal{F}\}$ is dense in $L^2(\Omega^{\mathbf{N}} \times M, \mathcal{B}^I, v^{\mathbf{N}} \times \mu)$. Then from (2.4) it follows that

$$L^2(\Lambda_{\mathcal{F}}, \mathcal{B}^I|_{\Lambda_{\mathcal{F}}}, v^{\mathbf{N}} \times \mu) \subset L^2(\Lambda_{\mathcal{F}}, \mathcal{B}^s|_{\Lambda_{\mathcal{F}}}, v^{\mathbf{N}} \times \mu)$$

which implies

$$\mathcal{B}^I \subset \mathcal{B}^s, v^{\mathbf{N}} \times \mu\text{-mod } 0$$

since $v^{\mathbf{N}} \times \mu(\Lambda_{\mathcal{F}}) = 1$. □

To conclude the preliminaries, we review some facts about local stable manifolds introduced in Chapter III.

Let $\mathcal{X}^+(M, v, \mu)$ be given. From (3.8) in Chapter III and the arguments at the beginning of Section III.5 we know that there exist a countable number of compact sets $\{\Lambda^i : \Lambda^i \subset \hat{\Lambda}_1, i \in \mathbf{N}\}$ such that $v^{\mathbf{N}} \times \mu(\hat{\Lambda}_1 \setminus \cup_{i=1}^{+\infty} \Lambda^i) = 0$ and each Λ^i is a set of the type Δ^l as considered in Section III.5 but with $E_0(\omega, x) = \cup_{\lambda^{(i)}(x) < 0} V_{(\omega, x)}^{(i)}$ for each $(\omega, x) \in \Lambda^i$.

Suppose that $\Lambda^i \in \{\Lambda^i : i \in \mathbf{N}\}$ and write $k_i = \dim E_0(\omega, x), (\omega, x) \in \Lambda^i$. Let $\{W_{\text{loc}}^s(\omega, x)\}_{(\omega, x) \in \Lambda^i}$ be a continuous family of C^1 embedded k_i-dimensional discs given by Theorem III. 3.1, corresponding to $n = 0$. Recall that this family of embedded discs have the following properties:

(I) There exist $\lambda_i > 0$ and $\gamma_i > 0$ such that for each $(\omega, x) \in \Lambda^i$, if $y, z \in W^s_{\text{loc}}(\omega, x)$, then for all $l \geq 0$

$$d^s(f^l_\omega y, f^l_\omega z) \leq \gamma_i e^{-\lambda_i l} d^s(y, z).$$

(II) Assume that $(\omega, x) \in \Lambda^i$. Let $\tilde{\Lambda}^i_\omega(x, q_{\Lambda^i})$ and its partition $\alpha^{(i)}$ be defined analogously to $\tilde{\Delta}^l_\omega(x, q_{\Delta^l})$ and the partition α considered in Section III.6. Denote $\tilde{\Lambda}^i_\omega(x, q_{\Lambda^i})$ by $[I^i_{(\omega,x)}]$ for simplicity, and let $\{\lambda^{[I^i_{(\omega,x)}]}_{\alpha^{(i)}(z)}\}_{z \in [I^i_{(\omega,x)}]}$ be a canonical system of conditional measures of $\lambda^{[I^i_{(\omega,x)}]}$ associated with the partition $\alpha^{(i)}$. Then, by Theorems III.5.1 and III.6.1, for λ-a.e. $z \in [I^i_{(\omega,x)}]$ the measure $\lambda^{[I^i_{(\omega,x)}]}_{\alpha^{(i)}(z)}$ is equivalent to $\lambda^{\alpha^{(i)}}_z$.

(III) For $(\omega, x) \in \Lambda^i$ and $r > 0$ we introduce

$$B_{\Lambda^i}((\omega, x), r) = \{(\omega', x') \in \Lambda^i : d(\omega, \omega') < r, d(x, x') < r\}$$

and, to repeat, define

$$B(x, r) = \{y \in M : d(x, y) < r\},$$
$$U_{\Lambda^i,\omega}(x, r) = \exp_x\{\zeta \in T_x M : \|\zeta\|_{(\omega,x),0} < r\}.$$

Then, owing to the compactness of Λ^i, there exist numbers $r_i, 0 < r_i < \rho_0/4$ (see Section II.1 for the definition of ρ_0), and $\varepsilon_i, 0 < \varepsilon_i < 1$, such that the following hold true:

(i) Let $(\omega, x) \in \Lambda^i$. If $(\omega', x') \in B_{\Lambda^i}((\omega, x), r_i)$, then

$$B(x, r_i) \subset U_{\Lambda^i,\omega'}(x', q_{\Lambda^i}/2);$$

(ii) For any $r \in [r_i/2, r_i]$ and each $(\omega, x) \in \Lambda^i$, if $(\omega', x') \in B_{\Lambda^i}((\omega, x), \varepsilon_i r)$, then $W^s_{\text{loc}}(\omega', x') \cap B(x, r)$ is connected and the map

$$(\omega', x') \mapsto W^s_{\text{loc}}(\omega', x') \cap B(x, r)$$

is a continuous map from $B_{\Lambda^i}((\omega, x), \varepsilon_i r)$ to the space of subsets of $B(x, r)$ (endowed with the Hausdorff topology);

(iii) Let $r \in [r_i/2, r_i]$ and $(\omega, x) \in \Lambda^i$. If $(\omega', x'), (\omega', x'') \in B_{\Lambda^i}((\omega, x), \varepsilon_i r)$, then either

$$W^s_{\text{loc}}(\omega', x') \cap B(x, r) = W^s_{\text{loc}}(\omega', x'') \cap B(x, r)$$

or the two terms in the above equation are disjoint. In the latter case, if it is assumed moreover that $x'' \in W^s(\omega', x')$, then

$$d^s(y, z) > 2r_i$$

for any $y \in W^s_{\text{loc}}(\omega', x') \cap B(x, r)$ and any $z \in W^s_{\text{loc}}(\omega', x'') \cap B(x, r)$;

(iv) There exists $R_i > 0$ such that for each $(\omega, x) \in \Lambda^i$, if $(\omega', x') \in B_{\Lambda^i}((\omega, x), r_i)$ and $y \in W^s_{\text{loc}}(\omega', x') \cap B(x, r_i)$, then $W^s_{\text{loc}}(\omega', x')$ contains the closed ball of centre y and d^s radius R_i in $W^s(\omega', x')$.

We are now prepared to prove Proposition 2.1.

Proof of Proposition 2.1. Let $\Lambda^i \in \{\Lambda^i : i \in \mathbf{N}\}$ be fixed arbitrarily. Since Λ^i is compact, the open cover $\{B_{\Lambda^i}((\omega,x),\varepsilon_i r_i/2)\}_{(\omega,x)\in\Lambda^i}$ has a finite subcover $\mathcal{U}_{\Lambda^i}$ of Λ^i. Fix arbitrarily $B_{\Lambda^i}((\omega_0,x_0),\varepsilon_i r_i/2) \in \mathcal{U}_{\Lambda^i}$. For each $r \in [r_i/2, r_i]$ put

$$S_r = \cup\{\{\omega\} \times [W^s_{\text{loc}}(\omega,x) \cap B(x_0,r)] : (\omega,x) \in B_{\Lambda^i}((\omega_0,x_0),\varepsilon_i r)\}.$$

Let ξ_r denote the partition of $\Omega^{\mathbf{N}} \times M$ into all the sets $\{\omega\} \times [W^s_{\text{loc}}(\omega,x) \cap B(x_0,r)]$, $(\omega,x) \in B_{\Lambda^i}((\omega_0,x_0),\varepsilon_i r)$ and the set $\Omega^{\mathbf{N}} \times M \backslash S_r$. From properties (III) (ii) and (III) (iii) it follows clearly that ξ_r is a measurable partition of $\Omega^{\mathbf{N}} \times M$. Define

$$\eta_r = \left(\bigvee_{n=0}^{+\infty} F^{-n}\xi_r \right) \vee \{\{\omega\} \times M : \omega \in \Omega^{\mathbf{N}}\}.$$

We claim that there exists $r \in [r_i/2, r_i]$ such that η_r has the following properties:

(1) $F^{-1}\eta_r \le \eta_r, \{\{\omega\} \times M : \omega \in \Omega^{\mathbf{N}}\} \le \eta_r$;

(2) Put $\hat{S}_r = \cup_{n=0}^{+\infty} F^{-n} S_r$. Then for $v^{\mathbf{N}} \times \mu$ -a.e.$(\omega,y) \in \hat{S}_r$, $(\eta_r)_\omega(y) \overset{\text{def.}}{=} \{z : (\omega,z) \in \eta_r(\omega,y)\} \subset W^s(\omega,y)$ and it contains an open neighbourhood of y in $W^s(\omega,y)$;

(3) For any $B \in \mathcal{B}(\Omega^{\mathbf{N}} \times M)$, the function

$$P_B(\omega,y) = \lambda^s_{(\omega,y)}((\eta_r)_\omega(y) \cap B_\omega)$$

is measurable and finite $v^{\mathbf{N}} \times \mu$ almost everywhere on $\hat{S}_r$;

(4) Let $\hat{\eta}_r = \eta_r|_{\hat{S}_r}$ and for $\omega \in \Omega^{\mathbf{N}}$ let $\{\mu_{(\hat{\eta}_r)_\omega(y)}\}_{y\in(\hat{S}_r)_\omega}$ be a canonical system of conditional measures of $\mu|_{(\hat{S}_r)_\omega}$ associated with the partition $(\hat{\eta}_r)_\omega$. If $\mu << \lambda$, then for $v^{\mathbf{N}}$- a.e. $\omega \in \Omega^{\mathbf{N}}$ it holds that

$$\mu_{(\hat{\eta}_r)_\omega(y)} << \lambda^s_{(\omega,y)}, \mu\text{-a.e. } y \in (\hat{S}_r)_\omega.$$

In fact, η_r satisfies (1)-(4) for Lebesgue almost every $r \in [r_i/2, r_i]$. We now prove this fact in four steps.

Step 1. From the definition of η_r it follows clearly that for each $r \in [r_i/2, r_i]$, η_r has property (1).

Step 2. Let $r \in [r_i/2, r_i]$. It is clear that for $v^{\mathbf{N}} \times \mu$ -a.e. $(\omega,y) \in \hat{S}_r$, $(\eta_r)_\omega(y) \subset W^s(\omega,y)$ since

$$(\eta_r)_\omega(y) \subset (f^n_\omega)^{-1} W^s_{\text{loc}}(\tau^n\omega,x) \tag{2.6}$$

for some $n \ge 0$ and some $(\tau^n\omega,x) \in B_{\Lambda^i}((\omega_0,x_0),\varepsilon_i r)$.

On the other hand, we first claim that there exists a function $\beta_r : S_r \to \mathbf{R}^+$ such that for each $(\omega,y) \in S_r$, $z \in W^s(\omega,y)$ and $d^s(y,z) \le \beta_r(\omega,y)$ imply that $z \in (\eta_r)_\omega(y)$. Indeed, define for $(\omega,y) \in S_r$

$$\beta_r(\omega,y) = \inf_{n\ge0} \left\{ R_i, \frac{1}{2\gamma_i} d(f^n_\omega y, \partial B(x_0,r)) e^{n\lambda_i}, \frac{r}{\gamma_i} \right\}.$$

Suppose that $(\omega, y) \in S_r$ and $z \in W^s(\omega, y)$ with $0 < d^s(y, z) \le \beta_r(\omega, y)$. We will check that for every $n \ge 0$

$$F^n(\omega, z) \in \xi_r(F^n(\omega, y)). \tag{2.7}$$

Since $d^s(y, z) \le R_i$, by property (III) (iv), there exists $(\omega, x) \in B_{\Lambda^i}((\omega_0, x_0), \varepsilon_i r)$ such that $y, z \in W^s_{\text{loc}}(\omega, x)$ and hence for every $n \ge 0$

$$d^s(f^n_\omega y, f^n_\omega z) \le \gamma_i e^{-n\lambda_i} d^s(y, z) \le \frac{1}{2} d(f^n_\omega y, \partial B(x_0, r)) \tag{2.8}$$

and

$$d^s(f^n_\omega y, f^n_\omega z) \le \gamma_i e^{-n\lambda_i} \frac{r}{\gamma_i} \le r. \tag{2.9}$$

We have now three cases to consider:

(a) If $F^n(\omega, y)$ and $F^n(\omega, z)$ both belong to S_r, we have (2.7) by property (III) (iii) and (2.9).

(b) If neither $F^n(\omega, y)$ nor $F^n(\omega, z)$ belong to S_r, we have (2.7) by the definition of ξ_r.

(c) If one of $F^n(\omega, y)$ and $F^n(\omega, z)$ belongs to S_r but the other does not, we should have

$$d^s(f^n_\omega y, f^n_\omega z) \ge d(f^n_\omega y, \partial B(x_0, r))$$

which would contradict (2.8). Thus case (c) is impossible. Our first claim is confirmed.

We next claim that $\beta_r > 0$ $\upsilon^{\mathbb{N}} \times \mu$ almost everywhere on S_r for Lebesgue almost every $r \in [r_i/2, r_i]$. In fact, let ν be the finite non-negative Borel measure on $[r_i/2, r_i]$ defined by

$$\nu(A) = \mu(\{x \in M : d(x, x_0) \in A\})$$

for each Borel subset A of $[r_i/2, r_i]$. According to Lemma 2.1, Lebesgue almost every $r \in [r_i/2, r_i]$ satisfies

$$\sum_{n=0}^{+\infty} \mu(\{x \in M : |d(x, x_0) - r| < e^{-n\lambda_i}\}) < +\infty. \tag{2.10}$$

Let $K_0 = \{r : r \in [r_i/2, r_i], r \text{ satisfies (2.10) and } \mu(\partial B(x_0, r)) = 0\}$. Clearly $|K_0| = r_i/2$. Let $r \in K_0$. Since $0 < r_i < \rho_0/4$, the standard knowledge about Riemannian metrics tells that there exists a constant $D > 0$ such that

$$d(x, \partial B(x_0, \rho)) < \tau$$

implies

$$|d(x, x_0) - \rho| < D\tau$$

for ρ and τ satisfying $0 < \tau < \rho \le r_i$. Thus from (2.10) we obtain

$$\sum_{n=0}^{+\infty} \mu(\{x : d(x, \partial B(x_0, r)) < D^{-1} e^{-n\lambda_i}\}) < +\infty. \tag{2.11}$$

Since for every $n \geq 0$

$$R_n : (\Omega^{\mathbf{N}} \times M, v^{\mathbf{N}} \times \mu) \to (M, \mu), (\omega, y) \mapsto f_\omega^n y$$

is a measure-preserving map, from (2.11) it follows that

$$\sum_{n=0}^{+\infty} v^{\mathbf{N}} \times \mu(\{(\omega, y) : d(f_\omega^n y, \partial B(x_0, r)) < D^{-1} e^{-n\lambda_i}\}) < +\infty.$$

Then, by Borel-Cantelli lemma, we know that $v^{\mathbf{N}} \times \mu$ -a.e. $(\omega, y) \in \Omega^{\mathbf{N}} \times M$ satisfies

$$d(f_\omega^n y, \partial B(x_0, r)) \geq D^{-1} e^{-n\lambda_i}$$

when n is sufficiently large. Therefore, $\beta_r(\omega, y) > 0$ for $v^{\mathbf{N}} \times \mu$ -a.e. $(\omega, y) \in S_r$. The second claim is proved.

Let $r \in K_0$. The two above claims together imply that for $v^{\mathbf{N}} \times \mu$ -a.e. $(\omega, y) \in S_r$, $(\eta_r)_\omega(y)$ contains an open neighbourhood of y in $W^s(\omega, y)$. Furthermore, for every $n \geq 0$ we have

$$\begin{aligned} \eta_r|_{F^{-n}S_r} &= \left[\left(\bigvee_{k=0}^{n-1} F^{-k}\xi_r \right) \vee (F^{-n}\eta_r) \right] \Bigg|_{F^{-n}S_r} \\ &= \left(\bigvee_{k=0}^{n-1} F^{-k}\xi_r \right) \Bigg|_{F^{-n}S_r} \vee (F^{-n}(\eta_r|_{S_r})). \end{aligned} \qquad (2.12)$$

Since $\mu(\partial B(x_0, r)) = 0$ implies that $v^{\mathbf{N}} \times \mu(\{(\omega, y) \in \Omega^{\mathbf{N}} \times M : f_\omega^l y \in \partial B(x_0, r)$ for some $l \geq 0\}) = 0$, from (2.12) it is easy to see that for $v^{\mathbf{N}} \times \mu$ -a.e. $(\omega, y) \in F^{-n}S_r$, $(\eta_r)_\omega(y)$ contains an open neighbourhood of y in $W^s(\omega, y)$. Thus η_r satisfies the requirements in (2).

Step 3. Let $r \in K_0$. From (2.6) it is easy to see that for each $B \in \mathcal{B}(\Omega^{\mathbf{N}} \times M)$ the function $P_B(\omega, y)$ is finite for $v^{\mathbf{N}} \times \mu$ -a.e. $(\omega, y) \in \hat{S}_r$.

Let $n \in \mathbf{Z}^+$, and put $\xi_r^n = \vee_{k=0}^{n-1} F^{-k}\xi_r$ and $\hat{S}_r^n = \cup_{k=0}^{n-1} F^{-k} S_r$. From the definition of S_r it is clear that, if U is an open ball in $\Omega^{\mathbf{N}} \times M$, the function

$$P_{U,n}(\omega, y) = \lambda^s_{(\omega,y)}((\xi_r^n)_\omega(y) \cap U_\omega)$$

is measurable on $\hat{S}_r^n$. Then the standard arguments from measure theory ensure that so is $P_{B,n}(\omega, y)$ for any $B \in \mathcal{B}(\Omega^{\mathbf{N}} \times M)$. Noticing that for any $B \in \mathcal{B}(\Omega^{\mathbf{N}} \times M)$

$$P_{B,n}(\omega, y) \geq P_{B,n+1}(\omega, y)$$

for each $(\omega, y) \in \hat{S}_r^n$, and

$$\lim_{n \to +\infty} P_{B,n}(\omega, y) = P_B(\omega, y)$$

for $v^{\mathbf{N}} \times \mu$ -a.e. $(\omega, y) \in \hat{S}_r$, we know for each $B \in \mathcal{B}(\Omega^{\mathbf{N}} \times M)$ the function $P_B(\omega, y)$ is measurable and finite $v^{\mathbf{N}} \times \mu$ almost everywhere on $\hat{S}_r$.

Step 4. Fix arbitrarily $r \in K_0$. Now assume that $\mu << \lambda$. Let $(\omega, x) \in B_{\Lambda^i}((\omega_0, x_0), \varepsilon_i r)$. From property (III) (i) and the definition of S_r we know that $(S_r)_\omega$ is a measurable subset of $\tilde{\Lambda}^i_\omega(x, q_{\Lambda^i}/2)$ and $(\xi_r|_{S_r})_\omega = \alpha^{(i)}|_{(S_r)_\omega}$. Then, denoting by $\{\lambda_{(\xi_r|_{S_r})_\omega(z)}\}_{z \in (S_r)_\omega}$ the conditional measures of $\lambda|_{(S_r)_\omega}$ associated with the partition $(\xi_r|_{S_r})_\omega$, we have

$$\lambda_{(\xi_r|_{S_r})_\omega(z)} << \lambda^s_{(\omega,z)}, \quad \lambda\text{-a.e. } z \in (S_r)_\omega$$

which implies, by Proposition III. 6.1,

$$\mu_{(\xi_r|_{S_r})_\omega(z)} << \lambda^s_{(\omega,z)}, \quad \mu\text{-a.e. } z \in (S_r)_\omega.$$

Since for μ-a.e.$z \in (S_r)_\omega, (\eta_r)_\omega|_{(\xi_r)_\omega(z)}$ is a countable partition, we have

$$\mu_{(\eta_r|_{S_r})_\omega(z)} << \lambda^s_{(\omega,z)}, \mu\text{-a.e. } z \in (S_r)_\omega.$$

Noticing that λ is quasi-invariant under the action of any C^1 diffeomorphism on M, i.e. $f\lambda$ is equivalent to λ for any $f \in \mathrm{Diff}^1(M)$, we see easily that for every $n \geq 0$ and $\omega \in \Omega^{\mathbf{N}}$

$$\mu_{(\eta_r|_{F^{-n}S_r})_\omega(z)} << \lambda^s_{(\omega,z)}, \mu\text{-a.e. } z \in (F^{-n}S_r)_\omega.$$

From this it follows clearly that η_r satisfies (4).

To finish the proof of Proposition 2.1, notice that the treatment above holds true for every element of $\cup_{i=1}^{+\infty}\mathcal{U}_{\Lambda^i} = \{U_1, U_2, \cdots\}$. For each U_n we denote by η_n the associated partition η_r satisfying (1)-(4) constructed above and denote by $\hat{S}_n$ the associated set $\hat{S}_r$. For each $n \in \mathbf{N}$ put $I_n = \cap_{l=0}^{+\infty} F^{-l}\hat{S}_n$. It is obvious that $F^{-1}I_n = I_n, n \geq 1$ and $v^{\mathbf{N}} \times \mu(\hat{\Lambda}_1 \setminus \cup_{n=1}^{+\infty} I_n) = 0$. By Lemma2.2 we may assume that $I_n \in \mathcal{B}^s$ for every $n \geq 1$ since otherwise we may find $I'_n \in \mathcal{B}^s$ such that $F^{-1}I'_n = I'_n$ and $v^{\mathbf{N}} \times \mu(I_n \Delta I'_n) = 0$ and we may restrict the procedure of constructing η_n to I'_n. Set $\hat{\eta}_n = \eta_n|_{I_n}$ for each $n \geq 1$ and define partition η of $\Omega^{\mathbf{N}} \times M$ by

$$\eta(\omega, x) = \begin{cases} \hat{\eta}_1(\omega, x), & \text{if } (\omega, x) \in I_1 \\ \hat{\eta}_n(\omega, x), & \text{if } (\omega, x) \in I_n \setminus \cup_{k=1}^{n-1} I_k \\ \{(\omega, x)\}, & \text{if } (\omega, x) \in \Omega^{\mathbf{N}} \times M \setminus \cup_{n=1}^{+\infty} I_n. \end{cases}$$

Then one can easily check that η satisfies the requirements of Proposition 2.1, completing the proof. □

The conclusion 3) of Proposition 2.1 allows us to define a Borel measure λ^* on $\Omega^{\mathbf{N}} \times M$ by

$$\lambda^*(K) = \int \lambda^s_{(\omega,x)}(\eta_\omega(x) \cap K_\omega) dv^{\mathbf{N}} \times \mu(\omega, x) \qquad (2.13)$$

for each Borel subset K of $\Omega^{\mathbf{N}} \times M$. It is easy to see that λ^* is σ-finite. Also, recall that by the definition of conditional measures we have

$$v^{\mathbf{N}} \times \mu(K) = \int \mu_x^{\eta_\omega}(\eta_\omega(x) \cap K_\omega) dv^{\mathbf{N}} \times \mu(\omega, x) \tag{2.14}$$

for each Borel subset K of $\Omega^{\mathbf{N}} \times M$. Since, by Proposition 2.1 4), $\mu_x^{\eta_\omega} << \lambda^s_{(\omega,x)}$ for $v^{\mathbf{N}} \times \mu$ -a.e.(ω, x), we have

$$v^{\mathbf{N}} \times \mu << \lambda^*.$$

Define

$$g = \frac{dv^{\mathbf{N}} \times \mu}{d\lambda^*}. \tag{2.15}$$

The next proposition follows from a measure-theoretic observation.

Proposition 2.2. *For $v^{\mathbf{N}} \times \mu$-a.e.$(\omega, x)$, we have*

$$g = \frac{d\mu_x^{\eta_\omega}}{d\lambda^s_{(\omega,x)}}$$

$\lambda^s_{(\omega,x)}$ *almost everywhere on $\eta_\omega(x)$.*

Proof. Let us first notice that (2.13) can be written equivalently as

$$\int \chi_K d\lambda^* = \int \left[\int_{\eta_\omega(x)} \chi_K(\omega, z) d\lambda^s_{(\omega,x)}(z) \right] dv^{\mathbf{N}} \times \mu(\omega, x)$$

for $K \in \mathcal{B}(\Omega^{\mathbf{N}} \times M)$. Then using standard methods of measure theory we easily obtain

$$\int h d\lambda^* = \int \left[\int_{\eta_\omega(x)} h(\omega, z) d\lambda^s_{(\omega,x)}(z) \right] dv^{\mathbf{N}} \times \mu(\omega, x) \tag{2.16}$$

for each $h \in L^1(\Omega^{\mathbf{N}} \times M, \mathcal{B}(\Omega^{\mathbf{N}} \times M), \lambda^*)$.

Let $A \in \mathcal{B}(\eta), B \in \mathcal{B}(\Omega^{\mathbf{N}} \times M)$ be two arbitrary sets. From (2.13)-(2.16) it follows that

$$\begin{aligned} \int_{A \cap B} g d\lambda^* &= \int_A \left[\int_{\eta_\omega(x) \cap B_\omega} g(\omega, z) d\lambda^s_{(\omega,x)}(z) \right] dv^{\mathbf{N}} \times \mu(\omega, x) \\ &= \int_{A \cap B} dv^{\mathbf{N}} \times \mu = \int_A \mu_x^{\eta_\omega}(\eta_\omega(x) \cap B_\omega) dv^{\mathbf{N}} \times \mu(\omega, x). \end{aligned} \tag{2.17}$$

Since $\Omega^{\mathbf{N}} \times M$ is a Borel subset of a Polish space (see Section I.1), by Theorem 0.1.3, the measure space $(\Omega^{\mathbf{N}} \times M, \mathcal{B}(\Omega^{\mathbf{N}} \times M), v^{\mathbf{N}} \times \mu)$ is separable. By Theorem 0.1.2, $\mathcal{B}(\Omega^{\mathbf{N}} \times M)$ can be generated ($v^{\mathbf{N}} \times \mu$- mod 0) by a countable subalgebra $\{B_j\}_{j=1}^{+\infty}$ of $\mathcal{B}(\Omega^{\mathbf{N}} \times M)$. Fixing $1 \leq j < +\infty$, we apply (2.17) to an arbitrary set $A \in \mathcal{B}(\eta)$ and to $B = B_j$. As A is arbitrary, (2.17) implies that there exists a measurable subset Z_j of $\Omega^{\mathbf{N}} \times M$ such that $v^{\mathbf{N}} \times \mu(Z_j) = 1$ and for each $(\omega, x) \in Z_j$ one has

$$\int_{\eta_\omega(x)\cap(B_j)_\omega} g(\omega, z) d\lambda^s_{(\omega,x)}(z) = \mu_x^{\eta_\omega}(\eta_\omega(x) \cap (B_j)_\omega).$$

Then, according to Theorem 0.1.1, we know that for $v^{\mathbf{N}} \times \mu$-a.e. (ω, x),

$$\int_{\eta_\omega(x)\cap B_\omega} g(\omega, z) d\lambda^s_{(\omega,x)}(z) = \mu_x^{\eta_\omega}(\eta_\omega(x) \cap B_\omega)$$

holds for any $B \in \mathcal{B}(\Omega^{\mathbf{N}} \times M)$, and therefore

$$g = \frac{d\mu_x^{\eta_\omega}}{d\lambda^s_{(\omega,x)}}$$

$\lambda^s_{(\omega,x)}$ almost everywhere on $\eta_\omega(x)$. □

§3 Estimation of the Entropy from Below

In this section we complete the proof of Theorem 1.1.

Proof of Theorem 1.1. Let $\mathcal{X}^+(M, v, \mu)$ be given with v also satisfying $\log|\det T_x f| \in L^1(\Omega \times M, v \times \mu)$, and assume that $\mu << \lambda$. In view of Ruelle's inequality (Theorem II.0.1) it remains to prove

$$h_\mu(\mathcal{X}^+(M, v)) \geq \int \sum_i \lambda^{(i)}(x)^+ m_i(x) d\mu. \tag{3.1}$$

In what follows we keep the notations in Chapter I. Let η be a partition of $\Omega^{\mathbf{N}} \times M$ of the type discussed in Proposition 2.1. Denote by η^+ the partition $P^{-1}\eta$ of $\Omega^{\mathbf{Z}} \times M$. For every integer $n \geq 1$, we now assume that

$$H_{v^{\mathbf{N}} \times \mu}(\eta | F^{-n}\eta \vee \sigma_0) < +\infty, \tag{3.2}$$

then from the general properties of conditional entropies (see Section 0.3) we have

$$\begin{aligned} &\frac{1}{n} H_{v^{\mathbf{N}} \times \mu}(\eta | F^{-n}\eta \vee \sigma_0) \\ =&\frac{1}{n} H_{\mu^*}(P^{-1}\eta | P^{-1}F^{-n}\eta \vee P^{-1}\sigma_0) \\ =&\frac{1}{n} H_{\mu^*}(\eta^+ | G^{-n}\eta^+ \vee \sigma^+) \\ =&\frac{1}{n} H_{\mu^*}(G^n\eta^+ | \eta^+ \vee G^n\sigma^+) \end{aligned}$$

$$=\frac{1}{n}\sum_{i=1}^{n} H_{\mu^*}(G^i\eta^+|G^{i-1}\eta^+ \vee G^n\sigma^+)$$

$$=\frac{1}{n}\sum_{i=0}^{n-1} H_{\mu^*}(\eta^+|G^{-1}\eta^+ \vee G^i\sigma^+). \tag{3.3}$$

Since $H_{\mu^*}(\eta^+|G^{-1}\eta^+ \vee \sigma^+) = H_{\upsilon^{\mathbb{N}}\times\mu}(\eta|F^{-1}\eta \vee \sigma_0) < +\infty$ and $G^i\sigma^+ \nearrow \sigma$ as $i \to +\infty$, from (3.3) we obtain

$$\begin{aligned}&\lim_{n\to+\infty}\frac{1}{n}H_{\upsilon^{\mathbb{N}}\times\mu}(\eta|F^{-n}\eta \vee \sigma_0)\\ =&H_{\mu^*}(\eta^+|G^{-1}\eta^+ \vee \sigma) = h^{\sigma}_{\mu^*}(G,\eta^+)\\ \le& h^{\sigma}_{\mu^*}(G) = h_\mu(\mathcal{X}^+(M,\upsilon)).\end{aligned}$$

Therefore, in order to prove (3.1) it suffices to prove (3.2) and

$$\frac{1}{n}H_{\upsilon^{\mathbb{N}}\times\mu}(\eta|F^{-n}\eta \vee \sigma_0) \ge \int \sum_i \lambda^{(i)}(x)^+ m_i(x)d\mu \tag{3.4}$$

for every $n \ge 1$.

Fix arbitrarily $n \ge 1$. We now begin to prove (3.2) and (3.4).

By the definition of conditional entropies we have

$$\begin{aligned}&H_{\upsilon^{\mathbb{N}}\times\mu}(\eta|F^{-n}\eta \vee \sigma_0)\\ =&-\int \log(\upsilon^{\mathbb{N}}\times\mu)^{F^{-n}\eta\vee\sigma_0}_{(\omega,x)}(\eta(\omega,x))d\upsilon^{\mathbb{N}}\times\mu\\ =&-\int_{\Omega^{\mathbb{N}}}\int_M \log\mu_x^{(f^n_\omega)^{-1}\eta_{\tau^n\omega}}(\eta_\omega(x))d\mu d\upsilon^{\mathbb{N}}.\end{aligned} \tag{3.5}$$

Let $\{I_l\}_{l=1}^{+\infty}$ be the sets introduced in the construction of η, i.e. in the proof of Proposition 2.1. Put $I = \bigcup_{l=1}^{+\infty} I_l$ and $I_0 = \Omega^{\mathbb{N}} \times M\backslash I$. Clearly, $F^{-1}I = I, F^{-1}I_0 = I_0$. Since η and $F^{-n}\eta \vee \sigma_0$ both refine the partition $\{I, I_0\}$ and their restrictions to I_0 are partitions into single points, we know that for each $(\omega, x) \in I_0$

$$\log\mu_x^{(f^n_\omega)^{-1}\eta_{\tau^n\omega}}(\eta_\omega(x)) = 0.$$

On the other hand, by Proposition I. 3.3,

$$\int_{I_0}\sum_i \lambda^{(i)}(x)^+ m_i(x)d\upsilon^{\mathbb{N}}\times\mu = 0.$$

Hence we may assume $\upsilon^{\mathbb{N}}\times\mu(I) = 1$ without any loss of generality.

Let $\varphi = d\mu/d\lambda$ be the Radon-Nikodym derivative. Put $A = \{x : \varphi(x) = 0\}$. Since

$$\int f^n_\omega\mu(A)d\upsilon^{\mathbb{N}}(\omega) = \mu(A) = 0,$$

we obtain $f_\omega^n\mu(A)=0$ for $v^{\mathbf{N}}$-a.e. $\omega\in\Omega^{\mathbf{N}}$. Let $B\subset M\backslash A$ be an arbitrary Borel set. If $\mu(B)=0$, then for any $\omega\in\Omega^{\mathbf{N}}$

$$\lambda(B)=0,\quad f_\omega^n\lambda(B)=0,\quad f_\omega^n\mu(B)=0.$$

It follows then that

$$f_\omega^n\mu<<\mu,\quad \mu<<(f_\omega^n)^{-1}\mu$$

for each point ω of a Borel subset Γ' of $\Omega^{\mathbf{N}}$ with $v^{\mathbf{N}}(\Gamma')=1$.

Let $\omega\in\Gamma'$. It is easy to verify that

$$\frac{d\mu}{d(f_\omega^n)^{-1}\mu}(z)=\frac{\varphi(z)}{\varphi(f_\omega^n z)}|\det T_z f_\omega^n|^{-1}\stackrel{\text{def.}}{=}\Phi_n(\omega,z).$$

Then, by Proposition III.6.1, we have

$$\frac{d\mu_x^{(f_\omega^n)^{-1}\eta_{\tau^n\omega}}}{d((f_\omega^n)^{-1}\mu)_x^{(f_\omega^n)^{-1}\eta_{\tau^n\omega}}}=\frac{\Phi_n(\omega,\cdot)|_{((f_\omega^n)^{-1}\eta_{\tau^n\omega})(x)}}{\displaystyle\int_{((f_\omega^n)^{-1}\eta_{\tau^n\omega})(x)}\Phi_n(\omega,z)d((f_\omega^n)^{-1}\mu)_x^{(f_\omega^n)^{-1}\eta_{\tau^n\omega}}}$$

for μ-a.e. $x\in M$. For $v^{\mathbf{N}}\times\mu$-a.e. $(\omega,y)\in\Omega^{\mathbf{N}}\times M$ we can define

$$\begin{aligned}
W_n(\omega,y)&=\mu_y^{(f_\omega^n)^{-1}\eta_{\tau^n\omega}}(\eta_\omega(y)),\\
X_n(\omega,y)&=\frac{\varphi(y)}{\varphi(f_\omega^n y)}\cdot\frac{g(F^n(\omega,y))}{g(\omega,y)},\\
Y_n(\omega,y)&=\frac{|\det(T_y f_\omega^n|_{E_0(\omega,y)})|}{|\det(T_y f_\omega^n)|},\\
Z_n(\omega,y)&=\int_{((f_\omega^n)^{-1}\eta_{\tau^n\omega})(y)}\Phi_n(\omega,z)d((f_\omega^n)^{-1}\mu)_y^{(f_\omega^n)^{-1}\eta_{\tau^n\omega}}.
\end{aligned}$$

It is easy to see that W_n, X_n, Y_n and Z_n are all measurable and $v^{\mathbf{N}}\times\mu$-a.e. finite functions on $\Omega^{\mathbf{N}}\times M$.

We now present several claims, whose proofs will be given a little later.

Claim 3.1. $W_n=\frac{X_nY_n}{Z_n}$, $v^{\mathbf{N}}\times\mu$ almost everywhere on $\Omega^{\mathbf{N}}\times M$.

Claim 3.2. (a) $-\log Y_n\in L^1(\Omega^{\mathbf{N}}\times M, v^{\mathbf{N}}\times\mu)$;
(b) $-\int\frac{1}{n}\log Y_n dv^{\mathbf{N}}\times\mu=\int\sum_i\lambda^{(i)}(x)^+m_i(x)d\mu$.

Claim 3.3.(a) $\log Z_n\in L^1(\Omega^{\mathbf{N}}\times M, v^{\mathbf{N}}\times\mu)$;
(b) $\int\log Z_n dv^{\mathbf{N}}\times\mu\geq 0$.

Claim 3.4. (a) $\log X_n\in L^1(\Omega^{\mathbf{N}}\times M, v^{\mathbf{N}}\times\mu)$;
(b) $\int\log X_n dv^{\mathbf{N}}\times\mu=0$.

Then we immediately obtain (3.2) and (3.4) from (3.5) and Claims 3.1-3.4. This proves Theorem 1.1.

Finally we give proofs of Claims 3.1-3.4.

Proof of Claim 3.1. It suffices to prove that for $v^{\mathbf{N}} \times \mu$-a.e. (ω, x) one has

$$W_n(\omega, y) = \frac{X_n(\omega, y)Y_n(\omega, y)}{Z_n(\omega, y)}, \quad \mu_x^{\eta_\omega}\text{-a.e.} y \in \eta_\omega(x).$$

In fact, for $v^{\mathbf{N}} \times \mu$-a.e.$(\omega, x)$ we have for any $B \in \mathcal{B}(M)$

$$\begin{aligned}
&\mu_x^{\eta_\omega}(B) \\
=&\frac{1}{W_n(\omega,x)}\int_{\eta_\omega(x)\cap B} d\mu_x^{(f_\omega^n)^{-1}\eta_{\tau^n\omega}} \\
=&\frac{1}{W_n(\omega,x)Z_n(\omega,x)}\int_{\eta_\omega(x)\cap B} \Phi_n(\omega,y)d((f_\omega^n)^{-1}\mu)_x^{(f_\omega^n)^{-1}\eta_{\tau^n\omega}}(y) \\
=&\frac{1}{W_n(\omega,x)Z_n(\omega,x)}\int_{\eta_\omega(x)\cap B} \Phi_n(\omega,y)d(f_\omega^n)^{-1}\mu_{f_\omega^n x}^{\eta_{\tau^n\omega}}(y) \\
=&\frac{1}{W_n(\omega,x)Z_n(\omega,x)}\int_{f_\omega^n(\eta_\omega(x)\cap B)} \Phi_n(\omega,(f_\omega^n)^{-1}y)d\mu_{f_\omega^n x}^{\eta_{\tau^n\omega}}(y) \\
=&\frac{1}{W_n(\omega,x)Z_n(\omega,x)}\int_{f_\omega^n(\eta_\omega(x)\cap B)} \Phi_n(\omega,(f_\omega^n)^{-1}y)g(\tau^n\omega,y)d\lambda_{F^n(\omega,x)}^s(y) \\
=&\frac{1}{W_n(\omega,x)Z_n(\omega,x)}\int_{\eta_\omega(x)\cap B} \Phi_n(\omega,y)g(F^n(\omega,y))|\det(T_yf_\omega^n|_{E_0(\omega,y)})|d\lambda_{(\omega,x)}^s(y)
\end{aligned}$$

and, on the other hand,

$$\mu_x^{\eta_\omega}(B) = \int_{\eta_\omega(x)\cap B} g(\omega,y)d\lambda_{(\omega,x)}^s(y).$$

Since B is arbitrary in $\mathcal{B}(M)$, one has

$$\frac{1}{W_n(\omega,x)Z_n(\omega,x)}\Phi_n(\omega,y)g(F^n(\omega,y))|\det(T_yf_\omega^n|_{E_0(\omega,y)})| = g(\omega,y)$$

for $\lambda_{(\omega,x)}^s$-a.e. $y \in \eta_\omega(x)$. Since $W_n(\omega,y) = W_n(\omega,x)$ and $Z_n(\omega,y) = Z_n(\omega,x)$ for any $y \in \eta_\omega(x)$, it follows then that

$$W_n(\omega, y) = \frac{X_n(\omega, y)Y_n(\omega, y)}{Z_n(\omega, y)}, \quad \mu_x^{\eta_\omega}\text{-a.e.} y \in \eta_\omega(x).$$

Claim 3.1 is proved. □

Proof of Claim 3.2. Noting that $\log^+ |f_\omega^n|_{C^1} \in L^1(\Omega^{\mathbf{N}}, v^{\mathbf{N}})$ and for $v^{\mathbf{N}} \times \mu$-a.e. $(\omega, y) \in \Omega^{\mathbf{N}} \times M$

$$|T_yf_\omega^n|_{E_0(\omega,y)}| \leq |f_\omega^n|_{C^1},$$

we know $\log^+ |T_y f_\omega^n|_{E_0(\omega,y)}| \in L^1(\Omega^{\mathbf{N}} \times M, v^{\mathbf{N}} \times \mu)$. By Oseledec multiplicative ergodic theorem we have

$$\frac{1}{n}\int \log|\det T_y f_\omega^n| dv^{\mathbf{N}} \times \mu = \int \sum_i \lambda^{(i)}(x) m_i(x) d\mu \tag{3.6}$$

and

$$\frac{1}{n}\int \log|\det(T_y f_\omega^n|_{E_0(\omega,y)})| dv^{\mathbf{N}} \times \mu = \int \sum_i \lambda^{(i)}(x)^- m_i(x) d\mu \tag{3.7}$$

(both sides of (3.6) and (3.7) may be $-\infty$). From $\log|\det T_y f| \in L^1(\Omega \times M, v \times \mu)$ it is easy to see that $\log|\det T_y f_\omega^n| \in L^1(\Omega^{\mathbf{N}} \times M, v^{\mathbf{N}} \times \mu)$. Hence, by (3.6),

$$\sum_i \lambda^{(i)}(x) m_i(x) \in L^1(M, \mu)$$

which together with (3.7) implies $\log|\det(T_y f_\omega^n|_{E_0(\omega,y)})| \in L^1(\Omega^{\mathbf{N}} \times M, v^{\mathbf{N}} \times \mu)$. Claim 3.2 follows then from (3.6) and (3.7). □

Proof of Claim 3.3. We first prove that $\log \Phi_n \in L^1(\Omega^{\mathbf{N}} \times M, v^{\mathbf{N}} \times \mu)$. Since

$$\begin{aligned}
&\iint \log^- \Phi_n(\omega, y) d\mu dv^{\mathbf{N}} \\
&= \iint [\log^- \Phi_n(\omega, y)] \Phi_n(\omega, y) d(f_\omega^n)^{-1} \mu dv^{\mathbf{N}} \\
&= \iint [\Phi_n(\omega, y) \log \Phi_n(\omega, y)]^- d(f_\omega^n)^{-1} \mu dv^{\mathbf{N}} \\
&> -\infty,
\end{aligned}$$

we obtain $\log^- \Phi_n \in L^1(\Omega^{\mathbf{N}} \times M, v^{\mathbf{N}} \times \mu)$. Noticing that for $v^{\mathbf{N}} \times \mu$-a.e. $(\omega, y) \in \Omega^{\mathbf{N}} \times M$

$$\log \Phi_n(\omega, y) = \log \frac{\varphi(y)}{\varphi(f_\omega^n y)} - \log|\det T_y f_\omega^n| \tag{3.8}$$

which implies

$$\log^- \frac{\varphi(y)}{\varphi(f_\omega^n y)} \geq \log^- \Phi_n(\omega, y) + \log^- |\det T_y f_\omega^n|,$$

we know that $\log^- \frac{\varphi}{\varphi o F^n} \in L^1(\Omega^{\mathbf{N}} \times M, v^{\mathbf{N}} \times \mu)$. Then, by Lemma I.3.1, $\log \frac{\varphi}{\varphi o F^n}$ is integrable with respect to $v^{\mathbf{N}} \times \mu$ and $\int \log \frac{\varphi}{\varphi o F^n} dv^{\mathbf{N}} \times \mu = 0$. Thus, from (3.8) the integrability of $\log \Phi_n$ follows.

We now remark that, given a probability space $(X, \mathcal{B}, \nu)$ and a sub-σ-algebra $\mathcal{A}$ of $\mathcal{B}$, we sometimes use $E_\nu(\cdot|\mathcal{A})$ to denote the conditional expectation operator.

Define $\theta(x) = \begin{cases} 0 & x = 0 \\ x \log x & x > 0 \end{cases}$, and define $\Phi_n^{(N)} = \min\{\Phi_n, N\}$ for each integer $N \geq 1$. Considering that for $v^{\mathbf{N}} \times \mu$ -a.e. $(\omega, y) \in \Omega^{\mathbf{N}} \times M$,

$$Z_n(\omega, y) = E_{(f_\omega^n)^{-1}\mu}(\Phi_n(\omega, \cdot)|\mathcal{B}((f_\omega^n)^{-1}\eta_{\tau^n\omega}))(y),$$

by the convexity of θ we have

$$\begin{aligned}
&\int \log \Phi_n(\omega, y) d v^{\mathbf{N}} \times \mu \\
=&\iint \theta(\Phi_n(\omega, y)) d(f_\omega^n)^{-1} \mu d v^{\mathbf{N}} \\
=&\iint E_{(f_\omega^n)^{-1}\mu}(\theta(\Phi_n(\omega, \cdot)) | \mathcal{B}((f_\omega^n)^{-1} \eta_{\tau^n\omega}))(y) d(f_\omega^n)^{-1} \mu d v^{\mathbf{N}} \\
\geq&\iint \theta(Z_n(\omega, y)) d(f_\omega^n)^{-1} \mu d v^{\mathbf{N}} \\
=&\lim_{N\to+\infty} \iint Z_n(\omega, y) \log E_{(f_\omega^n)^{-1}\mu}(\Phi_n^{(N)}(\omega, \cdot) | \mathcal{B}((f_\omega^n)^{-1} \eta_{\tau^n\omega}))(y) d(f_\omega^n)^{-1} \mu d v^{\mathbf{N}} \\
=&\lim_{N\to+\infty} \iint \Phi_n(\omega, y) \log E_{(f_\omega^n)^{-1}\mu}(\Phi_n^{(N)}(\omega, \cdot) | \mathcal{B}((f_\omega^n)^{-1} \eta_{\tau^n\omega}))(y) d(f_\omega^n)^{-1} \mu d v^{\mathbf{N}} \\
=&\iint \Phi_n(\omega, y) \log Z_n(\omega, y) d(f_\omega^n)^{-1} \mu d v^{\mathbf{N}} \\
=&\int \log Z_n d v^{\mathbf{N}} \times \mu
\end{aligned}$$

which proves $\log Z_n \in L^1(\Omega^{\mathbf{N}} \times M, v^{\mathbf{N}} \times \mu)$.

On the other hand, also by the convexity of θ, we have

$$\begin{aligned}
&\int \log Z_n d v^{\mathbf{N}} \times \mu \\
=&\iint \theta(Z_n(\omega, y)) d(f_\omega^n)^{-1} \mu d v^{\mathbf{N}} \\
\geq&\int \theta\left(\int Z_n(\omega, y) d(f_\omega^n)^{-1} \mu\right) d v^{\mathbf{N}} \\
=&0.
\end{aligned}$$

The proof is completed. □

Proof of Claim 3.4. By Claim 3.1, one has

$$\log W_n = \log X_n + \log Y_n - \log Z_n \leq 0, \quad v^{\mathbf{N}} \times \mu\text{-a.e.}$$

which yields

$$\log^+ X_n \leq -\log^- Y_n + \log^+ Z_n, \quad v^{\mathbf{N}} \times \mu\text{-a.e.}$$

Hence, by Claims 3.2 and 3.3, $\log^+ X_n \in L^1(\Omega^{\mathbf{N}} \times M, v^{\mathbf{N}} \times \mu)$. Then, according to Lemma I.3.1, we know that $\log X_n$ is integrable and $\int \log X_n d v^{\mathbf{N}} \times \mu = 0$. □

Chapter V Stochastic Flows of Diffeomorphisms

In the previous chapters we introduced the notions of entropy and Lyapunov exponents and presented some related ergodic-theoretic results for a random dynamical system generated by i.i.d. (independent and identically distributed) random diffeomorphisms. The main purpose of this chapter is to develop a generalization of the theory to the case of a (continuous time) stochastic flow of diffeomorphisms. Roughly speaking, all stochastic flows of diffeomorphisms are essentially solution flows of stochastic differential equations ($[\text{Kun}]_1$). The theory of stochastic differential equations was initiated by K. Itô in 1942. Since then the theory has been developed in various directions, of which an important one is the application to the study of diffusion processes associated with certain second order partial differential operators. A stochastic differential equation can also be used to describe a dynamical system disturbed by noise. In this chapter we adopt the latter point of view and we are mainly concerned with the random dynamical systems generated by solution flows of stochastic differential equations.

Here we assume that the reader has a reasonable background of random processes and stochastic differential equations. The notion of a stochastic flow in Section 1 needs a basic knowledge of random processes; Remarks 1.1, 1.2 and the proofs of Propositions 1.1 and 1.2 in Section 1 demand the reader being familiar with the standard machinery of Markov processes; when we deal with stochastic flows arising from stochastic differential equations, it is assumed that the reader has a reasonable acquaintance with the theory of stochastic differential equations (see the references therein).

§1 Preliminaries

Throughout this chapter, M is still a C^∞ compact connected Riemannian manifold without boundary. As usual, let $\text{Diff}^r(M)(r \geq 1)$ denote the space of C^r diffeomorphisms of M, equipped with the C^r topology (see $[\text{Hir}]_1$). We first mention the fact that $\text{Diff}^r(M)$ can be metrized in such a way that it becomes a Polish space. Indeed, let ρ be a metric on $C^r(M, M)$ such that $C^r(M, M)$ is separable and complete with respect to ρ. Define a metric $\hat{\rho}$ on $\text{Diff}^r(M)$ by $\hat{\rho}(f,g) = \rho(f,g) + \rho(f^{-1}, g^{-1})$. Then the topology of $\hat{\rho}$ is still the C^r topology, but now $\text{Diff}^r(M)$ is complete with respect to $\hat{\rho}$. This fact will be useful when we deal with random processes taking values in $\text{Diff}^r(M)$.

To begin with, we give the definition of a stochastic flow of diffeomorphisms on M.

Definition 1.1. *Let $(W, \mathcal{F}, P)$ be a probability space. A random process $\{\varphi_t : (W, \mathcal{F}, P) \to \mathrm{Diff}^r(M)\}_{t\geq 0}$ is called a stochastic flow of C^r diffeomorphisms if it has the following properties* (i)–(iv):

(i) *For any $0 \leq t_0 \leq t_1 \leq \cdots \leq t_n, \varphi_{t_i} \circ \varphi_{t_{i-1}}^{-1}, 1 \leq i \leq n$ are independent random variables;*

(ii) *For any $0 \leq s \leq t$, the distribution of $\varphi_t \circ \varphi_s^{-1}$ depends only on $t - s$;*

(iii) *With probability one $\{\varphi_t\}_{t\geq 0}$ has continuous sample paths, i.e. the map $\mathbf{R}^+ \to \mathrm{Diff}^r(M)$ given by $t \longmapsto \varphi_t(w)$ is continuous (with respect to the C^r topology on $\mathrm{Diff}^r(M)$) for P almost all $w \in W$;*

(iv) $\varphi_0 = id$, P–a.e.

Obviously, a stochastic flow of C^r diffeomorphisms is also a stochastic flow of $C^{r'}$ diffeomorphisms if $1 \leq r' \leq r$.

In this chapter we shall discuss stochastic flows of diffeomorphisms mainly from the point of view of ergodic theory of dynamical systems, that is, we shall mainly consider ergodic properties of dynamical systems generated by actions on M of diffeomorphisms of such flows.

Remark 1.1. A stochastic flow of diffeomorphisms can be characterized (up to equivalence of random processes with the same finite dimensional distributions) by a one-parameter convolution semigroup of Borel probability measures on the space of diffeomorphisms. First, let $\{\varphi_t : (W, \mathcal{F}, P) \to \mathrm{Diff}^r(M)\}_{t\geq 0}$ be a random process with $\varphi_0 = id$, a.e. such that properties (i) and (ii) above hold true. Let v_t be the distribution of φ_t, i.e. $v_t(\Gamma) = P\{w : \varphi_t(w) \in \Gamma\}$ for all Borel subsets Γ of $\mathrm{Diff}^r(M)$. One can easily prove that $\{\varphi_t\}_{t\geq 0}$ is a temporally homogeneous Markov process with transition probability functions $P(t, g, \Gamma) = v_t(\Gamma g^{-1}), t \geq 0, \Gamma \in \mathcal{B}(\mathrm{Diff}^r(M)), g \in \mathrm{Diff}^r(M)$. The property (i) and (ii) imply clearly the Chapman-Kolmogorov equation

$$P(s + t, g, \Gamma) = \int P(t, f, \Gamma) dP(s, g, \cdot)(f). \tag{1.1}$$

Then, by a standard argument from measure theory, from (1.1) it follows that $\{v_t\}_{t\geq 0}$ is a convolution semigroup, i.e. for all $s, t \geq 0$

$$v_t * v_s = v_{s+t} \tag{1.2}$$

which means

$$\int l(f \circ g) dv_t(f) dv_s(g) = \int l(h) dv_{s+t}(h)$$

for all bounded measurable functions $l : \mathrm{Diff}^r(M) \to \mathbf{R}$. Conversely, assume that $\{v_t\}_{t\geq 0}$ is a convolution semigroup of Borel probability measures on $\mathrm{Diff}^r(M)$ with $v_0(\{id\}) = 1$. Define $P(t, g, \Gamma) = v_t(\Gamma g^{-1}), t \geq 0, \Gamma \in \mathcal{B}(\mathrm{Diff}^r(M)), g \in \mathrm{Diff}^r(M)$. This is a family of transition probability functions since the convolution property (1.2) also imply the Chapman-Kolmogorov equation (1.1) for $\{P(t, g, \Gamma)\}$. We may use this family of transition probability functions to construct a temporally homogeneous Markov process $\{\varphi_t\}_{t\geq 0}$ with values in

$\text{Diff}^r(M)$ and with $\varphi_0 = id$, a.e. This process has then the properties (i) and (ii). Moreover, the process $\{\varphi_t\}_{t\geq 0}$ has a modification with continuous sample paths almost surely if and only if for all neighbourhood U of the identity of $\text{Diff}^r(M)$,

$$\frac{1}{t}v_t(\text{Diff}^r(M)\backslash U) \to 0$$

as $t \to 0$. We refer the reader to $[\text{Bax}]_2$ for a detailed treatment of this topic.

Remark 1.2. Properties (i) and (ii) above for a random process $\{\varphi_t\}_{t\geq 0}$ valued in $\text{Diff}^r(M)$ can also be characterized by Markov properties of all the n-point motions of $\{\varphi_t\}_{t\geq 0}$ in the following way. Suppose that $\{\varphi_t : (W, \mathcal{F}, P) \to \text{Diff}^r(M)\}_{t\geq 0}$ is a random process with $\varphi_0 = id$, a.e. Let $n \geq 1$ and let $x^{(n)} = (x_1, \cdots, x_n)$ be a point in M^n. Set $\varphi_t x^{(n)} = (\varphi_t x_1, \cdots, \varphi_t x_n)$. Then $\{\varphi_t x^{(n)} : (W, \mathcal{F}, P) \to M^n\}_{t\geq 0}$ is a random process starting at $x^{(n)}$ at time 0. It is called an *n-point motion* of the process $\{\varphi_t\}_{t\geq 0}$. If $\{\varphi_t\}_{t\geq 0}$ satisfies (i) and (ii) above, then it follows easily that for each $n \geq 1$ and $x^{(n)} = (x_1, \cdots, x_n) \in M^n$, $\{\varphi_t x^{(n)}\}_{t\geq 0}$ is a temporally homogeneous Markov process with transition probability functions $P_n(t, y^{(n)}, A) = P\{w : \varphi_t(w)y^{(n)} \in A\}, t \geq 0, A \in \mathcal{B}(M^n), y^{(n)} \in M^n$. Conversely, if for each $n \geq 1$ and $x^{(n)} = (x_1, \cdots, x_n) \in M^n$ the n-point motion $\{\varphi_t x^{(n)}\}_{t\geq 0}$ is a temporally homogeneous Markov process with transition probability functions $P_n(t, y^{(n)}, A) = P\{w : \varphi_t(w)y^{(n)} \in A\}, t \geq 0, A \in \mathcal{B}(M^n), y^{(n)} \in M^n$, then one can show that all the Chapman-Kolmogorov equations for $\{P_n(t, y^{(n)}, A) : t \geq 0, A \in \mathcal{B}(M^n), y^{(n)} \in M^n\}, n \geq 1$ put together lead to the convolution property (1.2) and consequently $\{\varphi_t\}_{t\geq 0}$ satisfies properties (i) and (ii).

Let $\{\varphi_t : (W, \mathcal{F}, P) \to \text{Diff}^r(M)\}_{t\geq 0}$ be a stochastic flow. We now introduce the notion of invariant measures for the flow.

Definition 1.2. *A Borel probability measure μ on M is called an invariant measure of $\{\varphi_t\}_{t\geq 0}$ if*

$$\int \varphi_t(w)\mu dP(w) = \mu$$

for all $t \geq 0$.

Let $P(t, x, \cdot), t \geq 0, x \in M$ be the transition probabilities of the one-point motions of $\{\varphi_t\}_{t\geq 0}$. Denote by $B(M)$ the space of all bounded measurable functions on M. We now introduce the linear operators $T_t : B(M) \to B(M), t \geq 0$ by the formula

$$(T_t g)(x) = \int g(y)dP(t, x, \cdot)(y) \tag{1.3}$$

for $g \in B(M)$ and $x \in M$. The family of linear operators $\{T_t\}_{t\geq 0}$ satisfies the semigroup property

$$T_t \circ T_s = T_{s+t}, \qquad s, t \geq 0 \tag{1.4}$$

because of the Markov property of the one-point motions, it is called the *semi-group of linear operators* for the one-point motions of $\{\varphi_t\}_{t\geq 0}$. Because of property (iii) in Definition 1.1 and the compactness of M one has $T_t C(M) \subset C(M)$ for all $t \geq 0$ and

$$\lim_{t\to 0}\sup_{x\in M}|(T_t g)(x) - g(x)| = 0 \tag{1.5}$$

for any $g \in C(M)$, where $C(M)$ denotes the space of all continuous functions on M.

Corresponding to T_t defined above, there is an adjoint operator $T_t^* : \mathcal{M}(M) \to \mathcal{M}(M)$ defined by

$$(T_t^*\rho)(A) = \int P(t,x,A)d\rho(x)$$

for $\rho \in \mathcal{M}(M)$ and $A \in \mathcal{B}(M)$. The family of operators $\{T_t^*\}_{t\geq 0}$ also has the semigroup property

$$T_t^* \circ T_s^* = T_{s+t}^*, \quad s,t \geq 0. \tag{1.6}$$

Clearly, a measure $\mu \in \mathcal{M}(M)$ is invariant for $\{\varphi_t\}_{t\geq 0}$ if and only if $T_t^*\mu = \mu$ for all $t \geq 0$.

Proposition 1.1. *Let $\{\varphi_t : (W,\mathcal{F},P) \to Diff(M)\}_{t\geq 0}$ be a stochastic flow. Then there is at least one invariant measure of the flow.*

Proof. Analogously as in the case of $\mathcal{X}^+(M,v)$ considered in Section I.1, for every positive integer n there exists a $T_{1/n}^*$-invariant measure $\mu_n \in \mathcal{M}(M)$. Now take a subsequence $\{\mu_{n_i}\}_{i\geq 1}$ of $\{\mu_n\}_{n\geq 1}$ such that μ_{n_i} weakly converges to some probability measure μ as $i \to +\infty$. For each fixed $t \geq 0$, if g is a continuous function on M, then using the semigroup properties (1.4) and (1.6) we have

$$\begin{aligned}
\int g dT_t^*\mu = & \int T_t g d\mu = \lim_{i\to+\infty}\int T_t g d\mu_{n_i} \\
= & \lim_{i\to+\infty}\int T_{t-[tn_i]\frac{1}{n_i}} g dT^*_{[tn_i]\frac{1}{n_i}}\mu_{n_i} \\
= & \lim_{i\to+\infty}\int T_{t-[tn_i]\frac{1}{n_i}} g d\mu_{n_i} \\
= & \int g d\mu
\end{aligned}$$

by the T_{1/n_i}^*-invariance of μ_{n_i} and (1.5). This implies that $T_t^*\mu = \mu$ for all $t \geq 0$.
□

In the sequel a stochastic flow $\{\varphi_t\}_{t\geq 0}$ will be denoted by $(\{\varphi_t\}_{t\geq 0}, \mu)$ when associated with an invariant measure μ.

In order to carry the notions and results presented in Chapters I–IV for a system $\mathcal{X}^+(M,v,\mu)$ over to the case of a stochastic flow of C^r diffeomorphisms, we shall prove a proposition (Proposition 1.2) which asserts that for a stochastic flow of $C^r (r = 1,2)$ diffeomorphisms the C^r-norms of the diffeomorphisms in the flow satisfy automatically an integrability condition. For this purpose we first present some preliminary facts.

Let $\{\varphi_t : (W, \mathcal{F}, P) \to \mathrm{Diff}^r(M)\}_{r\geq 1}$ be a stochastic flow. As we have said in Remark 1.1, it is a temporally homogeneous Markov process with transition probability functions $P(t, g, \Gamma) = v_t(\Gamma g^{-1}), t \geq 0, \Gamma \in \mathcal{B}(\mathrm{Diff}^r(M)), g \in \mathrm{Diff}^r(M)$, where v_t is as introduced in Remark 1.1. Let $\{T_t\}_{t\geq 0}$ be the semigroup of linear operators for $\{\varphi_t\}_{t\geq 0}$ defined analogously as in (1.3), i.e.

$$(T_t l)(g) = \int l(f) dP(t, g, \cdot)(f)$$

for each bounded measurable function $l : \mathrm{Diff}^r(M) \to \mathbf{R}$ and $g \in \mathrm{Diff}^r(M)$. It is easy to see that, if l is a bounded continuous function, so is $T_t l$. Hence $\{T_t\}_{t\geq 0}$ is a Feller semigroup. This together with property (iii) in Definition 1.1 implies that $\{\varphi_t\}_{t\geq 0}$ is a strong Markov process ([Dyn]).

Proposition 1.2. *Let $\{\varphi_t : (W, \mathcal{F}, P) \to \mathrm{Diff}^r(M)\}_{t\geq 0} (r = 1, 2)$ be a stochastic flow. Then*

$$\int [\sup_{0\leq t\leq T} \log^+ \|\varphi_t(w)\|_{C^r} + \sup_{0\leq t\leq T} \log^+ \|\varphi_t(w)^{-1}\|_{C^r}] dP(w) < +\infty \qquad (1.7)$$

for any $T \geq 0$, where $\|f\|_{C^r} = \sup_{x\in M} |T_x f| = |f|_{C^1}$ if $r = 1$, and $\|f\|_{C^r}$ is as defined in Section II. 1 if $r = 2$.

Proof. For any $f, g \in \mathrm{Diff}^r(M)$, by Proposition II. 1.1 (if $r = 2$),

$$\|f \circ g\|_{C^r} \leq C \|f\|_{C^r} \|g\|_{C^r} \max\{\|g\|_{C^1}, 1\} \qquad (1.8)$$

where $C \geq 1$ is a constant. Define

$$U = \{f \in \mathrm{Diff}^r(M) : \max\{\|f\|_{C^r}, \|f^{-1}\|_{C^r}, \|f\|_{C^1}, \|f^{-1}\|_{C^1}\} < K\}$$

where $K = \|id\|_{C^r} + 2$. By properties (iii) and (iv) in Definition 1.1, for any given $\delta > 0$ there exists $t(\delta) > 0$ such that

$$P\{\tau_1 \leq t(\delta)\} \leq \delta \qquad (1.9)$$

where τ_1 is defined to be the stopping (Markov) time $\inf\{t : \varphi_t \notin U\}$. For positive integer $n > 1$, put

$$U^n = \{f \in \mathrm{Diff}^r(M) : f = f_n \circ f_{n-1} \circ \cdots \circ f_1, f_i \in U, 1 \leq i \leq n\}.$$

Then, by (1.8),

$$\max_{1\leq k\leq r} \{\|f\|_{C^k}, \|f^{-1}\|_{C^k}\} \leq C^n K^{2n} \qquad (1.10)$$

for any $f \in U^n$. Define $\tau_n = \inf\{t : \varphi_t \notin U^n\}$, then, using the strong Markov property of the process $\{\varphi_t\}_{t\geq 0}$ at the stopping time τ_n, we have for any given

$t_0 > 0$

$$\begin{aligned}
& P\{\tau_n \le t_0\} \\
= \;& P\{\varphi_t \notin U^n \text{ for some } t \le t_0\} \\
\le \;& P\{\tau_{n-1} \le t_0 \text{ and } \varphi_t \circ \varphi_{\tau_{n-1}}^{-1} \notin U \text{ for some } t \text{ satisfying } \tau_{n-1} \le t \le t_0\} \\
\le \;& P\{\tau_{n-1} \le t_0\} P\{\tau_1 \le t_0\}.
\end{aligned}$$

Thus, by induction,

$$P\{\tau_n \le t_0\} \le (P\{\tau_1 \le t_0\})^n. \tag{1.11}$$

Taking $t_0 = t(\delta)$ and using (1.9) we obtain

$$P\{\tau_n \le t(\delta)\} \le \delta^n. \tag{1.12}$$

Now, by (1.10) and (1.12),

$$\begin{aligned}
& \int \max_{1\le k\le r} \left\{ \sup_{0\le t\le t(\delta)} \log^+ \|\varphi_t\|_{C^k}, \sup_{0\le t\le t(\delta)} \log^+ \|\varphi_t^{-1}\|_{C^k} \right\} dP \\
\le \;& \sum_{n=1}^{+\infty} n \log(CK^2) P\{\tau_{n-1} \le t(\delta) < \tau_n\} \qquad (1.13) \\
\le \;& \delta^{-1} \log(CK^2) \sum_{n=1}^{\infty} n\delta^n < +\infty
\end{aligned}$$

if $\delta < 1$, where $\tau_0 = 0$. This clearly imply (1.7) for $T \le t(\delta)$. In order to prove (1.7) for all $T > 0$, notice that for any $n \ge 1$, by (1.8),

$$\begin{aligned}
& \sup_{0\le t\le (n+1)t(\delta)} \log^+ \|\varphi_t\|_{C^r} \\
\le \;& \sup_{0\le t\le t(\delta)} \log^+ \|\varphi_t\|_{C^r} + \sup_{t(\delta)\le t\le (n+1)t(\delta)} \log^+ \|\varphi_t \circ \varphi_{t(\delta)}^{-1} \circ \varphi_{t(\delta)}\|_{C^r} \\
\le \;& 2 \sup_{0\le t\le t(\delta)} \log^+ \|\varphi_t\|_{C^r} + \log C + \log^+ \|\varphi_{t(\delta)}\|_{C^1} \qquad (1.14) \\
& + \sup_{t(\delta)\le t\le (n+1)t(\delta)} \log^+ \|\varphi_t \circ \varphi_{t(\delta)}^{-1}\|_{C^r}.
\end{aligned}$$

Notice that if we define $\psi_s = \varphi_{t(\delta)+s}\varphi_{t(\delta)}^{-1}$ then $\{\psi_s\}_{s\ge 0}$ is also a stochastic flow. It is a temporally homogeneous Markov process with transition probability functions $P(s, g, \Gamma) = v_s'(\Gamma g^{-1}), s \ge 0, \Gamma \in \mathcal{B}(\mathrm{Diff}^r(M)), g \in \mathrm{Diff}^r(M)$, where v_s' is the distribution of ψ_s. Because of property (ii) in Definition 1.1 we have $v_s' = v_s$ for all $s \ge 0$. Hence $\{\varphi_t\}_{t\ge 0}$ and $\{\psi_s\}_{s\ge 0}$ have the same finite dimensional distributions. This together with property (iii) in Definition 1.1 implies that

$$\int \sup_{t(\delta)\le t\le (n+1)t(\delta)} \log^+ \|\varphi_t \circ \varphi_{t(\delta)}^{-1}\|_{C^r} dP = \int \sup_{0\le t\le nt(\delta)} \log^+ \|\varphi_t\|_{C^r} dP.$$

Then by (1.13), (1.14) and an inductive argument we get

$$\int \sup_{0\le t\le (n+1)t(\delta)} \log^+ \|\varphi_t\|_{C^r} dP < +\infty$$

for all $n \geq 1$, proving the integrability of the first expression in (1.7). We now prove the integrability of the second one. For any $n \geq 1$, by (1.8),

$$\begin{aligned}
&\sup_{0\le t\le (n+1)t(\delta)} \max_{1\le k\le r} \log^+ \|\varphi_t^{-1}\|_{C^k} \\
\le\ & \sup_{0\le t\le t(\delta)} \max_{1\le k\le r} \log^+ \|\varphi_t^{-1}\|_{C^k} \\
&+ \sup_{t(\delta)\le t\le (n+1)t(\delta)} \max_{1\le k\le r} \log^+ \|\varphi_{t(\delta)}^{-1} \circ (\varphi_t \circ \varphi_{t(\delta)}^{-1})^{-1}\|_{C^k} \\
\le\ & 2 \sup_{0\le t\le t(\delta)} \max_{1\le k\le r} \log^+ \|\varphi_t^{-1}\|_{C^k} + \log C \\
&+2 \sup_{t(\delta)\le t\le (n+1)t(\delta)} \max_{1\le k\le r} \log^+ \|(\varphi_t \circ \varphi_{t(\delta)}^{-1})^{-1}\|_{C^k}
\end{aligned}$$

Then, for the same reason as above, we obtain

$$\int \sup_{0\le t\le (n+1)t(\delta)} \max_{1\le k\le r} \log^+ \|\varphi_t^{-1}\|_{C^k} dP < +\infty$$

for all $n \geq 1$. The proof of the proposition is completed. □

Remark 1.3. It can be shown that Proposition 1.2 holds true for all $r \geq 1$. In fact, let $|f|_{C^r}$ be the usual C^r-norm for $f \in \mathrm{Diff}^r(M)(r \geq 1)$ (see [Fra] for the definition). By Lemma 3.2 of [Fra] one has

$$\begin{aligned}
|f \circ g|_{C^r} \le\ & C_r |f|_{C^r} (|g|_{C^r}^r + 1) \\
\le\ & 2C_r |f|_{C^r} \max\{|g|_{C^r}^r, 1\}
\end{aligned} \tag{1.15}$$

for any $f, g \in \mathrm{Diff}^r(M)$, where C_r is a constant depending only on r. Using (1.15) in place of (1.8) one can prove

$$\int \left[\sup_{0\le t\le T} \log^+ |\varphi_t|_{C^r} + \sup_{0\le t\le T} \log^+ |\varphi_t^{-1}|_{C^r}\right] dP < +\infty$$

for all $T \geq 0$ by almost the same argument as the proof of Proposition 1.2.

Remark 1.4. Y. Kifer presented in $[\mathrm{Kif}]_3$ an even stronger result which says that, if $\{\varphi_t : (W, \mathcal{F}, P) \to \mathrm{Diff}^r(M)\}_{t\ge 0} (r \geq 1)$ is a stochastic flow, then

$$\int \left[\sup_{0\le t\le T} |\varphi_t|_{C^r}^l + \sup_{0\le t\le T} |\varphi_t^{-1}|_{C^r}^l\right] dP < +\infty$$

for all $T \geq 0$ and $l \in \mathbf{N}$. But we are not sure if this result in [Kif]$_3$ and the proof provided there are correct. Actually, the proof there is based on the following two statements:

(1) $|\varphi_{t(\delta)}|_{C^r}$ and $\sup_{t(\delta)\leq t\leq(n+1)t(\delta)} |\varphi_t \circ \varphi_{t(\delta)}^{-1}|_{C^r} (n \geq 1)$ are independent;

(2) For all $l \in \mathbf{N}$

$$\int \max\left\{\sup_{0\leq t\leq t(\delta)} |\varphi_t|_{C^r}^l, \sup_{0\leq t\leq t(\delta)} |\varphi_t^{-1}|_{C^r}^l\right\} dP < +\infty.$$

On the one hand, we are dubious if statement (1) is correct. On the other hand, Y. Kifer did not give a proper justification of statement (2) since he only proved that for each fixed $l \in \mathbf{N}$ there exists $\delta_l > 0$ such that

$$\int \max\left\{\sup_{0\leq t\leq t(\delta_l)} |\varphi_t|_{C^r}^l, \sup_{0\leq t\leq t(\delta_l)} |\varphi_t^{-1}|_{C^r}^l\right\} dP < +\infty.$$

We now pass to stochastic flows of diffeomorphisms arising from solutions of stochastic differential equations. This subject has been discussed in many recent papers and books from various points of view. Here we shall neither attempt to touch on all the topics of the subject nor attempt to give the full bibliography on them. We shall only give the basic idea of how stochastic flows of diffeomorphisms are related to solutions of stochastic differential equations. Now we assume that the reader has a standard knowledge of stochastic differential equations.

For the sake of presentation we first review briefly some basic knowledge about stochastic differential equations defined on M, which can be found in [Ike], [Elw]$_2$ or [Kun]$_2$. Suppose now we are given vector fields $X_0, X_1, \cdots, X_d$ on M. We assume that $X_1, \cdots, X_d$ are of class C^3, i.e. with a local coordinate $x = (x_1, \cdots, x_{m_0})$ (where $m_0 = \dim M$ as before), these vector fields are expressed as

$$X_k(x) = \sum_{i=1}^{m_0} X_k^i(x) \frac{\partial}{\partial x_i}$$

where $X_k^i(x), i = 1, \cdots, m_0, k = 1, \cdots, d$ are C^3 functions of x. As to vector field X_0, we assume it is of class C^2. Let $\{B_t = (B_t^1, \cdots, B_t^d)\}_{t\geq 0}$ be a d-dimensional standard Brownian motion defined on a probability space $(W, \mathcal{F}, P)$ and let $\{\mathcal{F}_t\}_{t\geq 0}$ be the proper reference family of $\{B_t\}_{t\geq 0}$, i.e. $\mathcal{F}_t = \bigcap_{\varepsilon>0} \sigma\{B_u : u \leq t + \varepsilon\}$ where $\sigma\{B_u : u \leq t + \varepsilon\}$ is the smallest sub-σ-algebra of $\mathcal{F}$ with respect to which B_u is measurable for each $u \leq t + \varepsilon$. We now consider a Stratonovich SDE (stochastic differential equation) on the manifold M

$$d\xi_t = X_0(\xi_t)dt + \sum_{k=1}^{d} X_k(\xi_t) \circ dB_t^k. \tag{1.16}$$

By a solution to SDE (1.16) starting at x at time s we mean a sample continuous, $\{\mathcal{F}_t\}_{t\geq 0}$-adapted random process $\{\varphi_{s,t}(x) : (W,\mathcal{F},P) \to M\}_{t\geq s}$ with $\varphi_{s,s}(x) = x$, P-a.e. such that for any C^3 map $F : M \to \mathbf{R}$ there holds the following *Itô* formula

$$F(\varphi_{s,t}(x)) = F(x) + \int_s^t (X_0F)(\varphi_{s,u}(x))du + \sum_{k=1}^d \int_s^t (X_kF)(\varphi_{s,u}(x)) \circ dB_u^k$$

for all $t \geq s$.

For any given $x \in M$ and $s \in \mathbf{R}^+$, SDE (1.16) always has a unique solution starting at x at time s. The uniqueness holds in the sense that, if $\{\varphi_{s,t}(x)\}_{t\geq s}$ and $\{\varphi'_{s,t}(x)\}_{t\geq s}$ are both solutions to SDE (1.16) starting at x at time s, then $P\{\varphi_{s,t}(x) = \varphi'_{s,t}(x)$ for all $t \geq s\} = 1$. It is remarkable that, in this special case and with specific modifications, there exists a system of random processes $\{\varphi_{s,t}(x) : (W,\mathcal{F},P) \to M\}_{t\geq s}, s \geq 0, x \in M$ such that the following (1) and (2) hold true:

(1) For each $x \in M$ and $s \geq 0, \{\varphi_{s,t}(x)\}_{t\geq s}$ is a solution to SDE (1.16) starting at x at time s;

(2) We write $\varphi_{s,t}(x) : (W,\mathcal{F},P) \to M$ as $\varphi_{s,t}(x,\cdot)$ for the sake of presentation. Then there exists a measurable set $N \in \mathcal{F}$ with $P(N) = 1$ such that for each fixed $w \in N, \varphi_{s,t}(\cdot, w) : M \to M, x \longmapsto \varphi_{s,t}(x,w)$ defines a C^1 diffeomorphism on M for all $0 \leq s \leq t$.

More crucially, the Markov property of the system of solutions $\{\varphi_{s,t}(x)\}_{t\geq s}, s \geq 0, x \in M$ manifests itself as the independence property (the property (d) that follows) of the family of C^1 diffeomorphisms $\{\varphi_{s,t}(\cdot,w) : 0 \leq s \leq t, w \in N\}$. The detailed properties of the family of diffeomorphisms are summarized in the following paragraph.

We write $\varphi_{s,t}(\cdot,w)$ simply as $\varphi_{s,t}(w)$. Then the set N of full P measure can be chosen such that the family of C^1 diffeomorphisms $\{\varphi_{s,t}(w)\}$ has the following properties:

(a) $\varphi_{s,t}(w) = \varphi_{u,t}(w) \circ \varphi_{s,u}(w), s \leq u \leq t$ for each $w \in N$;

(b) $\varphi_{s,s}(w) = id$ for each $w \in N$;

(c) $\varphi_{s,t}(w)$ depends continuously on (s,t) (with respect to the C^1 topology of $\mathrm{Diff}^1(M)$) for each $w \in N$;

(d) $\varphi_{t_i,t_{i+1}}(\cdot) : N \to \mathrm{Diff}^1(M), i = 0,\cdots,n-1$ are independent for any $0 \leq t_0 \leq \cdots \leq t_n$;

(e) The distribution of $\varphi_{s,t}(\cdot) : N \to \mathrm{Diff}^1(M)$ coincides with that of $\varphi_{s+h,t+h}(\cdot) : N \to \mathrm{Diff}^1(M)$ for any $0 \leq s \leq t$ and $h \geq 0$.

The proof of the above facts is far from being easy as compared to the case of an ordinary differential equation. It requires a lot of careful arguments about null sets. We refer the reader to the important treatise [Kun]$_1$ of Kunita for a detailed proof.

Let $\{\varphi_{s,t}(w) : 0 \leq s \leq t, w \in N\}$ be as obtained above. Define $\varphi_t = \varphi_{0,t}$ for $t \geq 0$. Then $\{\varphi_t\}_{t\geq 0}$ is clearly a stochastic flow of C^1 diffeomorphisms and

$\varphi_{s,t}(w) = \varphi_t(w) \circ \varphi_s(w)^{-1}$ for all $0 \leq s \leq t$ and $w \in N$. We call $\{\varphi_t\}_{t \geq 0}$ a stochastic flow generated by solutions of SDE (1.16).

When the coefficients $X_1, \cdots, X_d$ are vector fields of class $C^{r+2}(r \geq 1)$ and X_0 is of class C^{r+1}, the stochastic flow $\{\varphi_t\}_{t \geq 0}$ can be chosen such that it is a stochastic flow of C^r diffeomorphisms (see [Kun]$_2$).

Conversely, for a given stochastic flow of $C^r(r \geq 1)$ diffeomorphisms with a suitable regularity condition, one can write down an SDE such that its solutions generate the given flow. In the general case, however, such an SDE has to be based on an infinite number of Brownian motions, or equivalently a Brownian motion with values in vector fields. In this sense, there exists essentially a one-to-one correspondence between stochastic flows of diffeomorphisms and stochastic differential equations. For a detailed treatment of the precise relationship we refer the reader to [Kun]$_1$.

Let $\{\varphi_t\}_{t \geq 0}$ be a stochastic flow generated by solutions of SDE (1.16). The transition probability functions of the solutions of SDE (1.16) are clearly given by $P(t, x, A) = P\{w : \varphi_t(w)x \in A\}, t \geq 0, A \in \mathcal{B}(M), x \in M$. They are uniquely determined by the generator of SDE (1.16)

$$L = \frac{1}{2}\sum_{k=1}^{d} X_k^2 + X_0. \tag{1.17}$$

Hence L determines all the invariant measures of $\{\varphi_t\}_{t \geq 0}$. If L is elliptic, or equivalently SDE (1.16) is non-degenerate, i.e. $\{X_1(x), \cdots, X_d(x)\}$ spans $T_x M$ for each $x \in M$, then $\{\varphi_t\}_{t \geq 0}$ has a unique invariant measure μ. The measure μ has a smooth density with respect to the Lebesgue measure λ on M induced by the Riemannian metric, i.e. $d\mu/d\lambda = \rho$ for some smooth function $\rho : M \longrightarrow \mathbf{R}^+$. This is because ρ is a solution to the adjoint elliptic partial differential equation and Schwartz-Weyl lemma holds true (see, e.g. Chapter V of [Ike]).

Remark 1.5. Since we regard in this chapter stochastic differential equations as dynamical systems disturbed by noise, we consider here a Stratonovich SDE rather than an Itô SDE. If we want to consider an Itô SDE on M, then we should not regard the coefficients of the equation as vector fields (see Section II. 8 of [Kun]$_2$).

§2 Lyapunov Exponents and Stable Manifolds of Stochastic Flows of Diffeomorphisms

In this section we present a version of Oseledec multiplicative ergodic theorem for a stochastic flow of diffeomorphisms. This theorem establishes the existence of Lyapunov exponents for the flow, which describe the almost-sure limiting exponential growth rates of tangent vectors under the flow. Then we establish the existence of (global and local) stable manifolds for the flow associated with the Lyapunov exponents.

Now suppose that we are given a probability space $(W, \mathcal{F}, P)$, a stochastic flow $\{\varphi_t : (W, \mathcal{F}, P) \to \mathrm{Diff}^r(M)\}_{t\geq 0} (r \geq 1)$ and an invariant measure μ of the flow. Let $(W \times M, \mathcal{F} \times \mathcal{B}(M), P \times \mu)$ be the product space of $(W, \mathcal{F}, P)$ and $(M, \mathcal{B}(M), \mu)$.

Theorem 2.1. *Let $(\{\varphi_t\}_{t\geq 0}, \mu)$ be as given above with $r \geq 1$. Then we have a measurable set $\Lambda_0 \subset W \times M$ of full $P \times \mu$ measure such that for each $(w, x) \in \Lambda_0$, there exist numbers depending only on x*

$$-\infty < \lambda^{(1)}(x) < \lambda^{(2)}(x) < \cdots < \lambda^{(r(x))}(x) < +\infty \tag{2.1}$$

and an associated filtration by linear subspaces of $T_x M$

$$\{0\} = V^{(0)}_{(w,x)} \subset V^{(1)}_{(w,x)} \subset \cdots \subset V^{(r(x))}_{(w,x)} = T_x M \tag{2.2}$$

such that

$$\lim_{t\to+\infty} \frac{1}{t} \log |T_x \varphi_t(w)\xi| = \lambda^{(i)}(x) \tag{2.3}$$

for each $\xi \in V^{(i)}_{(w,x)} \backslash V^{(i-1)}_{(w,x)}, 1 \leq i \leq r(x)$ and

$$\lim_{t\to+\infty} \frac{1}{t} \log |T_x \varphi_t(w)| = \lambda^{(r(x))}(x), \tag{2.4}$$

$$\lim_{t\to+\infty} \frac{1}{t} \log |det\ T_x \varphi_t(w)| = \sum_{i=1}^{r(x)} \lambda^{(i)}(x) m_i(x) \tag{2.5}$$

where $m_i(x) = \dim V^{(i)}_{(w,x)} - \dim V^{(i-1)}_{(w,x)}$, which also depends only on x. In addition, the numbers $r(x), \lambda^{(i)}(x)$ and the subspaces $V^{(i)}_{(w,x)}$ all depend measurably on (w, x).

Remark 2.1. According to Fubini theorem, Theorem 2.1 implies that for μ almost every $x \in M$ there exist the numbers in (2.1) such that for P almost all $w \in W$ there exists a filtration (2.2) satisfying (2.3)–(2.5). We shall call the numbers $\lambda^{(1)}(x) < \cdots < \lambda^{(r(x))}(x)$ the *Lyapunov exponents* of $(\{\varphi_t\}_{t\geq 0}, \mu)$ at point $x \in M$ and call $m_i(x)$ the *multiplicity* of $\lambda^{(i)}(x)$.

Proof. We prove the theorem first for discrete time steps of length 1. Denote $\mathrm{Diff}^1(M)$ by X for simplicity of notation. Let v_1 be the distribution of φ_1. Then by Proposition 1.2 it holds that

$$\int_X [\log^+ |f|_{C^1} + \log^+ |f^{-1}|_{C^1}] dv_1(f) < +\infty. \tag{2.6}$$

Put

$$(X^{\mathbf{N}}, \mathcal{B}(X)^{\mathbf{N}}, v_1^{\mathbf{N}}) = \prod_1^{+\infty} (X, \mathcal{B}(X), v_1) \tag{2.7}$$

and define a map

$$\Sigma_1 : (W, \mathcal{F}, P) \to (X^{\mathbf{N}}, \mathcal{B}(X)^{\mathbf{N}}, v_1^{\mathbf{N}}),$$
$$w \longmapsto (\varphi_1(w) \circ \varphi_0(w)^{-1}, \varphi_2(w) \circ \varphi_1(w)^{-1}, \cdots). \tag{2.8}$$

By Properties (i) and (ii) in Definition 1.1 we know that Σ_1 is a measure-preserving map. Now note that Theorem 3.2 in Chapter I also holds true if $\Omega = \mathrm{Diff}^2(M)$ is replaced by $X = \mathrm{Diff}^1(M)$ and $\int_X \log^+ |f|_{C^1} dv_1(f) < +\infty$. This together with (2.6) shows immediately that Theorem 2.1 holds true if we consider only $t \in \mathbf{Z}^+$.

In order to obtain the full (continuous time) result, note that if $(w, x) \in W \times M$ and $\xi \in T_xM$ then for all $n \in \mathbf{Z}^+$ and $t \in [n, n+1]$ we have

$$\begin{aligned} &\log |T_x\varphi_{n+1}(w)\xi| - \log^+ |T_{\varphi_t(w)x}(\varphi_{n+1}(w) \circ \varphi_t(w)^{-1})| \\ \leq\ &\log |T_x\varphi_t(w)\xi| \\ \leq\ &\log |T_x\varphi_n(w)\xi| + \log^+ |T_{\varphi_n(w)x}(\varphi_t(w) \circ \varphi_n(w)^{-1})|, \end{aligned} \tag{2.9}$$

$$\begin{aligned} &\log |T_x\varphi_{n+1}(w)| - \log^+ |T_{\varphi_t(w)x}(\varphi_{n+1}(w) \circ \varphi_t(w)^{-1})| \\ \leq\ &\log |T_x\varphi_t(w)| \\ \leq\ &\log |T_x\varphi_n(w)| + \log^+ |T_{\varphi_n(w)x}(\varphi_t(w) \circ \varphi_n(w)^{-1})|, \end{aligned} \tag{2.10}$$

and

$$\begin{aligned} &\log |\det T_x\varphi_{n+1}(w)| - \log^+ |T_{\varphi_t(w)x}(\varphi_{n+1}(w) \circ \varphi_t(w)^{-1})|^{m_0} \\ \leq\ &\log |\det T_x\varphi_t(w)| \\ \leq\ &\log |\det T_x\varphi_n(w)| + \log^+ |T_{\varphi_n(w)x}(\varphi_t(w) \circ \varphi_n(w)^{-1})|^{m_0}. \end{aligned} \tag{2.11}$$

Thus, if we put

$$\Phi_n(w) = \sup_{n \leq t \leq n+1} \log^+ |\varphi_t(w) \circ \varphi_n(w)^{-1}|_{C^1} \tag{2.12}$$

and

$$\Psi_n(w) = \sup_{n \leq t \leq n+1} \log^+ |\varphi_{n+1}(w) \circ \varphi_t(w)^{-1}|_{C^1}, \tag{2.13}$$

then the desired full result follows from (2.9)–(2.11) provided that for P almost all $w \in W$

$$\lim_{n \to +\infty} \frac{1}{n} \Phi_n(w) = 0 \tag{2.14}$$

and

$$\lim_{n \to +\infty} \frac{1}{n} \Psi_n(w) = 0. \tag{2.15}$$

So it remains to prove (2.14) and (2.15). Put $\tilde{X} = \{\tilde{f} : \tilde{f} = \{\tilde{f}_t : \tilde{f}_t \in X, t \geq 0\}$ such that $\tilde{f}_0 = id$ and $t \longmapsto \tilde{f}_t$ is a continuous map from $\mathbf{R}^+$ to X (with respect to the C^1 topology on X) $\}$. Let $\mathcal{B}(\tilde{X})$ be the smallest σ-algebra containing all the cylinder sets $\{\tilde{f} : \tilde{f}_{t_i} \in \Gamma_i, 1 \leq i \leq n\}, \Gamma_1, \cdots, \Gamma_n \in \mathcal{B}(X), 0 \leq t_1 < \cdots < t_n, n \geq 1$ and let $\tilde{P}$ be the measure on $(\tilde{X}, \mathcal{B}(\tilde{X}))$ which on a cylinder set $\{\tilde{f} : \tilde{f}_{t_i} \in \Gamma_i, 1 \leq i \leq n\}$ is given by

$$\int_{\Gamma_1} P(t_1, id, dg_1) \int_{\Gamma_2} P(t_2 - t_1, g_1, dg_2) \cdots \int_{\Gamma_n} P(t_n - t_{n-1}, g_{n-1}, dg_n),$$

where $P(t, g, \Gamma) = v_t(\Gamma g^{-1}), t \geq 0, \Gamma \in \mathcal{B}(X), g \in X$ and v_t is the distribution of φ_t. For each $s \geq 0$ we define a map $\theta_s : \tilde{X} \to \tilde{X}$ by $(\theta_s \tilde{f})_t = \tilde{f}_{s+t} \circ \tilde{f}_s^{-1}, t \geq 0, \tilde{f} \in \tilde{X}$. By properties (i)–(iv) in Definition 1.1 it is easy to verify that θ_s preserves $\tilde{P}$ for each $s \geq 0$ and the map

$$\Sigma_1' : (W, \mathcal{F}, P) \to (\tilde{X}, \mathcal{B}(\tilde{X}), \tilde{P}), \quad w \longmapsto \{\varphi_t(w) : t \geq 0\}$$

is measure-preserving. We now define $\Phi, \Psi : \tilde{X} \longmapsto \mathbf{R}^+$ by

$$\Phi(\tilde{f}) = \sup_{0 \leq t \leq 1} \log^+ |\tilde{f}_t|_{C^1},$$

$$\Psi(\tilde{f}) = \sup_{0 \leq t \leq 1} \log^+ |\tilde{f}_1 \circ \tilde{f}_t^{-1}|_{C^1}.$$

From Proposition 1.2 it follows that $\Phi, \Psi \in L^1(\tilde{X}, \mathcal{B}(\tilde{X}), \tilde{P})$. Then, by Birkhoff ergodic theorem, one bas

$$\lim_{n \to +\infty} \frac{1}{n} \Phi(\theta_1^n \tilde{f}) = 0$$

and

$$\lim_{n \to +\infty} \frac{1}{n} \Psi(\theta_1^n \tilde{f}) = 0$$

for $\tilde{P}$ almost all $\tilde{f} \in \tilde{X}$. This together with the measure-preserving property of Σ_1' proves (2.14) and (2.15). □

In what follows we show that, to the negative elements of the Lyapunov exponents (which indicate the exponential shrinking rates of the tangent vectors under the flow), there correspond stable manifolds in M on which the trajectories of the flow cluster together at exponential rates.

We first have the following theorem concerning the global stable manifolds of the flow.

Theorem 2.2. *Let $(\{\varphi_t\}_{t \geq 0}, \mu)$ be as given above with $r \geq 2$. For $(w, x) \in \Lambda_0$ (see Theorem 2.1), if $\lambda^{(1)}(x) < 0$ and $\lambda^{(1)}(x) < \cdots < \lambda^{(p)}(x)$ are the strictly negative Lyapunov exponents at x, we define $W^{s,1}(w, x) \subset \cdots \subset W^{s,p}(w, x)$ by*

$$W^{s,i}(w, x) = \left\{ y \in M : \limsup_{t \to +\infty} \frac{1}{t} \log d(\varphi_t(w)x, \varphi_t(w)y) \leq \lambda^{(i)}(x) \right\},$$

$1 \le i \le p$. Then for $P \times \mu$ almost every $(w, x), W^{s,i}(w, x)$ is the image of $V^{(i)}_{(w,x)}$ under an injective immersion of class $C^{1,1}$ and is tangent to $V^{(i)}_{(w,x)}$ at point x. In addition, if $y \in W^{s,i}(w, x)$, then

$$\limsup_{t\to+\infty} \frac{1}{t} \log d^s(\varphi_t(w)x, \varphi_t(w)y) \le \lambda^{(i)}(x)$$

where $d^s(\ ,\ )$ is the distance along the submanifold $\varphi_t(w)W^{s,i}(w, x)$.

Proof. We also prove the result first for discrete time steps of length 1. Denote $\Omega = \mathrm{Diff}^2(M)$ and let v_1 be the distribution of φ_1. By Proposition 1.2 one has

$$\int_\Omega \log^+ |f|_{C^2} dv_1(f) < +\infty.$$

Put

$$(\Omega^{\mathbf{N}}, \mathcal{B}(\Omega)^{\mathbf{N}}, v_1^{\mathbf{N}}) = \prod_1^{+\infty} (\Omega, \mathcal{B}(\Omega), v_1)$$

and, similarly as in (2.8), define a map

$$\begin{aligned} \Sigma_2 :\ & (W, \mathcal{F}, P) \to (\Omega^{\mathbf{N}} . \mathcal{B}(\Omega)^{\mathbf{N}}, v_1^{\mathbf{N}}), \\ & w \longmapsto (\varphi_1(w) \circ \varphi_0(w)^{-1}, \varphi_2(w) \circ \varphi_1(w)^{-1}, \cdots). \end{aligned} \tag{2.16}$$

Then Σ_2 is also a measure-preserving map. Thus, by applying Theorem III. 3.2, the theorem is proved if we consider only $t \in \mathbf{Z}^+$.

In order to prove the full (continuous time) result, note that, if $y, z \in W^{s,i}(w, x)$, then for all $n \ge 0$ and $t \in [n, n+1]$

$$\begin{aligned} & d^s(\varphi_t(w)y, \varphi_t(w)z) \\ = \ & d^s(\varphi_t(w) \circ \varphi_n(w)^{-1} \circ \varphi_n(w)y, \varphi_t(w) \circ \varphi_n(w)^{-1} \circ \varphi_n(w)z) \\ \le \ & \sup_{n \le t \le n+1} |\varphi_t(w) \circ \varphi_n(w)^{-1}|_{C^1} d^s(\varphi_n(w)y, \varphi(w)z). \end{aligned} \tag{2.17}$$

Put

$$C_n(w) = \sup_{n \le t \le n+1} |\varphi_t(w) \circ \varphi_n(w)^{-1}|_{C^1}. \tag{2.18}$$

From (2.14) it follows that

$$\lim_{n\to+\infty} \frac{1}{n} \log^+ C_n(w) = 0, \quad P-\text{a.e.} w. \tag{2.19}$$

This together with (2.17) proves the theorem. □

As for the local properties of the stable manifolds $W^{s,i}(w, x)$ we have the following local stable manifold theorem.

Theorem 2.3. *Let $(\{\varphi_t\}_{t\geq 0}, \mu)$ be as given above with $r \geq 2$. Given $\lambda < 0$, we put $\Lambda_0^\lambda = \{(w,x) \in \Lambda_0 : \lambda^{(1)}(x) < \lambda$ and $\lambda^{(i)}(x) \neq \lambda, 1 \leq i \leq r(x)\}$ (Λ_0 is given in Theorem 2.1). Then we have a measurable set $\Lambda^\lambda \subset \Lambda_0^\lambda$ with $P \times \mu(\Lambda_0^\lambda \backslash \Lambda^\lambda) = 0$, and measurable functions $\alpha, \beta, \gamma : \Lambda^\lambda \to (0, +\infty)$ such that for each $(w,x) \in \Lambda^\lambda$ there exists a C^1 embedded disc $W_{\text{loc}}^{s,\lambda}(w,x)$ which contains x and has the following properties:*

1) $W_{\text{loc}}^{s,\lambda}(w,x)$ is tangent to $V_{(w,x)}^{(i)}$ at point x, where i is such that $\lambda^{(i)}(x) < \lambda < \lambda^{(i+1)}(x)$, and $W_{\text{loc}}^{s,\lambda}(w,x) = \exp_x$ Graph $(h_{(w,x)})$ where $h_{(w,x)} : \{\xi \in V_{(w,x)}^{(i)} : |\xi| < \alpha(w,x)\} \to (V_{(w,x)}^{(i)})^\perp$ is a $C^{1,1}$ map with $h_{(w,x)}(0) = 0$ and max $\{Lip(h_{(w,x)}), Lip(T.h_{(w,x)})\} < \beta(w,x)$;

2) If $y, z \in W_{\text{loc}}^{s,\lambda}(w,x)$, then for all $t \geq 0$

$$d^s(\varphi_t(w)y, \varphi_t(w)z) \leq \gamma(w,x) d^s(y,z) e^{\lambda t}$$

where $d^s(\ ,\ )$ is the distance along the submanifold $\varphi_t(w)W_{\text{loc}}^{s,\lambda}(w,x)$.

Proof. Put $\Lambda_0^{\lambda,1} = \{(w,x) \in \Lambda_0^\lambda : \lambda^{(i)}(x) \in (-\infty, \lambda - 1)\}$ and $\Lambda_0^{\lambda,l} = \{(w,x) \in \Lambda_0^\lambda : \lambda^{(i)}(x) \in [\lambda - (l-1)^{-1}, \lambda - l^{-1})\}, l \geq 2$ where i is such that $\lambda^{(i)}(x) < \lambda < \lambda^{(i+1)}(x)$ ($\lambda^{(i+1)}(x) = +\infty$ if $i = r(x)$). In a way analogous to that in the proof of Theorem 2.2, it follows from Theorem III. 3.1 that for each $\Lambda_0^{\lambda,l}$ we have a measurable set $\Lambda^{\lambda,l} \subset \Lambda_0^{\lambda,l}$ with $P \times \mu(\Lambda_0^{\lambda,l} \backslash \Lambda^{\lambda,l}) = 0$, and measurable functions $\alpha_l, \beta_l, \gamma_l : \Lambda^{\lambda,l} \to (0, +\infty)$ such that for each $(w,x) \in \Lambda^{\lambda,l}$ there exists a C^1 embedded disc $W_{\text{loc}}^{s,\lambda}(w,x)$ which has the following properties:

1)′ $W_{\text{loc}}^{s,\lambda}(w,x) = \exp_x$ Graph $(h_{(w,x)}^l)$ where $h_{(w,x)}^l : \{\xi \in V_{(w,x)}^{(i)} : |\xi| < \alpha_l(w,x)\} \to (V_{(w,x)}^{(i)})^\perp$ is a $C^{1,1}$ map such that $h_{(w,x)}^l(0) = 0, T_0 h_{(w,x)}^l = 0$ and max $\{\text{Lip}(h_{(w,x)}^l), \text{Lip}(T.h_{(w,x)}^l)\} < \beta_l(w,x)$;

2)′ For any $y, z \in W_{\text{loc}}^{s,\lambda}(w,x)$ and $n \in \mathbf{Z}^+$,

$$d^s(\varphi_n(w)y, \varphi_n(w)z) \leq \gamma_l(w,x) d^s(y,z) e^{\lambda_l n}$$

where $\lambda_l = \lambda - (2l)^{-1}$.

Now note that by (2.19) we may define a measurable and P almost everywhere finite function $K : W \to (0, +\infty]$ by

$$K(w) = \sup_{k \geq 0} \left\{ C_k(w) e^{-\frac{1}{2l}k} \right\},$$

where $C_k(w)$ is defined in (2.18). Choose a measurable set $\Gamma \subset W$ with $P(\Gamma) = 1$ such that $K(w) < +\infty$ for each $w \in \Gamma$. Then for each $l \geq 1$, if $(w,x) \in \Lambda^{\lambda,l} \cap (\Gamma \times M)$ and $y, z \in W_{\text{loc}}^{s,\lambda}(w,x)$, we have

$$\begin{aligned} d^s(\varphi_t(w)y, \varphi_t(w)z) &\leq C_n(w) d^s(\varphi_n(w)y, \varphi_n(w)z) \\ &\leq K(w) e^{\frac{1}{2l}n} \gamma_l(w,x) d^s(y,z) e^{\lambda_l n} \leq e^{-\lambda} K(w) \gamma_l(w,x) d^s(y,z) e^{\lambda t} \end{aligned}$$

for all $n \in \mathbf{Z}^+$ and $t \in [n, n+1]$.

Setting $\Lambda^\lambda = \cup_{l\geq 1}[\Lambda^{\lambda,l} \cap (\Gamma \times M)]$ and defining $\alpha(w,x) = \alpha_l(w,x), \beta(w,x) = \beta_l(w,x)$, $\gamma(w,x) = e^{-\lambda} K(w)\gamma_l(w,x)$ and $h_{(w,x)} = h^l_{(w,x)}$ if $(w,x) \in \Lambda^{\lambda,l} \cap (\Gamma \times M)$, we get the proof completed. □

The results above of this section are actually an extension of the work of Ruelle $[\text{Rue}]_2$ to the case of stochastic flows of diffeomorphisms, though our techniques used in dealing with the local stable manifolds are rather different from those of Ruelle. The programme of extending Ruelle's ergodic theory of (deterministic) differentiable dynamical systems to the stochastic case was suggested by L. Arnold at Les Houches, June 1980 and was fulfilled by A. Carverhill [Car] for stochastic flows of diffeomorphisms generated by stochastic differential equations. By the present time, there have been many papers and books concerning various applications as well as estimation and calculation of Lyapunov exponents of random dynamical systems. For further information we refer the reader to the recent volume [Arn] and $[\text{Elw}]_1$ and the references therein.

§3 Entropy of Stochastic Flows of Diffeomorphisms

We now turn to the notion of (measure-theoretic) entropy of a stochastic flow of diffeomorphisms. This is an extension to the continuous time case of what was introduced in Section I.2 for a random dynamical system generated by i.i.d. random diffeomorphisms.

Now let $\{\varphi_t : (W, \mathcal{F}, P) \to \text{Diff}^r(M)\}_{t\geq 0} (r \geq 1)$ be a stochastic flow and μ an invariant measure of the flow.

Definition 3.1. *Let $t_0 > 0$ be given. If ξ is a finite measurable partition of M, then the limit*

$$h_\mu^{t_0}(\{\varphi_t\}_{t\geq 0}, \xi) \overset{\text{def}}{=} \lim_{n\to+\infty} \frac{1}{n} \int H_\mu \left(\bigvee_{k=0}^{n-1} \varphi_{kt_0}(w)^{-1}\xi \right) dP(w) \tag{3.1}$$

is called the t_0-time-step entropy of $(\{\varphi_t\}_{t\geq 0}, \mu)$ with respect to ξ.

The limit (3.1) does exist and it holds that

$$h_\mu^{t_0}(\{\varphi_t\}_{t\geq 0}, \xi) = \inf_{n\geq 1} \frac{1}{n} \int H_\mu \left(\bigvee_{k=0}^{n-1} \varphi_{kt_0}(w)^{-1}\xi \right) dP(w). \tag{3.2}$$

In fact, we put

$$a_n(t_0, \xi) = \int H_\mu \left(\bigvee_{k=0}^{n-1} \varphi_{kt_0}(w)^{-1}\xi \right) dP(w). \tag{3.3}$$

Since for any $s > 0$ the map

$$\begin{aligned} \Sigma_1^s : \quad & (W, \mathcal{F}, P) \to (X^{\mathbf{N}}, \mathcal{B}(X)^{\mathbf{N}}, v_s^{\mathbf{N}}) \\ & w \longmapsto (\varphi_s(w) \circ \varphi_0(w)^{-1}, \varphi_{2s}(w) \circ \varphi_s(w)^{-1}, \cdots) \end{aligned} \tag{3.4}$$

(where $X = \mathrm{Diff}^1(M)$ and v_s is the distribution of φ_s) is a measure-preserving transformation, by almost the same argument as in the proof of Theorem I. 2.1 one has

$$a_{n+m}(t_0, \xi) \le a_n(t_0, \xi) + a_m(t_0, \xi)$$

for all $n, m \ge 1$. Thus, according to the proof of Theorem 0.4.1,

$$\lim_{n \to +\infty} \frac{1}{n} a_n(t_0, \xi) = \inf_{n \ge 1} \frac{1}{n} a_n(t_0, \xi).$$

This proves (3.2).

Definition 3.2. *Let $(\{\varphi_t\}_{t \ge 0}, \mu)$ be as given above and let $t_0 > 0$ be given. Then*

$$h_\mu^{t_0}(\{\varphi_t\}_{t \ge 0}) \stackrel{\text{def}}{=} \sup h_\mu^{t_0}(\{\varphi_t\}_{t \ge 0}, \xi),$$

where the supremum is taken over the set of all finite measurable partitions of M, is called the t_0-time-step entropy of $(\{\varphi_t\}_{t \ge 0}, \mu)$.

As in the case of a deterministic flow of diffeomorphisms, we have the following result, which is adopted from [Kif]$_1$.

Proposition 3.1. *For any $t_0 > 0$,*

$$h_\mu^{t_0}(\{\varphi_t\}_{t \ge 0}) = t_0 h_\mu^1(\{\varphi_t\}_{t \ge 0}). \tag{3.5}$$

Proof. Still write $X = \mathrm{Diff}^1(M)$ for simplicity of notation. For any given $s > 0$ let v_s be the distribution of φ_s. Define

$$\begin{aligned} F_s : \quad & (X^{\mathbf{N}} \times M, \mathcal{B}(X)^{\mathbf{N}} \times \mathcal{B}(M), v_s^{\mathbf{N}} \times \mu) \hookleftarrow \\ & (\omega, x) \longmapsto (\tau\omega, f_0(\omega)x) \end{aligned}$$

where we write $\omega = (f_0(\omega), f_1(\omega), \cdots)$ and τ is the left shift operator on $X^{\mathbf{N}}$. Analogously to Proposition I.1.1, $v_s^{\mathbf{N}} \times \mu$ is F_s-invariant. Since Σ_1^s defined by (3.4) is a measure-preserving map, in the same way as the proof of Theorem I.2.2 we have

$$h_\mu^s(\{\varphi_t\}_{t \ge 0}) = h_{v_s^{\mathbf{N}} \times \mu}^{\sigma_0}(F_s), \tag{3.6}$$

where σ_0 is the σ-algebra $\{\Gamma \times M : \Gamma \in \mathcal{B}(X)^{\mathbf{N}}\}$ on $X^{\mathbf{N}} \times M$. Also, by Theorem 0.4.3, for any $k \geq 1$

$$\begin{aligned} h_\mu^{ks}(\{\varphi_t\}_{t\geq 0}) = & \ h^{\sigma_0}_{v_{ks}^{\mathbf{N}} \times \mu}(F_{ks}) = h^{\sigma_0}_{v_s^{\mathbf{N}} \times \mu}(F_s^k) \\ = & \ k h^{\sigma_0}_{v_s^{\mathbf{N}} \times \mu}(F_s) = k h_\mu^s(\{\varphi_t\}_{t\geq 0}). \end{aligned} \tag{3.7}$$

Form this it follows that (3.5) holds true for all rational numbers $t_0 > 0$.

In order to prove (3.5) for all real number $t_0 > 0$, choose an increasing sequence of finite measurable partitions $\xi_1 \leq \xi_2 \leq \cdots$ of M such that the union of the boundaries $\partial A_k^i = A_k^i \backslash \mathrm{int}(A_k^i)$ of elements A_k^i of the partitions ξ_k has μ measure zero and

$$\lim_{k \to +\infty} \mathrm{diam}(\xi_k) = 0.$$

Note that Theorem I.2.5 also holds true if Ω is replaced by $X = \mathrm{Diff}^1(M)$. Then for any $s > 0$ we have

$$\begin{aligned} h_\mu^s(\{\varphi_t\}_{t\geq 0}) = & \lim_{k\to+\infty} h_\mu^s(\{\varphi_t\}_{t\geq 0}, \xi_k) \\ = & \lim_{k\to+\infty} \lim_{n\to+\infty} \frac{1}{n} a_n(s, \xi_k) \\ = & \lim_{k\to+\infty} \inf_{n\geq 1} \frac{1}{n} a_n(s, \xi_k) \end{aligned} \tag{3.8}$$

by (3.2), where $a_n(s, \xi_k)$ is defined by (3.3).

Since the union of the boundaries of all elements of the partitions ξ_k has μ measure zero, we know that for any given $s > 0$ the union of the boundaries of all elements of the partitions $\varphi_s(w)^{-1}\xi_k$ has μ measure zero for P almost all $w \in W$. Indeed, if $A \in \mathcal{B}(M)$ and $\mu(A) = 0$, then $\mu(A) = \int \mu(\varphi_t(w)^{-1}A)dP(w) = 0$ implies $\mu(\varphi_t(w)^{-1}A) = 0$ for P almost all w. Thus the above assertion holds true. This together with property (iii) in Definition 1.1 yields that $a_n(t, \xi_k)$ is continuous in t. Then, by (3.8),

$$\begin{aligned} \lim_{s\to s_0} h_\mu^s(\{\varphi_t\}_{t\geq 0}) = & \lim_{s\to s_0} \lim_{k\to+\infty} \inf_n \frac{1}{n} a_n(s, \xi_k) \\ \leq & \lim_{k\to+\infty} \inf_n \lim_{s\to s_0} \frac{1}{n} a_n(s, \xi_k) \\ = & \lim_{k\to+\infty} \inf_n \frac{1}{n} a_n(s_0, \xi_k) \\ = & \ h_\mu^{s_0}(\{\varphi_t\}_{t\geq 0}) \end{aligned} \tag{3.9}$$

for any $s_0 > 0$.

Consider the function $\psi(s) = \frac{1}{s} h_\mu^s(\{\varphi_t\}_{t\geq 0})$. By (3.7) we know that $\psi(r) = \psi(1)$ for any rational number $r > 0$. On the other hand, from (3.9) it follows that $\psi(s)$ is upper semi-continuous. Since the rational numbers are dense in $\mathbf{R}$, these two conditions together imply that $\psi(s) = \psi(1)$ for all $s > 0$. The proof is completed. □

Ruelle's inequality and Pesin's entropy formula respectively worked out in Chapter II and Chapter IV for i.i.d. random diffeomorphisms can be easily carried over to the case of a stochastic flow of C^2 diffeomorphisms. In fact, assume that $(\{\varphi_t\}_{t\geq 0}, \mu)$ is a stochastic flow of C^2 diffeomorphisms. Let $\Omega = \mathrm{Diff}^2(M)$ and let v_1 be the distribution of φ_1. Then, as we have seen in the previous discussions in this and the last sections, from the measure-preserving property of the map Σ_2 defined by (2.16) it follows that for μ-a.e. x the Lyapunov exponents $\lambda^{(1)}(x) < \cdots < \lambda^{(r(x))}(x)$ together with their respective multiplicities $m_i(x), 1 \leq i \leq r(x)$ of $(\{\varphi_t\}_{t\geq 0}, \mu)$ coincide with those of $\mathcal{X}^+(M, v_1, \mu)$ (see Section I. 1) at point x and $h^1_\mu(\{\varphi_t\}_{t\geq 0}) = h_\mu(\mathcal{X}^+(M, v_1))$. In addition, Proposition 1.2 asserts that $\int \log^+ |f|_{C^2} dv_1(f) < +\infty$ and $\log |\det T_x f|$ is integrable in (f, x) with respect to $v_1 \times \mu$ (see Remark IV. 1.1). Thus, by Theorem II. 0.1 and Theorem IV. 1.1, we obtain

Theorem 3.1. *Assume that $\{\varphi_t\}_{t\geq 0}$ is a stochastic flow of C^2 diffeomorphisms and μ is an invariant measure of the flow. Then*

$$h^1_\mu(\{\varphi_t\}_{t\geq 0}) \leq \int \sum_i \lambda^{(i)}(x)^+ m_i(x) d\mu. \tag{3.10}$$

Moreover, if $\mu \ll$ Leb., then

$$h^1_\mu(\{\varphi_t\}_{t\geq 0}) = \int \sum_i \lambda^{(i)}(x)^+ m_i(x) d\mu. \tag{3.11}$$

We shall call respectively (3.10) and (3.11) *Ruelle's inequality* and *Pesin's formula* for the stochastic flow $(\{\varphi_t\}_{t\geq 0}, \mu)$.

If $\{\varphi_t\}_{t\geq 0}$ is a stochastic flow of C^2 diffeomorphisms arising from a non-degenerate SDE of the form (1.16) and μ is the unique invariant measure of the flow (see Section 1), then Pesin's formula (3.11) holds true for $(\{\varphi_t\}_{t\geq 0}, \mu)$. It is remarkable that (3.11) is valid for arbitrary SDE's satisfying only the smooth and non-degenerate conditions, no other conditions like hyperbolicity or its like are required. This is in sharp contradistinction with the deterministic case.

Chapter VI Characterization of Measures Satisfying Entropy Formula

We consider in this chapter systems generated by two-sided compositions of random diffeomorphisms. Our main purpose here is to prove that Pesin's entropy formula holds true in this random case if and only if the sample measures, i.e. the natural invariant family of measures associated with individual realizations of the random process have Sinai-Bowen-Ruelle (SBR) property. Roughly speaking, we say that the sample measures have SBR property if their conditional measures on unstable manifolds are absolutely continuous with respect to Lebesgue measures on these manifolds. The idea of the above result goes back to the ergodic theory of Axiom-A attractors. Recall that, if f is a twice differentiable diffeomorphism on a compact manifold N and Λ is an Axiom-A attractor of f with besin of attraction U, then there is a unique f-invariant measure ρ with support in Λ that is characterized by each of the following properties: (a) ρ has absolutely continuous conditional measures on unstable manifolds; (b) Pesin's entropy formula holds true for the system (N, f, ρ); (c) There exists a set $S \subset U$ such that $U \backslash S$ has Lebesgue measure zero and $\lim_{n\to+\infty} \frac{1}{n}\sum_{k=0}^{n-1}\delta_{f^k x} = \rho$ whenever $x \in S$. The measures with the above properties were first shown to exist by Sinai ([Sin]) for Anosov diffeomorphisms and this result was later extended to Axiom-A attractors by Bowen and Ruelle ([Bow]$_2$ and [Rue]$_3$). These measures are then called SBR measures. Let us emphasize here that each one of properties (a)-(c) has been shown to be significant in its own right, but it is also striking that they are equivalent to one another. In addition, we remark that Y.Kifer gives another equivalent characterization of such measures via their stochastic perturbations (see [Kif]$_4$). Some of these results for deterministic uniformly hyperbolic systems have been shown to remain valid in more general frameworks. A well-known theorem of Ledrappier and Young (Theorem A of [Led]$_2$) asserts that properties (a) and (b) remain equivalent for all C^2 diffeomorphisms on compact manifolds. What this means is that, if N is a compact Riemannian manifold without boundary and f is a C^2 diffeomorphism on N preserving a Borel probability measure m, then a sufficient and necessary condition for the validity of entropy formula $h_m(f) = \int \sum_i \lambda^{(i)}(x)^+ m_i(x) dm$ is that m has absolutely continuous conditional measures on unstable manifolds (That (a) implies (b) for this case is proved by Ledrappier and Strelcyn in [Led]$_3$). The main result of this chapter thus turns out to be a generalization of the above theorem to the random case. This generalization was actually first mentioned by Ledrappier and Young themselves, though not clearly stated (see [Led]$_1$ for the idea). We present here a (first) detailed treatment. Although the technical details are quite different, our proof here follows the main ideas in the deterministic case given by [Led]$_3$ and [Led]$_2$.

This chapter is organized as follows. In the first part (Section 1) we introduce the relevant concepts of ergodic theory of systems generated by two-sided compositions of random diffeomorphisms, then we formulate the main result (Theorem 1.1) of this chapter and give an important consequence (Corollary 1.2) of this result. The second part (Section 2) is devoted to the proof of the " if " part of Theorem 1.1. The third part (Sections 3-8) consists of a detailed proof of the " only if " part of Theorem 1.1. The first part is basic. The second part is fairly easy for readers who are familiar with Chapter IV, and the conclusion and arguments of this part will be very useful when we deal with in Chapter VII hyperbolic attractors subjected to random perturbations. The third part is not essential for those readers who are not interested in technical details of the proof of the " only if " part of Theorem 1.1. Such readers can omit this part on the first reading.

§1 Basic Concepts and Formulation of the Main Result

For the sake of clearity of presentation, we divide this section into several subsections.

A. The General Setting

As in the previous chapters, let M be a C^∞ connected compact Riemannian manifold without boundary, and write $m_0 = \dim M$ and $\Omega = \mathrm{Diff}^2(M)$. Suppose that v is a Borel probability measure on Ω satisfying

$$\begin{cases} \int_\Omega \log^+ |f|_{C^2} dv(f) < +\infty \\ \int_\Omega \log^+ |f^{-1}|_{C^2} dv(f) < +\infty. \end{cases} \tag{1.1}$$

In this chapter we consider the evolution process generated by forward and backward successive applications of randomly chosen maps from Ω, these maps being independent and identically distributed with law v. More precisely, let

$$(\Omega^{\mathbf{Z}}, \mathcal{B}(\Omega)^{\mathbf{Z}}, v^{\mathbf{Z}}) = \prod_{-\infty}^{+\infty} (\Omega, \mathcal{B}(\Omega), v)$$

be the bi-infinite product of copies of $(\Omega, \mathcal{B}(\Omega), v)$. For each $w = (\cdots, f_{-1}(w), f_0(w), f_1(w), \cdots) \in \Omega^{\mathbf{Z}}$ and $n > 0$, define

$$\begin{aligned} f_w^0 &= id, \\ f_w^n &= f_{n-1}(w) \circ f_{n-2}(w) \circ \cdots \circ f_0(w), \\ f_w^{-n} &= f_{-n}(w)^{-1} \circ f_{-n+1}(w)^{-1} \circ \cdots \circ f_{-1}(w)^{-1}. \end{aligned}$$

We are here concerned with the random system generated by actions on M of $\{f_w^n : n \in \mathbf{Z}, w \in (\Omega^{\mathbf{Z}}, \mathcal{B}(\Omega)^{\mathbf{Z}}, v^{\mathbf{Z}})\}$ and we denote this set-up by $\mathcal{X}(M, v)$. Let us notice that, when dealing with a system $\mathcal{X}(M, v)$, one can consider simultaneously the associated forward system $\mathcal{X}^+(M, v)$ as discussed in previous chapters. Relationship between these two systems will play an important role in this chapter.

Throughout this chapter, the spaces $\Omega^{\mathbf{Z}}$ and $\Omega^{\mathbf{Z}} \times M$ are always endowed with the product topology. Recall that

$$\mathcal{B}(\Omega)^{\mathbf{Z}} = \mathcal{B}(\Omega^{\mathbf{Z}})$$

and

$$\mathcal{B}(\Omega)^{\mathbf{Z}} \times \mathcal{B}(M) = \mathcal{B}(\Omega^{\mathbf{Z}} \times M)$$

(see Section I.1). Also, to repeat, put for each $w = (\cdots, f_{-1}(w), f_0(w), f_1(w), \cdots) \in \Omega^{\mathbf{Z}}$

$$w^+ = (f_0(w), f_1(w), \cdots), \quad w^- = (\cdots, f_{-2}(w), f_{-1}(w))$$

and define maps

$$\begin{aligned} P_1 &: \Omega^{\mathbf{Z}} \times M \to \Omega^{\mathbf{Z}}, \quad (w, x) \mapsto w, \\ P_2 &: \Omega^{\mathbf{Z}} \times M \to M, \quad (w, x) \mapsto x, \\ P &: \Omega^{\mathbf{Z}} \times M \to \Omega^{\mathbf{N}} \times M, \quad (w, x) \mapsto (w^+, x) \end{aligned}$$

and

$$G : \Omega^{\mathbf{Z}} \times M \to \Omega^{\mathbf{Z}} \times M, \quad (w, x) \mapsto (\tau w, f_0(w)x),$$

where τ is the shift operator on $\Omega^{\mathbf{Z}}$.

B. Invariant Measures, Sample Measures

Definition 1.1. *A Borel probability measure μ on M is called an invariant measure of $\mathcal{X}(M, v)$ if*

$$\int_{\Omega} f\mu dv(f) = \mu$$

where $(f\mu)(E) = \mu(f^{-1}E)$ for all $E \in \mathcal{B}(M)$ and $f \in \Omega$.

We denote by $\mathcal{M}(\mathcal{X}(M, v))$ the set of all invariant measures of $\mathcal{X}(M, v)$. Obviously, $\mathcal{M}(\mathcal{X}(M, v)) = \mathcal{M}(\mathcal{X}^+(M, v))$. When associated with an invariant measure μ, $\mathcal{X}(M, v)$ will be referred to as $\mathcal{X}(M, v, \mu)$.

Given $\mathcal{X}(M, v, \mu)$, by Proposition I.1.2 we know that there exists a unique Borel probability measure μ^* on $\Omega^{\mathbf{Z}} \times M$ which satisfies $G\mu^* = \mu^*$ and $P\mu^* = v^{\mathbf{N}} \times \mu$. Henceforth, unless indicated otherwise, the σ-algebra associated with $(\Omega^{\mathbf{Z}} \times M, \mu^*)$ is always understood to be the completion $\mathcal{B}_{\mu^*}(\Omega^{\mathbf{Z}} \times M)$ of $\mathcal{B}(\Omega^{\mathbf{Z}} \times M)$ with respect to μ^*. Since $\Omega^{\mathbf{Z}} \times M$ is a Borel subset of the Polish space $\Pi_{-\infty}^{+\infty} C^2(M, M) \times M$, by Theorem 0.1.10 we know that $(\Omega^{\mathbf{Z}} \times M, \mu^*)$ is a Lebesgue space. Let $\{\mu^*_{\{w\} \times M} : w \in \Omega^{\mathbf{Z}}\}$ be a (essentially unique) canonical system of conditional measures of μ^* associated with the measurable partition $\sigma = \{\{w\} \times M : w \in \Omega^{\mathbf{Z}}\}$. Identifying any $\{w\} \times M$ with M and denoting $\mu^*_{\{w\} \times M}$ simply by μ_w, we obtain a family of Borel probability measures $\{\mu_w\}_{w \in \Omega^{\mathbf{Z}}}$ on M.

Definition 1.2. *$\{\mu_w\}_{w\in\Omega^{\mathbf{Z}}}$ is called the family of sample measures of $\mathcal{X}(M,v,\mu)$.*

The following proposition says that $\{\mu_w\}_{w\in\Omega^{\mathbf{Z}}}$ is intuitively a natural invariant family of measures associated with individual realizations of the random process $\mathcal{X}(M,v,\mu)$.

Proposition 1.1. *Let $\mathcal{X}(M,v,\mu)$ be given. Then the family of sample measures $\{\mu_w\}_{w\in\Omega^{\mathbf{Z}}}$ of $\mathcal{X}(M,v,\mu)$ is the $v^{\mathbf{Z}}$-mod 0 unique family of Borel probability measures on M such that the following 1)-4) hold true:*

1) $w \mapsto \mu_w(\Lambda_w)$ is a measurable function on $(\Omega^{\mathbf{Z}}, \mathcal{B}_{v^{\mathbf{Z}}}(\Omega^{\mathbf{Z}}))$ for any $\Lambda \in \mathcal{B}(\Omega^{\mathbf{Z}} \times M)$, where $\mathcal{B}_{v^{\mathbf{Z}}}(\Omega^{\mathbf{Z}})$ is the completion of $\mathcal{B}(\Omega^{\mathbf{Z}})$ with respect to $v^{\mathbf{Z}}$;

2) $f_0(w)\mu_w = \mu_{\tau w}$, $v^{\mathbf{Z}}$ -a.e. w;

3) μ_w depends only on w^- for $v^{\mathbf{Z}}$ -a.e. w;

4) $\int \mu_w dv^{\mathbf{Z}}(w) = \mu$.

Moreover, for $v^{\mathbf{Z}}$ -a.e. w we have $f^n_{\tau^{-n}w}\mu \to \mu_w$ as $n \mapsto +\infty$.

Proof. Let $\{\mu_w\}_{w\in\Omega^{\mathbf{Z}}}$ be the family of sample measures of $\mathcal{X}(M,v,\mu)$. By the definition of $\{\mu_w\}_{w\in\Omega^{\mathbf{Z}}}$ and Proposition I.1.2, 1),2) and 4) follow immediately from the general properties of conditional measures. We now prove the last conclusion which implies 3) clearly.

Recall that we use σ^+ in Chapter I to denote the partition $P^{-1}\{\{\omega\}\times M : \omega \in \Omega^{\mathbf{N}}\}$ of $\Omega^{\mathbf{Z}} \times M$. Now for each $n \geq 0$ we put $\sigma_n^+ = G^n\sigma^+$ and denote also by σ_n^+ the σ-algebra consisting of all measurable σ_n^+-sets. Let $\{g_i : i \in \mathbf{N}\}$ be a dense subset of $C(M)$ (the space of all continuous functions $g : M \to \mathbf{R}$). For each $i \in \mathbf{N}$ we define $\hat{g}_i : \Omega^{\mathbf{Z}} \times M \to \mathbf{R}, (w,x) \mapsto g_i(x)$. Since $\sigma_n^+ \nearrow \sigma$, one has for each $i \in \mathbf{N}$

$$\lim_{n\to+\infty} E(\hat{g}_i|\sigma_n^+) = E(\hat{g}_i|\sigma),\ \mu^* - \text{a.e.}$$

which implies

$$\lim_{n\to+\infty} \int g_i df^n_{\tau^{-n}w}\mu = \int g_i d\mu_w,\ v^{\mathbf{Z}} - \text{a.e.}w.$$

Since $\{g_i : i \in \mathbf{N}\}$ is dense in $C(M)$, then for $v^{\mathbf{Z}}$ − a.e.$w \in \Omega^{\mathbf{Z}}$ we have

$$\lim_{n\to+\infty} \int g df^n_{\tau^{-n}w}\mu = \int g d\mu_w$$

for all $g \in C(M)$. This means that

$$f^n_{\tau^{-n}w}\mu \to \mu_w$$

as $n \to +\infty$ for $v^{\mathbf{Z}}$ − a.e.w.

On the other hand, let $\{\mu_w\}_{w\in\Omega^{\mathbf{Z}}}$ be a family of Borel probability measures on M with properties 1)-4). Define a Borel probability measure $\hat{\mu}^*$ on $\Omega^{\mathbf{Z}} \times M$ by

$$\hat{\mu}^*(\Lambda) = \int\int \chi_\Lambda(w,x)d\mu_w(x)dv^{\mathbf{Z}}(w), \Lambda \in \mathcal{B}(\Omega^{\mathbf{Z}} \times M).$$

From Properties 2)-4) it is easy to see that $\hat{\mu}^*$ is G-invariant and $P\hat{\mu}^* = v^{\mathbf{N}} \times \mu$. By Proposition I.1.2 we then obtain $\hat{\mu}^* = \mu^*$. Thus, by the essential uniqueness of conditional measures, $\{\mu_w\}_{w \in \Omega^{\mathbf{Z}}}$ is the family of sample measures of $\mathcal{X}(M, v, \mu)$. □

Definition 1.3. *Let $\rho \in \mathcal{M}(\mathcal{X}(M, v))$. We say that ρ is ergodic if ρ is an extreme point of $\mathcal{M}(\mathcal{X}(M, v))$. In this case we also say that $\mathcal{X}(M, v, \rho)$ is ergodic.*

From Proposition I.1.3 we see that the ergodicity of $\rho \in \mathcal{M}(\mathcal{X}(M, v))$ coincides with that of ρ as an invariant measure of $\mathcal{X}^+(M, v)$. Recall that if $\rho \in \mathcal{M}(\mathcal{X}(M, v))$ then the following three conditions are equivalent: (1) ρ is ergodic; (2) $F : (\Omega^{\mathbf{N}} \times M, v^{\mathbf{N}} \times \rho) \hookleftarrow$ is ergodic; (3) $G : (\Omega^{\mathbf{Z}} \times M, \rho^*) \hookleftarrow$ is ergodic. We now denote by $\mathcal{M}_e(\mathcal{X}(M, v))$ the set of all ergodic invariant measures of $\mathcal{X}(M, v)$. In what follows we discuss briefly ergodic decompositions of invariant measures of $\mathcal{X}(M, v)$.

Consider now $\mathcal{X}^+(M, v)$ simultaneously. Note that the transition probabilities $P(x, \cdot)$, $x \in M$ of $\mathcal{X}^+(M, v)$ as defined in Section IV.1 can be viewed as transition probabilities of a Markov chain $\{X_n\}_{n \geq 0}$, i.e. X_{k+1} has the distribution $P(x, \cdot)$ provided $X_k = x, k \geq 0$. Employing ergodic decompositions of Markov processes (see Appendix A.1 of [Kif]$_1$ and see also Chapter 13 of [Yos]), we know that there exists a measurable partition η_0 of M into a collection of disjoint sets $\{C_\alpha\}_{\alpha \in \mathcal{A}}$ and $M \setminus \cup_{\alpha \in \mathcal{A}} C_\alpha$ such that the following hold true:

i) $\cup_{\alpha \in \mathcal{A}} C_\alpha$ has full measure with respect to any $\mu \in \mathcal{M}(\mathcal{X}(M, v))$;

ii) For each $\alpha \in \mathcal{A}$ and $x \in C_\alpha, P(x, C_\alpha) = 1$;

iii) For each $\alpha \in \mathcal{A}$ there exists a unique invariant measure ρ_α of $\mathcal{X}(M, v)$ concentrated on C_α and this measure is ergodic;

iv) For any $\mu \in \mathcal{M}(\mathcal{X}(M, v))$, $\{\rho_\alpha\}_{\alpha \in \mathcal{A}}$ is a canonical system of conditional measures of μ associated with the partition η_0.

In view of Proposition I.1.2 and Rohlin's ergodic decomposition theorem about measure -preserving transformations (see the arguments at the end of Section 0.5 or [Roh]$_3$), one can show that, corresponding to the partition η_0, there exists a measurable partition $\hat{\eta}_0$ of $\Omega^{\mathbf{Z}} \times M$ into a collection of disjoint sets $\{\widehat{C}_\alpha\}_{\alpha \in \mathcal{A}}$ and $\Omega^{\mathbf{Z}} \times M \setminus \cup_{\alpha \in \mathcal{A}} \widehat{C}_\alpha$ such that:

i)′ $\cup_{\alpha \in \mathcal{A}} \widehat{C}_\alpha$ has full μ^* measure for any $\mu \in \mathcal{M}(\mathcal{X}(M, v))$;

ii)′ For each $\alpha \in \mathcal{A}, G^{-1}\widehat{C}_\alpha = \widehat{C}_\alpha, \rho_\alpha^*$ is concentrated on $\widehat{C}_\alpha$ and $G : (\widehat{C}_\alpha, \rho_\alpha^*) \hookleftarrow$ is ergodic.

iii)′ For any $\mu \in \mathcal{M}(\mathcal{X}(M, v)), \{\rho_\alpha^*\}_{\alpha \in \mathcal{A}}$ is a canonical system of conditional measures of μ^* associated with $\hat{\eta}_0$;

iv)′ For any $\mu \in \mathcal{M}(\mathcal{X}(M, v))$, the σ-algebra consisting of all measurable $\hat{\eta}_0$-sets equals μ^*-mod 0 the σ-algebra $\{A \in \mathcal{B}_{\mu^*}(\Omega^{\mathbf{Z}} \times M) : G^{-1}A = A\}$.

As an easy consequence of properties i)′-iv)′ we have

Corollary 1.1. *Let $\mu \in \mathcal{M}(\mathcal{X}(M, v))$. Then for each measurable function $H : \Omega^{\mathbf{Z}} \times M \to \mathbf{R} \cup \{\infty\}$ satisfying*

$$H \circ G = H, \mu^* - a.e.$$

there exists a measurable function $h: M \to \mathbf{R} \cup \{\infty\}$ *such that*

$$H(w,x) = h(x), \mu^* - a.e.(w,x).$$

C. Entropy, Lyapunov Exponents

Let $\mathcal{X}(M,v,\mu)$ be given. From Theorem I.2.1 it follows clearly that for each finite measurable partition ξ of M the following limit exists

$$h_\mu(\mathcal{X}(M,v),\xi) \overset{\text{def.}}{=} \lim_{n\to+\infty} \frac{1}{n} \int H_\mu\left(\bigvee_{k=0}^{n-1} (f_w^k)^{-1}\xi\right) dv^{\mathbf{Z}}(w).$$

Definition 1.4. $h_\mu(\mathcal{X}(M,v)) \overset{\text{def.}}{=} \sup\{h_\mu(\mathcal{X}(M,v),\xi) : \xi$ *is a finite measurable partition of* $M\}$ *is called the (measure-theoretic) entropy of* $\mathcal{X}(M,v,\mu)$.

By this definition and Theorems I.2.2 and I.2.3 one has

$$h_\mu(\mathcal{X}(M,v)) = h_\mu(\mathcal{X}^+(M,v)) = h^\sigma_{\mu^*}(G). \tag{1.2}$$

We now turn to Lyapunov exponents of $\mathcal{X}(M,v,\mu)$. Condition (1.1) implies that

$$\int [\log^+ |f|_{C^1} + \log^+ |f^{-1}|_{C^1}] dv(f) < +\infty.$$

Then applying Oseledec multiplicative ergodic theorem about invertible systems (see Appendix 2 of [Kat]) in a way completely analogous to Theorem I.3.2, we obtain

Proposition 1.2. *Let* $\mathcal{X}(M,v,\mu)$ *be given. Then there exists a Borel set* $\Delta_0 \subset \Omega^{\mathbf{Z}} \times M$ *satisfying* $\mu^*(\Delta_0) = 1, G\Delta_0 = \Delta_0$ *and having the following properties:*

1) For each $(w,x) \in \Delta_0$ *there exist a decomposition*

$$T_xM = E_1(w,x) \oplus E_2(w,x) \oplus \cdots \oplus E_{r(w,x)}(w,x) \tag{1.3}$$

and numbers

$$-\infty < \lambda^{(1)}(w,x) < \lambda^{(2)}(w,x) < \cdots < \lambda^{(r(w,x))}(w,x) < +\infty \tag{1.4}$$

such that

$$\lim_{n\to\pm\infty} \frac{1}{n} \log |T_x f_w^n \xi| = \lambda^{(i)}(w,x)$$

for any $\xi \in E_i(w,x)$ *with* $\xi \neq 0, 1 \leq i \leq r(w,x)$. *Moreover, the decomposition (1.3) and the numbers in (1.4) all depend measurably on* $(w,x) \in \Delta_0$. *They also satisfy for each* $(w,x) \in \Delta_0$

$$r(G(w,x)) = r(w,x)$$

and

$$\lambda^{(i)}(G(w,x)) = \lambda^{(i)}(w,x), T_x f_0(w) E_i(w,x) = E_i(G(w,x)), 1 \le i \le r(w,x).$$

2) Let $(w,x) \in \Delta_0$ and let $\rho^{(1)}(w,x) \le \cdots \le \rho^{(m_0)}(w,x)$ denote $\lambda^{(1)}(w,x) \le \cdots \le \lambda^{(1)}(w,x) \le \cdots \le \lambda^{(r(w,x))}(w,x) \le \cdots \le \lambda^{(r(w,x))}(w,x)$ with $\lambda^{(i)}(w,x)$ being repeated $m_i(w,x) \stackrel{\text{def.}}{=} \dim E_i(w,x)$ times. If $\{\xi_1, \cdots, \xi_{m_0}\}$ is a basis of $T_x M$ satisfying

$$\lim_{n \to \pm\infty} \frac{1}{n} \log |T_x f_w^n \xi_i| = \rho^{(i)}(w,x), \quad 1 \le i \le m_0,$$

then for any two non-empty disjoint subsets $P, Q \subset \{1, \cdots, m_0\}$ we have

$$\lim_{n \to \pm\infty} \frac{1}{n} \log \gamma(T_x f_w^n E_P, T_x f_w^n E_Q) = 0$$

where E_P and E_Q respectively denote the subspaces of $T_x M$ spanned by $\{\xi_i\}_{i \in P}$ and $\{\xi_j\}_{j \in Q}$.

Definition 1.5. *The numbers $\lambda^{(i)}(w,x), 1 \le i \le r(w,x)$ introduced above are called the Lyapunov exponents of $\mathcal{X}(M,v,\mu)$ at point (w,x), $m_i(w,x)$ is called the multiplicity of $\lambda^{(i)}(w,x)$.*

Remark 1.1. Given $\mathcal{X}(M,v,\mu)$, let $\lambda^{(1)}(x) < \lambda^{(2)}(x) < \cdots < \lambda^{(r(x))}(x)$ be the Lyapunov exponents of $\mathcal{X}^+(M,v,\mu)$ at point $x \in M$, as introduced by Theorem I.3.2, and let $m_i(x)$ be the multiplicity of $\lambda^{(i)}(x)$. From 1) of Proposition 1.2 and Proposition I.1.2 it is easy to see that for μ^*-a.e. $(w,x) \in \Omega^{\mathbf{Z}} \times M$

$$r(w,x) = r(x)$$

and

$$\lambda^{(i)}(w,x) = \lambda^{(i)}(x), m_i(w,x) = m_i(x), \quad 1 \le i \le r(x).$$

Moreover, the following equation holds clearly true:

$$\int \sum_i \lambda^{(i)}(w,x)^+ m_i(w,x) d\mu^* = \int \sum_i \lambda^{(i)}(x)^+ m_i(x) d\mu. \tag{1.5}$$

D. Unstable Manifolds

Let $\mathcal{X}(M,v,\mu)$ be given. By an argument completely analogous to Lemma III.1.4, one can find a Borel set $\widehat{\Gamma}_0 \subset \Omega^{\mathbf{Z}}$ with $v^{\mathbf{Z}}(\widehat{\Gamma}_0) = 1$ and $\tau\widehat{\Gamma}_0 = \widehat{\Gamma}_0$ such that for any given $\varepsilon > 0$ there exists a Borel function $r : \widehat{\Gamma}_0 \to (0, +\infty)$ with the following properties: For each $w \in \widehat{\Gamma}_0$ and $x \in M$, the map

$$F_{(w,x)}^{-1} \stackrel{\text{def.}}{=} \exp_{f_w^{-1}x}^{-1} \circ f_w^{-1} \circ \exp_x : T_x M(r(w)^{-1}) \to T_{f_w^{-1}x} M$$

is well defined, $\mathrm{Lip}(T.F^{-1}_{(w,x)}) \leq r(w)$ and $r(\tau^{-n}w) \leq r(w)e^{\varepsilon n}$ for all $n \in \mathbf{Z}^+$. Let Δ_0 be as introduced in Proposition 1.2. Then we put

$$\widehat{\Delta}_0 = \Delta_0 \cap (\widehat{\Gamma}_0 \times M)$$

and

$$\widehat{\Delta}_1 = \{(w,x) \in \widehat{\Delta}_0 : \lambda^{(i)}(w,x) > 0 \quad \text{for some} \quad i\}. \tag{1.6}$$

Clearly, $\mu^*(\widehat{\Delta}_0) = 1, G\widehat{\Delta}_0 = \widehat{\Delta}_0$ and $G\widehat{\Delta}_1 = \widehat{\Delta}_1$.

Let $[b,c], 0 \leq b < c$, be a closed interval of $\mathbf{R}$. Denote by $\Delta_{b,c}$ the subset of $\widehat{\Delta}_1$ consisting of points (w,x) such that $\lambda^{(i)}(w,x) \notin [b,c]$ for all $1 \leq i \leq r(w,x)$. If $(w,x) \in \Delta_{b,c}$, we put

$$E(w,x) = \sum_{\lambda^{(i)}(w,x)<b} \oplus E_i(w,x), H(w,x) = \sum_{\lambda^{(i)}(w,x)>c} \oplus E_i(w,x).$$

Let $k \in \{1, \cdots, m_0\}$ and suppose that

$$\Delta_{b,c,k} \stackrel{\text{def.}}{=} \{(w,x) \in \Delta_{b,c} : \dim H(w,x) = k\} \neq \phi.$$

Substituting $G^{-1} : (\Omega^{\mathbf{Z}} \times M, \mu^*) \hookleftarrow$ for $F : (\Omega^{\mathbf{N}} \times M, v^{\mathbf{N}} \times \mu) \hookleftarrow$ and the decompositions $T_xM = H(w,x) \oplus E(w,x), (w,x) \in \Delta_{b,c}$ for $T_xM = E_0(\omega,x) \oplus H_0(\omega,x), (\omega,x) \in \Lambda_{a,b}$, one can easily adapt the arguments in Sections III.1-III.3 to the case of $\mathcal{X}(M,v,\mu)$ to obtain the following two results.

Proposition 1.3. *Let $\Delta_{b,c,k}$ be given as above and let $0 < \varepsilon < \min\{1,(c-b)/200\}$. Then $\Delta_{b,c,k}$ can be divided into a countable number of Borel subsets $\{\Delta^i_{b,c,k}\}_{i \geq 1}$ with each $\Delta^i_{b,c,k}$ having the following properties:*

1) $E(w,x)$ and $H(w,x)$ depend continuously on $(w,x) \in \Delta^i_{b,c,k}$;

2) There exists a continuous family of C^1 embedded k-dimensional discs $\{W(w,x)\}_{(w,x)\in\Delta^i_{b,c,k}}$ in M together with numbers α_i, β_i and γ_i such that for each $(w,x) \in \Delta^i_{b,c,k}$, the following hold true:

i) $W(w,x) = \exp_x Graph(h_{(w,x)} : U(w,x) \to E(w,x))$ where $U(w,x)$ is an open subset of $H(w,x)$ that contains $\{\xi \in H(w,x) : |\xi| < \alpha_i\}$ and $h_{(w,x)}$ is of class $C^{1,1}$ and satisfies $h_{(w,x)}(0) = 0, T_0h_{(w,x)} = 0$, $Lip(h_{(w,x)}) \leq \beta_i$ and $Lip(T.h_{(w,x)}) \leq \beta_i$;

ii) $d^u(f_w^{-l}y, f_w^{-l}z) \leq \gamma_i e^{-(c-4\varepsilon)l} d^u(y,z)$ for all $y,z \in W(w,x)$ and $l \in \mathbf{Z}^+$, where $d^u(\cdot,\cdot)$ denotes the distance along $f_w^{-l}W(w,x)$ for each $l \in Z^+$.

Proposition 1.4. *Let $(w,x) \in \widehat{\Delta}_1$ and let $\lambda^{(q)}(w,x) < \cdots < \lambda^{(r(w,x))}(w,x)$ be the strictly positive Lyapunov exponents of $\mathcal{X}(M,v,\mu)$ at (w,x). Define $W^{u,r(w,x)}(w,x) \subset \cdots \subset W^{u,q}(w,x)$ by*

$$W^{u,i}(w,x) = \left\{ y \in M : \limsup_{n\to+\infty} \frac{1}{n} \log d(f_w^{-n}x, f_w^{-n}y) \leq -\lambda^{(i)}(w,x) \right\}$$

for $q \le i \le r(w,x)$. *Then* $W^{u,i}(w,x)$ *is the image of* $\sum_{j\ge i} \oplus E_j(w,x)$ *under an injective immersion of class* $C^{1,1}$. *In addition, if* $y \in W^{u,i}(w,x)$, *then*

$$\limsup_{n\to+\infty} \frac{1}{n} \log d^u(f_w^{-n}x, f_w^{-n}y) \le -\lambda^{(i)}(w,x),$$

where $d^u(\cdot,\cdot)$ *is the distance along the submanifold* $f_w^{-n} W^{u,i}(w,x) = W^{u,i}(G^{-n}(w,x))$ *for each* $n \in \mathbf{Z}^+$.

Remark 1.2. Given $\mathcal{X}(M,v,\mu)$ and $(w,x) \in \Omega^{\mathbf{Z}} \times M$, the *global unstable manifold* $W^u(w,x)$ is defined by

$$W^u(w,x) = \left\{ y \in M : \limsup_{n\to+\infty} \frac{1}{n} \log d(f_w^{-n}x, f_w^{-n}y) < 0 \right\}.$$

In a way similar to Remark III.3.2 we know that, if $(w,x) \in \widehat{\Delta}_1$ and $\lambda^{(q)}(w,x) < \cdots < \lambda^{(r(w,x))}(w,x)$ are the strictly positive exponents at (w,x), then

$$W^u(w,x) = W^{u,q}(w,x)$$

and hence $W^u(w,x)$ is the image of $\sum_{\lambda^{(i)}(w,x)>0} \oplus E_i(w,x)$ under an injective immersion of class $C^{1,1}$, and in addition, $W^u(w,x)$ is tangent to this subspace of T_xM at point x.

E. SBR Sample Measures, Formulation of the Main Result

Let $\mathcal{X}(M,v,\mu)$ be given and let $\widehat{\Delta}_1$ be as introduced in (1.6). For each $(w,x) \in \Omega^{\mathbf{Z}} \times M$, if $(w,x) \notin \widehat{\Delta}_1$, we define $W^u(w,x) = \{x\}$.

Definition 1.6. *A measurable partition* η *of* $\Omega^{\mathbf{Z}} \times M$ *with* $\eta \ge \sigma$ *is said to be subordinate to* W^u*-manifolds of* $\mathcal{X}(M,v,\mu)$ *if for* μ^**-a.e.* (w,x), $\eta_w(x) \overset{\text{def.}}{=} \{y : (w,y) \in \eta(w,x)\} \subset W^u(w,x)$ *and contains an open neighbourhood of* x *in* $W^u(w,x)$, *this neighbourhood being taken in the topology of* $W^u(w,x)$ *as a submanifold of* M.

Now Suppose that η is a partition of $\Omega^{\mathbf{Z}} \times M$ subordinate to W^u-manifolds of $\mathcal{X}(M,v,\mu)$. For each $w \in \Omega^{\mathbf{Z}}$ we denote by η_w the partition $\{\eta_w(x) : x \in M\}$ of M. Let $\{(\mu^*)^\eta_{(w,x)}\}_{(w,x)\in\Omega^{\mathbf{Z}}\times M}$ be a canonical system of conditional measures of μ^* associated with η and let $\{(\mu_w)^{\eta_w}_x\}_{x\in M}, w \in \Omega^{\mathbf{Z}}$ be defined analogously. Identifying $\{w\} \times \eta_w(x)$ with $\eta_w(x)$, by the transitivity of conditional measures we have $(\mu^*)^\eta_{(w,x)} = (\mu_w)^{\eta_w}_x$ for μ^* -a.e. (w,x).

Definition 1.7. *We say that the sample measures* $\mu_w, w \in \Omega^{\mathbf{Z}}$ *of* $\mathcal{X}(M,v,\mu)$ *have absolutely continuous conditional measures on* W^u*-manifolds, or equivalently that the family of sample measures* $\{\mu_w\}_{w\in\Omega^{\mathbf{Z}}}$ *has SBR property, if for every measurable partition* η *subordinate to* W^u*-manifolds of* $\mathcal{X}(M,v,\mu)$ *we have* *for* $v^{\mathbf{Z}}$ *-a.e.* $w \in \Omega^{\mathbf{Z}}$,

$$(\mu_w)_x^{\eta_w} << \lambda^u_{(w,x)}, \quad \mu_w\text{-a.e.} x \in M$$

where $\lambda^u_{(w,x)}$ *denotes the Lebesgue measure on* $W^u(w,x)$ *induced by its inherited Riemannian structure as a submanifold of* M ($\lambda^u_{(w,x)} = \delta_x$ *if* $(w,x) \notin \widehat{\Delta}_1$).

The main purpose of this chapter is to prove the following

Theorem 1.1. *For any given* $\mathcal{X}(M,v,\mu)$, *the following two conditions are equivalent:*

1) The family of sample measures $\{\mu_w\}_{w\in\Omega^{\mathbf{Z}}}$ *has SBR property;*

2) Pesin's entropy formula holds true, i.e.

$$h_\mu(\mathcal{X}(M,v)) = \int \sum_i \lambda^{(i)}(w,x)^+ m_i(w,x) d\mu^*.$$

Remark 1.3. We show in fact that, if the entropy formula in Theorem 1.1 is satisfied, then for μ^*-a.e. (w,x) the density $d(\mu_w)_x^{\eta_w}/d\lambda^u_{(w,x)}$ is a strictly positive function that is locally Lipschitz along $W^u(w,x)$ (see Corollary 8.2).

Theorem 1.1 together with Theorem IV.1.1, (1.2) and (1.5) implies the following

Corollary 1.2. *Let* $\mathcal{X}(M,v,\mu)$ *be given and suppose that* $\mu << Leb$. *Then the family of sample measures* $\{\mu_w\}_{w\in\Omega^{\mathbf{Z}}}$ *has SBR property.*

The remainning part of this chapter is devoted to the proof of Theorem 1.1.

§2 SBR Sample Measures: Sufficiency for Entropy Formula

Our purpose in this section is to prove that Theorem 1.1 1) implies Theorem 1.1 2). Let $\mathcal{X}(M,v,\mu)$ be given. First let us notice that Theorem II.0.1 together with (1.2) and (1.5) implies

$$h_\mu(\mathcal{X}(M,v)) \le \int \sum_i \lambda^{(i)}(w,x)^+ m_i(w,x) d\mu^*.$$

Thus, the point is to prove that, if the family of sample measures $\{\mu_w\}_{w\in\Omega^{\mathbf{Z}}}$ has SBR property, we have

$$h_\mu(\mathcal{X}(M,v)) \ge \int \sum_i \lambda^{(i)}(w,x)^+ m_i(w,x) d\mu^*. \tag{2.1}$$

Analogously to the proof of Theorem IV.1.1, we need to construct a measurable partition of $\Omega^{\mathbf{Z}} \times M$ subordinate to W^u-manifolds of $\mathcal{X}(M,v,\mu)$, by means of which (2.1) will be achieved. The construction is accomplished by means of local unstable manifolds. We present below the necessary arguments.

First notice that the partition of $\Omega^{\mathbf{Z}} \times M$ into global unstable manifolds $\{w\} \times W^u(w,x)$, $(w,x) \in \Omega^{\mathbf{Z}} \times M$ is in general not measurable, but we may consider the σ-algebra consisting of measurable subsets of $\Omega^{\mathbf{Z}} \times M$ which are unions of some global unstable manifolds, i.e. the σ-algebra

$$\mathcal{B}^u(\mathcal{X}(M,v,\mu)) = \left\{ B \in \mathcal{B}_{\mu^*}(\Omega^{\mathbf{Z}} \times M) : B = \bigcup_{(w,x)\in B} \{w\} \times W^u(w,x) \right\}.$$

In addition, put

$$\mathcal{B}^I(\mathcal{X}(M,v,\mu)) = \{A \in \mathcal{B}_{\mu^*}(\Omega^{\mathbf{Z}} \times M) : G^{-1}A = A\}.$$

We then have the following useful fact:

Proposition 2.1. $\mathcal{B}^I(\mathcal{X}(M,v,\mu)) \subset \mathcal{B}^u(\mathcal{X}(M,v,\mu)), \mu^*-$*mod 0.*

Keeping Corollary 1.1 in mind, one can easily adapt the proof of Lemma IV.2.2 to the present case to prove Proposition 2.1. Since the arguments are completely analogous, they are omitted here.

Proposition 2.2. *There exists a measurable partition η of $\Omega^{\mathbf{Z}} \times M$ with the following properties:*

1) $\eta \leq G^{-1}\eta, \sigma \leq \eta$;

2) η is subordinate to W^u-manifolds of $\mathcal{X}(M,v,\mu)$;

3) For every $B \in \mathcal{B}(\Omega^{\mathbf{Z}} \times M)$ the function

$$P_B(w,x) = \lambda^u_{(w,x)}(\eta_w(x) \cap B_w)$$

is measurable and μ^* almost everywhere finite, where B_w is the section $\{y : (w,y) \in B\}$.

Considering $G^{-1} : (\Omega^{\mathbf{Z}} \times M, \mu^*) \hookleftarrow$ instead of $F : (\Omega^{\mathbf{N}} \times M, v^{\mathbf{N}} \times \mu) \hookleftarrow$ and applying Proposition 1.3 and Proposition 2.1 instead of Theorem III.3.1 and Lemma IV.2.2 one can easily adapt those arguments in Section IV.2 concerning the existence of a partition satisfying 1)-3) of Proposition IV.2.1 to the present case to accomplish the proof of Proposition 2.2. Details are left to the reader.

Proposition 2.3. *Suppose that the family of sample measures $\{\mu_w\}_{w\in\Omega^{\mathbf{Z}}}$ has SBR property. Let η be a partition of the type as introduced in Proposition 2.2. Then there exists a Borel function $g : \Omega^{\mathbf{Z}} \times M \to \mathbf{R}^+$ such that for μ^* -a.e. $(w,x) \in \Omega^{\mathbf{Z}} \times M$,*

$$g(w,z) = \frac{d(\mu^*)^\eta_{(w,x)}}{d\lambda^u_{(w,x)}}(z), \quad \lambda^u_{(w,x)}\text{-a.e.} z \in \eta_w(x)$$

where $(\mu^*)^\eta_{(w,x)}$ is regarded as a measure on $\eta_w(x)$.

Proof of this proposition is completely analogous to that of Proposition IV.2.2.

Now we turn to the main part of this section.

Proof of Theorem 1.1 1)⇒ 2). It is sufficient to prove (2.1). Let η be as introduced in Proposition 2.2. By (1.3) and Theorem 0.4.3, we have

$$\begin{aligned} h_\mu(\mathcal{X}(M,v)) &= h^\sigma_{\mu^*}(G) = h^\sigma_{\mu^*}(G^{-1}) \\ &\geq h^\sigma_{\mu^*}(G^{-1},\eta) = H_{\mu^*}(\eta|G\eta) = H_{\mu^*}(G^{-1}\eta|\eta) \\ &= -\int \log(\mu^*)^\eta_{(w,x)}((G^{-1}\eta)(w,x))d\mu^*. \end{aligned} \tag{2.2}$$

Let $\widehat{\Delta}_1$ be as introduced above. Put $\widehat{\Delta}_2 = \Omega^{\mathbf{Z}} \times M \backslash \widehat{\Delta}_1$. Clearly $G^{-1}\widehat{\Delta}_i = \widehat{\Delta}_i$, $i = 1,2$. According to Proposition 2.1, η and $G^{-1}\eta$ refine (μ^*-mod 0) the partition $\{\widehat{\Delta}_1, \widehat{\Delta}_2\}$. Since their restrictions to $\widehat{\Delta}_2$ are the partition into single points, one has for μ^* -a.e. $(w,x) \in \widehat{\Delta}_2$

$$\log(\mu^*)^\eta_{(w,x)}((G^{-1}\eta)(w,x)) = 0.$$

On the other hand,

$$\int_{\widehat{\Delta}_2} \sum_i \lambda^{(i)}(w,x)^+ m_i(w,x)d\mu^* = 0.$$

Therefore, we may assume $\mu^*(\widehat{\Delta}_1) = 1$ without loss of generality.

For μ^*-a.e. $(w,x) \in \Omega^{\mathbf{Z}} \times M$ we may define

$$X(w,x) = (\mu^*)^\eta_{(w,x)}((G^{-1}\eta)(w,x)),$$

$$Y(w,x) = \frac{g(w,x)}{g(G(w,x))},$$

$$Z(w,x) = |\det(T_x f_0(w)|_{E^u_{(w,x)}})|, \text{ where } \quad E^u_{(w,x)} = \sum_{\lambda^{(i)}(w,x)>0} \oplus E_i(w,x).$$

It is easy to see that X, Y and Z are all measurable and μ^*-a.e. finite functions on $\Omega^{\mathbf{Z}} \times M$. We first claim the following results, whose proofs will be given a little later.

Claim 2.1. $X = YZ^{-1}, \mu^*$-a.e.

Claim 2.2. (a) $\log Z \in L^1(\Omega^{\mathbf{Z}} \times M, \mu^*)$;
(b) $\int \log Z d\mu^* = \int \sum_i \lambda^{(i)}(w,x)^+ m_i(w,x)d\mu^*$.

Claim 2.3. (a) $\log Y \in L^1(\Omega^{\mathbf{Z}} \times M, \mu^*)$;
(b) $\int \log Y d\mu^* = 0$.

From these and (2.2) one immediately obtains (2.1) and completes the proof of Theorem 1.1 1)$\Rightarrow$ 2). $\square$

Proof of Claim 2.1. First notice that for μ^*-a.e. (w', x'), $(G^{-1}\eta)|_{\eta(w',x')}$ is $((\mu^*)^{\eta}_{(w',x')}$-mod 0) a countable partition. Then for μ^*-a.e.(w, x), identifying $\{w\} \times \eta_w(x)$ with $\eta_w(x)$ and $\{w\} \times (G^{-1}\eta)_w(x)$ with $(G^{-1}\eta)_w(x)$, we have for any $B \in \mathcal{B}(M)$

$$
\begin{aligned}
&(\mu^*)^{G^{-1}\eta}_{(w,x)}(B) \\
=&\frac{1}{X(w,x)}(\mu^*)^{\eta}_{(w,x)}((G^{-1}\eta)_w(x) \cap B) \\
=&\frac{1}{X(w,x)}\int_{(G^{-1}\eta)_w(x)\cap B} g(w,z)d\lambda^u_{(w,x)}(z);
\end{aligned}
$$

on the other hand,

$$
\begin{aligned}
&(\mu^*)^{G^{-1}\eta}_{(w,x)}(B) \\
=&(\mu^*)^{\eta}_{G(w,x)}(f_0(w)B) \\
=&\int_{\eta_{\tau w}(f_0(w)x)\cap(f_0(w)B)} g(\tau w, z)d\lambda^u_{G(w,x)}(z) \\
=&\int_{(G^{-1}\eta)_w(x)\cap B} g(\tau w, f_0(w)z)|\det(T_z f_0(w)|_{E^u_{(w,z)}})|d\lambda^u_{(w,x)}(z).
\end{aligned}
$$

Since B is arbitrarily chosen from $\mathcal{B}(M)$, we obtain for μ^*-a.e. (w, x)

$$\frac{1}{X(w,x)}g(w,z) = g(\tau w, f_0(w)z)|\det(T_z f_0(w)|_{E^u_{(w,z)}})|, \lambda^u_{(w,x)}\text{-a.e.}z \in (G^{-1}\eta)_w(x).$$

This implies that for μ^*-a.e. (w, x) one has

$$X(w,z) = Y(w,z)Z(w,z)^{-1}, (\mu^*)^{G^{-1}\eta}_{(w,x)} \text{-a.e.}z \in (G^{-1}\eta)_w(x)$$

since $X(w, x) = X(w, z)$ for each $z \in (G^{-1}\eta)_w(x)$. From this Claim 2.1 follows clearly. $\square$

Proof of Claim 2.2. Since for μ^*-a.e. $(w, x) \in \Omega^{\mathbf{Z}} \times M$

$$|T_x f_0(w)|_{E^u_{(w,x)}}| \le |f_0(w)|_{C^1}$$

and $\log^+ |f_0(w)|_{C^1} \in L^1(\Omega^{\mathbf{Z}}, \nu^{\mathbf{Z}})$, by Oseledec multiplicative ergodic theorem we know that $\log Z \in L^1(\Omega^{\mathbf{Z}} \times M, \mu^*)$ and

$$\int \log Z d\mu^* = \int \sum_i \lambda^{(i)}(w,x)^+ m_i(w,x)d\mu^*,$$

completing the proof. $\square$

Proof of Claim 2.3. By Claim 2.1, one has

$$\log X = \log Y - \log Z \leq 0, \mu^*\text{-a.e.}$$

which implies

$$\log^+ Y \leq \log^+ Z, \quad \mu^*\text{-a.e.}$$

Hence, by Claim 2.2, $\log^+ Y \in L^1(\Omega^{\mathbf{Z}} \times M, \mu^*)$. This claim follows then from Lemma I.3.1. □

§3 Lyapunov Charts

In this and the subsequent sections we address ourselves to proving Theorem 1.1 2)⇒ 1). The idea and outline of the proof are as follows. Qualitatively, negative Lyapunov exponents indicate stability in the sence that certain points converge towards one another asymptotically under repeated applications of random diffeomorphisms. On the other hand, positive Lyapunov exponents are associated with chaotic instability due to "sensitive dependence on initial conditions". Since entropy measures the degree of chaotic instability, it is mainly determined by positive Lyapunov exponents. In fact, we shall first prove in Section 7 that for a certain class of measurable partitions ξ's subordinate to W^u-manifolds of $\mathcal{X}(M, v, \mu)$ it holds true that $h_\mu(\mathcal{X}(M, v)) = H_{\mu^*}(\xi | G\xi)$. Then we shall prove in Section 8 that this together with entropy formula implies $(\mu_w)_x^{\xi_w} << \lambda^u_{(w,x)}$ for μ^*-a.e. (w, x). To carry out the first part of the proof is not easy. It is necessary to consider explicitly the role played by the zero exponent as well as by the positive exponents. So we shall introduce in Section 4 some nonlinear constructions, i.e. unstable manifolds and center unstable sets related to these exponents. These constructions are worked out by means of Lyapunov charts which are treated in this section. Two classes of needed measurable partitions connected with these constructions are introduced in Section 5. Some averaging results in Euclidean spaces are given in Section 6. Finally, in view of Proposition 2.1, we shall complete the proof by reducing the problem to ergodic case. While not at all essential, this line of approach simplifies the presentation, especially where notation is concerned. Thus we now declare the following

Hypothesis for Sections 3-7:
$\mathcal{X}(M, v, \mu)$ is given ergodic.

Under this hypothesis we know that there exists a Borel set $\Delta_0' \subset \Delta_0$ (see Subsection 1. C) with $\mu^*(\Delta_0') = 1$ and $G\Delta_0' = \Delta_0'$ such that for each $(w, x) \in \Delta_0'$, $r(w, x), \lambda^{(i)}(w, x)$ and $m_i(w, x)$ equal respectively constants r_0, $\lambda^{(i)}$ and m_i, $1 \leq i \leq r_0$.

For $(w,x) \in \Delta'_0$, let

$$E^u_{(w,x)} = \sum_{\lambda^{(i)}>0} \oplus E_i(w,x), \qquad u = \dim E^u_{(w,x)};$$

$$E^c_{(w,x)} = E_{i_0}(w,x) \quad \text{where} \quad \lambda^{(i_0)} = 0, c = \dim E^c_{(w,x)};$$

$$E^s_{(w,x)} = \sum_{\lambda^{(i)}<0} \oplus E_i(w,x), \qquad s = \dim E^s_{(w,x)};$$

$$\lambda^+ = \min\{\lambda^{(i)} : \lambda^{(i)} > 0\}, \quad \lambda^- = \max\{\lambda^{(i)} : \lambda^{(i)} < 0\}.$$

It is easy to see that Theorem 1.1 2) $\Rightarrow$ 1) is completely trivial if $u = 0$. So we assume that $u > 0$.

We now begin the proof by constructing in this section Lyapunov charts, which will be particularly useful in the subsequent sections. The construction needs the following two lemmas.

Lemma 3.1. *There exists a Borel set $\widehat{\Gamma}'_0 \subset \Omega^{\mathbf{Z}}$ with $v^{\mathbf{Z}}(\widehat{\Gamma}'_0) = 1$ and $\tau\widehat{\Gamma}'_0 = \widehat{\Gamma}'_0$ such that for any given $\delta > 0$ one can define a Borel function $B : \widehat{\Gamma}'_0 \to [1, +\infty)$ with the following properties:*

1) For each $w \in \widehat{\Gamma}'_0$ and $x \in M$, the maps

$$F_{(w,x)} \stackrel{\text{def.}}{=} \exp^{-1}_{f^1_w x} \circ f^1_w \circ \exp_x : T_xM(B(w)^{-1}) \to T_{f^1_w x}M,$$

$$F^{-1}_{(w,x)} \stackrel{\text{def.}}{=} \exp^{-1}_{f^{-1}_w x} \circ f^{-1}_w \circ \exp_x : T_xM(B(w)^{-1}) \to T_{f^{-1}_w x}M$$

are well defined and

$$Lip(T.F_{(w,x)}) \leq B(w),$$

$$Lip(T.F^{-1}_{(w,x)}) \leq B(w);$$

2) $B(\tau^{\pm 1}w) \leq B(w)e^{\delta}$ for all $w \in \widehat{\Gamma}'_0$.

Proof. By Condition (1.1) and by arguments completely analogous to the proof of Lemma III.1.4 we know that there exists a Borel function $\widehat{B} : \Omega^{\mathbf{Z}} \to [1, +\infty)$ such that for each $w \in \Omega^{\mathbf{Z}}$ and $x \in M$ the maps

$$\widehat{F}_{(w,x)} \stackrel{\text{def.}}{=} \exp^{-1}_{f^1_w x} \circ f^1_w \circ \exp_x : T_xM(\widehat{B}(w)^{-1}) \to T_{f^1_w x}M,$$

$$\widehat{F}^{-1}_{(w,x)} \stackrel{\text{def.}}{=} \exp^{-1}_{f^{-1}_w x} \circ f^{-1}_w \circ \exp_x : T_xM(\widehat{B}(w)^{-1}) \to T_{f^{-1}_w x}M$$

are well defined, $\max\{\text{Lip}(T.\widehat{F}_{(w,x)}), \text{Lip}(T.\widehat{F}^{-1}_{(w,x)})\} \leq \widehat{B}(w)$ and $\log \widehat{B} \in L^1(\Omega^{\mathbf{Z}}, v^{\mathbf{Z}})$. According to Birkhoff ergodic theorem,

$$\lim_{n\to\pm\infty} \frac{1}{n} \log \widehat{B}(\tau^n w) = 0, \quad v^{\mathbf{Z}}\text{-a.e.}w.$$

Then there exists a Borel set $\widehat{\Gamma}_0' \subset \Omega^{\mathbf{Z}}$ such that $v^{\mathbf{Z}}(\widehat{\Gamma}_0') = 1, \tau\widehat{\Gamma}_0' = \widehat{\Gamma}_0'$ and for each $w \in \widehat{\Gamma}_0'$

$$\lim_{n\to\pm\infty} \frac{1}{n} \log \widehat{B}(\tau^n w) = 0.$$

Given $\delta > 0$, define $B : \widehat{\Gamma}_0' \to [1, +\infty)$ by the formula

$$B(w) = \sup\{\widehat{B}(\tau^n w)e^{-|n|\delta} : n \in \mathbf{Z}\}.$$

Then one can easily verify that $\widehat{\Gamma}_0'$ and the function B satisfy the requirements of the lemma. □

Lemma 3.2. *For any given $\delta > 0$ there exists a Borel function $C : \Delta_0' \to [1, +\infty)$ such that :*

1) For each $(w, x) \in \Delta_0'$ we have

$$|T_x f_w^n \xi| \le C(w, x)e^{(\lambda^{(i)}+\delta)n}|\xi|,$$

$$|T_x f_w^{-n} \xi| \le C(w, x)e^{(-\lambda^{(i)}+\delta)n}|\xi|$$

for all $n \ge 0$ and all $\xi \in E_i(w, x), 1 \le i \le r_0$;

2) $\gamma(E_i(w, x), \sum_{j\neq i} \oplus E_j(w, x)) \ge C(w, x)^{-1}, 1 \le i \le r_0$ for all $(w, x) \in \Delta_0'$;

3) $C(G^{\pm 1}(w, x)) \le C(w, x)e^{\delta}$ for all $(w, x) \in \Delta_0'$.

Proof. By Proposition 1.2 and by arguments analogous to the proof of Lemma III.1.1 we know that the functions $C_i^+, C_i^-, C_i' : \Delta_0' \to [0, +\infty), 1 \le i \le r_0$ defined by

$$C_i^+(w, x) = \sup\left\{\frac{|T_x f_w^{k+n}\xi|}{|T_x f_w^k \xi|}e^{-(\lambda^{(i)}+\delta)n-\delta|k|} : n \ge 0, k \in \mathbf{Z}, 0 \ne \xi \in E_i(w, x)\right\},$$

$$C_i^-(w, x) = \sup\left\{\frac{|T_x f_w^{k+n}\xi|}{|T_x f_w^k \xi|}e^{-(\lambda^{(i)}-\delta)n-\delta|k|} : n \le 0, k \in \mathbf{Z}, 0 \ne \xi \in E_i(w, x)\right\},$$

$$C_i'(w, x) = \inf\left\{\gamma\left(E_i(G^n(w, x)), \sum_{j\neq i} \oplus E_j(G^n(w, x))\right) e^{|n|\delta} : n \in \mathbf{Z}\right\}$$

are all measurable, everywhere finite, and positive functions. Define $C : \Delta_0' \to [1, +\infty)$ by

$$C(w, x) = \max_i\{C_i^+(w, x), C_i^-(w, x), C_i'(w, x)^{-1}\}.$$

Then one can easily check that the function C satisfies the requirements of this lemma. □

For $(\xi, \eta, \zeta) \in \mathbf{R}^u \times \mathbf{R}^c \times \mathbf{R}^s$ we define

$$||(\xi, \eta, \zeta)|| = \max\{||\xi||_u, ||\eta||_c, ||\zeta||_s\}$$

where $\|\cdot\|_u, \|\cdot\|_c$ and $\|\cdot\|_s$ are the Euclidean norms on $\mathbf{R}^u, \mathbf{R}^c$ and $\mathbf{R}^s$ respectively. And for $r > 0$ we put

$$\bar{\mathbf{R}}(r) = \bar{\mathbf{R}}^u(r) \times \bar{\mathbf{R}}^c(r) \times \bar{\mathbf{R}}^s(r)$$

where $\bar{\mathbf{R}}^u(r), \bar{\mathbf{R}}^c(r)$ and $\bar{\mathbf{R}}^s(r)$ denote respectively the closed discs in $\mathbf{R}^u, \mathbf{R}^c$ and $\mathbf{R}^s$ of radius r centered at 0.

Let $0 < \varepsilon < \min\{1, \lambda^+/100m_0, -\lambda^-/100m_0\}$ be fixed arbitrarily, and let ρ_0 be a number as introduced at the beginning of Section II.1. Put

$$\Delta_0'' = \Delta_0' \cap (\widehat{\Gamma}_0' \times M)$$

which satisfies clearly $\mu^*(\Delta_0'') = 1$ and $G\Delta_0'' = \Delta_0''$. In what follows we define for each $(w, x) \in \Delta_0''$ a change of coordinates in some neighbourhood of x in M. The size of the neighbourhood, the local chart and the related estimates will vary with $(w, x) \in \Delta_0''$. This is the following

Proposition 3.1. *There exists a measurable function $l : \Delta_0'' \to [1, +\infty)$ satisfying $l(G^{\pm 1}(w, x)) \le l(w, x)e^{\varepsilon}$ for all $(w, x) \in \Delta_0''$, and for each $(w, x) \in \Delta_0''$ there is a C^∞ embedding $\Phi_{(w,x)} : \bar{\mathbf{R}}(l(w, x)^{-1}) \to M$ with the following properties:*

1) $\Phi_{(w,x)}(0) = x, T_0\Phi_{(w,x)}$ maps $\mathbf{R}^u, \mathbf{R}^c$ and $\mathbf{R}^s$ onto $E^u_{(w,x)}, E^c_{(w,x)}$ and $E^s_{(w,x)}$ respectively.

2) Let

$$H_{(w,x)} = \Phi^{-1}_{G(w,x)} \circ f^1_w \circ \Phi_{(w,x)},$$

$$H^{-1}_{(w,x)} = \Phi^{-1}_{G^{-1}(w,x)} \circ f^{-1}_w \circ \Phi_{(w,x)},$$

defined wherever they make sense. Then

i)

$$e^{\lambda^+ - \varepsilon}\|\xi\| \le \|T_0 H_{(w,x)}\xi\| \quad \textit{for } \xi \in \mathbf{R}^u,$$

$$e^{-\varepsilon}\|\xi\| \le \|T_0 H_{(w,x)}\xi\| \le e^{\varepsilon}\|\xi\| \quad \textit{for } \xi \in \mathbf{R}^c,$$

$$\|T_0 H_{(w,x)}\xi\| \le e^{\lambda^- + \varepsilon}\|\xi\| \quad \textit{for } \xi \in \mathbf{R}^s;$$

ii)

$$Lip(H_{(w,x)} - T_0 H_{(w,x)}) \le \varepsilon,$$

$$Lip(H^{-1}_{(w,x)} - T_0 H^{-1}_{(w,x)}) \le \varepsilon,$$

$$Lip(T.H_{(w,x)}) \le \varepsilon l(w, x),$$

$$Lip(T.H^{-1}_{(w,x)}) \le \varepsilon l(w, x);$$

iii)

$$\max\{\|T_\xi H_{(w,x)}\|, \|T_\xi H^{-1}_{(w,x)}\|\} \le e^{\lambda_0}$$

for all $\xi \in \bar{\mathbf{R}}(e^{-\lambda_0 - \varepsilon} l(w, x)^{-1})$, where $\lambda_0 > 0$ is a number depending only on ε and the exponents.

3) For any $\xi, \eta \in \bar{\mathbf{R}}(l(w, x)^{-1})$ one has

$$K_0^{-1} d(\Phi_{(w,x)}\xi, \Phi_{(w,x)}\eta) \le \|\xi - \eta\| \le l(w, x) d(\Phi_{(w,x)}\xi, \Phi_{(w,x)}\eta)$$

for some universal constant $K_0 > 0$.

Proof. Let $B : \widehat{\Gamma}_0' \to [1, +\infty)$ and $C : \Delta_0' \to [1, +\infty)$ be functions of the types as discussed in Lemmas 3.1 and 3.2 respectively, corresponding to $\delta = \varepsilon/2(r_0 + 1)$.

We now define a new inner product $\langle\ ,\ \rangle'_{(w,x)}$ on T_xM for each $(w,x) \in \Delta_0''$. Let $(w,x) \in \Delta_0''$. First we define

$$\langle \xi, \eta \rangle'_{(w,x)} = \sum_{n=0}^{+\infty} \frac{\langle T_x f_w^n \xi, T_x f_w^n \eta \rangle}{e^{2(\lambda^{(i)}+\varepsilon)n}} + \sum_{n=-1}^{-\infty} \frac{\langle T_x f_w^n \xi, T_x f_w^n \eta \rangle}{e^{2(\lambda^{(i)}-\varepsilon)n}}$$

for $\xi, \eta \in E_i(w,x), 1 \le i \le r_0$ where $\langle\ ,\ \rangle$ is the Riemannian metric on M. Then we extend $\langle\ ,\ \rangle'_{(w,x)}$ to all of T_xM by demanding that the subspaces $E_i(w,x), 1 \le i \le r_0$ are orthogonal to one another with respect to $\langle\ ,\ \rangle'_{(w,x)}$. Let $|\cdot|'_{(w,x)}$ be the norm on T_xM defined by

$$|\xi|'_{(w,x)} = \max\{(\langle \xi_a, \xi_a \rangle'_{(w,x)})^{\frac{1}{2}} : a = u, c, s\}$$

for each $\xi = \xi_u + \xi_c + \xi_s \in E^u_{(w,x)} \oplus E^c_{(w,x)} \oplus E^s_{(w,x)}$. Then it can be verified that

$$e^{\lambda^{(i)}-\varepsilon} |\xi|'_{(w,x)} \le |T_x f_w^1 \xi|'_{G(w,x)} \le e^{\lambda^{(i)}+\varepsilon} |\xi|'_{(w,x)}$$

for all $\xi \in E_i(w,x), 1 \le i \le r_0$ and

$$\frac{1}{3\sqrt{r_0}} |\xi| \le |\xi|'_{(w,x)} \le \widehat{C}(w,x) |\xi|$$

for all $\xi \in T_xM$, where $\widehat{C}(w,x) = 4^{r_0} C_0 C(w,x)^{r_0+1}$ and $C_0 = (\sum_{n=-\infty}^{+\infty} e^{-\varepsilon|n|})^{\frac{1}{2}}$, the last estimate coming from Lemma 3.2 and a consideration analogous to (1.4) in Chapter III.

Let $b(\rho_0/2)$ be a number introduced by Lemma II.1.1. Define for each $(w,x) \in \Delta_0''$

$$l(w,x) = \max\{6\sqrt{r_0}\rho_0^{-1}, \varepsilon^{-1}(3\sqrt{r_0})^2 b(\rho_0/2) e^{\varepsilon/2} \widehat{C}(w,x) B(w)\}$$

and let

$$\lambda_0 = \max\{|\lambda^{(i)}| : 1 \le i \le r_0\} + 2\varepsilon,$$
$$K_0 = 3\sqrt{r_0} b(\rho_0/2).$$

Next, for each $(w,x) \in \Delta_0''$ take a linear map $L_{(w,x)} : T_xM \to \mathbf{R}^u \times \mathbf{R}^c \times \mathbf{R}^s$ such that it takes $E^u_{(w,x)}, E^c_{(w,x)}$ and $E^s_{(w,x)}$ onto $\mathbf{R}^u \times \{0\} \times \{0\}, \{0\} \times \mathbf{R}^c \times \{0\}$ and $\{0\} \times \{0\} \times \mathbf{R}^s$ respectively and satisfies

$$\langle\langle L_{(w,x)}\xi, L_{(w,x)}\eta \rangle\rangle = \langle \xi, \eta \rangle'_{(w,x)}$$

for all $\xi, \eta \in T_xM$, where $\langle\langle\ ,\ \rangle\rangle$ is the usual Euclidean inner product. Then we set for each $(w,x) \in \Delta_0''$

$$\Phi_{(w,x)} = \exp_x \circ L^{-1}_{(w,x)}|_{\bar{\mathbf{R}}(l(w,x)^{-1})}.$$

With the entries defined above, one can easily check that 1)-3) of the proposition are satisfied. □

From here on, we shall refer to any system of local charts $\{\Phi_{(w,x)}\}_{(w,x)\in\Delta_0''}$ satisfying 1)-3) of Proposition 3.1 as a system of (ε, l)-charts, and λ_0 and K_0 will be defined as above.

§4 Local Unstable Manifolds and Center Unstable Sets

As we have said in Section 3, in order to prove Theorem 1.1 2)⇒1) it is necessary to consider explicitly the role played by the zero exponent as well as by the positive Lyapunov exponents. In this section we use Lyapunov charts described in Section 3 to introduce some nonlinear constructions related to these exponents. These constructions will be used in the next section to deal with some measurable partitions of $\Omega^{\mathbf{Z}} \times M$ which play central roles in the whole proof. For the sake of clearity of presentation, we divide this section into three subsections.

A. Local Unstable Manifolds and Center Unstable Sets

Let $\{\Phi_{(w,x)}\}_{(w,x)\in\Delta_0''}$ be a system of (ε, l)-charts. Sometimes it is necessary to reduce the size of the charts. Let $0 < \delta \leq 1$ be a reduction factor. For $(w,x) \in \Delta_0''$ we define

$$S_\delta^{cu}(w,x) = \{\zeta \in \bar{\mathbf{R}}(l(w,x)^{-1}) : \|\Phi^{-1}_{G^{-n}(w,x)} \circ f_w^{-n} \circ \Phi_{(w,x)}\| \leq \delta l(G^{-n}(w,x))^{-1}, n \in \mathbf{Z}^+\},$$

that is, $\Phi_{(w,x)} S_\delta^{cu}(w,x)$ consists of those points in M whose backward orbit under actions $f_w^{-n}, n \geq 0$ stays inside the domains of the charts at $G^{-n}(w,x)$ for all $n \geq 0$. It is called the *center unstable set* of $\mathcal{X}(M, v, \mu)$ at (w,x) associated with $(\{\Phi_{(w,x)}\}_{(w,x)\in\Delta_0''}, \delta)$. On $S_\delta^{cu}(w,x)$ and for all $n \geq 1$, one clearly has

$$H^{-n}_{(w,x)} \overset{\text{def.}}{=} \Phi^{-1}_{G^{-n}(w,x)} \circ f_w^{-n} \circ \Phi_{(w,x)} = H^{-1}_{G^{-n+1}(w,x)} \circ \cdots \circ H^{-1}_{G^{-1}(w,x)} \circ H^{-1}_{(w,x)}.$$

We next introduce the *local unstable manifold* of $\mathcal{X}(M, v, \mu)$ at $(w,x) \in \Delta_0''$ associated with $(\{\Phi_{(w,x)}\}_{(w,x)\in\Delta_0''}, \delta)$. It is defined to be the component of $W^u(w,x) \cap \Phi_{(w,x)} \bar{\mathbf{R}}(\delta l(w,x)^{-1})$ that contains x. The $\Phi^{-1}_{(w,x)}$ -image of this set is denoted by $W^u_{(w,x),\delta}(x)$.

Lemma 4.1. *Let $\{\Phi_{(w,x)}\}_{(w,x)\in\Delta_0''}$ be a system of (ε, l)-charts.*

1) If $0 < \delta \leq e^{-(\lambda_0+\varepsilon)}$ and $(w,x) \in \Delta_0''$, then

i) $W^u_{(w,x),\delta}(x)$ is the graph of a C^1 function

$$g_{(w,x),x} : \bar{\mathbf{R}}^u(\delta l(w,x)^{-1}) \to \bar{\mathbf{R}}^{c+s}(\delta l(w,x)^{-1})$$

with $g_{(w,x),x}(0) = 0$ and $Lip(g_{(w,x),x}) < 1$;

ii) $W^u_{(w,x),\delta}(x) \subset S_\delta^{cu}(w,x)$;

2) If $0 < \delta \leq e^{-2(\lambda_0+\varepsilon)}$ and $(w,x) \in \Delta_0''$, then

$$H_{(w,x)} W^u_{(w,x),\delta}(x) \cap \bar{\mathbf{R}}(\delta l(G(w,x))^{-1}) = W^u_{G(w,x),\delta}(f_w^1 x).$$

Proof. Let $0 < \delta \le e^{-(\lambda_0+\varepsilon)}$ and $(w,x) \in \Delta_0''$. By Proposition 3.1 we have

$$\|T_0 H_{(w,x)}^{-1}\xi\| \le e^{-\lambda^+ + \varepsilon}\|\xi\| \qquad \text{for} \quad \xi \in \mathbf{R}^u,$$
$$\|T_0 H_{(w,x)}^{-1}\eta\| \ge e^{-\varepsilon}\|\eta\| \qquad \text{for} \quad \eta \in \mathbf{R}^{c+s}$$

and

$$\begin{aligned}\mathrm{Lip}(H_{(w,x)}^{-1} - T_0 H_{(w,x)}^{-1}) &\le \varepsilon l(w,x) \cdot \delta l(w,x)^{-1} = \varepsilon\delta \\ &< \min\{e^{-\lambda^+ + 2\varepsilon} - e^{-\lambda^+ + \varepsilon}, e^{-\varepsilon} - e^{-\lambda^+ + 2\varepsilon}\},\end{aligned}$$

where $H_{(w,x)}^{-1}$ is restricted to $\bar{\mathbf{R}}(\delta l(w,x)^{-1})$. Then by arguments analogous to those in Steps 1-6 in the proof of Theorem III.3.1 one easily see that there is a C^1 function

$$g_{(w,x),x} : \bar{\mathbf{R}}^u(\delta l(w,x)^{-1}) \to \bar{\mathbf{R}}^{c+s}(\delta l(w,x)^{-1})$$

with $g_{(w,x),x}(0) = 0$ and $\mathrm{Lip}(g_{(w,x),x}) < 1$ such that

$$\Phi_{(w,x)}\mathrm{Graph}(g_{(w,x),x}) \subset W^u(w,x)$$

and moreover

$$\begin{aligned}\mathrm{Graph}(g_{(w,x),x}) = \{\zeta \in \bar{\mathbf{R}}(\delta l(w,x)^{-1}) : &\|H_{(w,x)}^{-n}\zeta\| \\ &\le \delta l(w,x)^{-1} e^{-(\lambda^+ - 2\varepsilon)n}, n \in \mathbf{Z}^+\}.\end{aligned}$$

Then 1) follows immediately. If $0 < \delta \le e^{-2(\lambda_0+\varepsilon)}$ and $(w,x) \in \Delta_0''$, then

$$W^u_{G(w,x),\delta}(f_w^1 x) \subset H_{(w,x)} W^u_{(w,x),\delta}(x) \subset W^u_{G(w,x),\delta'}(f_w^1 x)$$

where $\delta' = e^{-(\lambda_0+\varepsilon)}$. This together with 1) yields 2). □

Lemma 4.2. *If* $0 < \delta \le e^{-2(\lambda_0+\varepsilon)}$*, then for* μ^**-a.e.*$(w,x) \in \Delta_0''$ *one has*

$$S_\delta^{cu}(w,x) \cap \Phi_{(w,x)}^{-1} W^u(w,x) = W^u_{(w,x),\delta}(x).$$

Proof. In view of ii) in Lemma 4.1, it suffices to prove that for μ^*-a.e.$(w,x) \in \Delta_0''$ one has

$$S_\delta^{cu}(w,x) \cap \Phi_{(w,x)}^{-1} W^u(w,x) \subset W^u_{(w,x),\delta}(x).$$

Let $\zeta \in S_\delta^{cu}(w,x) \cap \Phi_{(w,x)}^{-1} W^u(w,x)$ and let $d^u(\ ,\ )$ denote the Riemannian distance along W^u-manifolds of $\mathcal{X}(M,v,\mu)$. Since $\Phi_{(w,x)}\zeta \in W^u(w,x)$, by Proposition 1.4 and Remark 1.2 one has $d^u(f_w^{-n}\Phi_{(w,x)}\zeta, f_w^{-n}x) \to 0$ as $n \to +\infty$. But, according to Poincaré Recurrence Theorem, $l(G^{-n}(w,x))^{-1}$ does not tend to 0 as $n \to +\infty$ for μ^*-a.e. $(w,x) \in \Delta_0''$. This implies that for μ^*-a.e. $(w,x) \in \Delta_0''$ there is some $k \ge 0$ such that $H_{(w,x)}^{-k}\zeta \in W^u_{G^{-k}(w,x),\delta}(f_w^{-k}x)$. Let $k = k(w,x)$ be the smallest one among such nonnegative integers. If $k > 0$, then by Lemma 4.1 2), $H_{(w,x)}^{-k+1}\zeta \notin \bar{\mathbf{R}}(\delta l(G^{-k+1}(w,x))^{-1})$, which contradicts $\zeta \in S_\delta^{cu}(w,x)$. So

$k = 0$, or equivalently, $\zeta \in W^u_{(w,x),\delta}(x)$. The conclusion in Lemma 4.2 is thus obtained. □

Let $(w,x) \in \Delta_0''$. Consider now $y \in \Phi_{(w,x)} S^{cu}_\delta(w,x)$ with $(w,y) \in \Delta_0''$, where $0 < \delta \leq 1/4$. Let $W^u_{(w,x),2\delta}(y)$ be the $\Phi^{-1}_{(w,x)}$-image of the component of $W^u(w,y) \cap \Phi_{(w,x)}[\bar{\mathbf{R}}^u(2\delta l(w,x)^{-1}) \times \bar{\mathbf{R}}^{c+s}(4\delta l(w,x)^{-1})]$ that contains y. Then $\Phi_{(w,x)} W^u_{(w,x),2\delta}(y)$ contains an open neighbourhood of y in $W^u(w,y)$ and is also referred to as a local unstable manifold of $\mathcal{X}(M,v,\mu)$ at (w,y) (although in general $\Phi_{(w,y)} W^u_{(w,y),2\delta}(y) \neq \Phi_{(w,x)} W^u_{(w,x),2\delta}(y)$). A reduction factor $0 < \delta \leq 1/4$ is taken because when working in $G^{-n}(w,x)$-charts we cannot control the unstable manifolds of points whose backward orbits under the actions f_w^{-n} come too close to the boundaries of $\Phi_{G^{-n}(w,x)} \bar{\mathbf{R}}(l(G^{-n}(w,x))^{-1})$. Another technical nuisance is that $\|\Phi^{-1}_{G^{-n}(w,x)} f_w^{-n} y\| \neq 0$. Aside from these, we have the following analogue of Lemmas 4.1 and 4.2.

Lemma 4.3. *Let $\{\Phi_{(w,x)}\}_{(w,x)\in\Delta_0''}$ be a system of (ε,l)-charts.*

1) Let $0 < \delta \leq \frac{1}{4} e^{-(\lambda_0+\varepsilon)}$ and $(w,x) \in \Delta_0''$. If $y \in \Phi_{(w,x)} S^{cu}_\delta(w,x)$ with $(w,y) \in \Delta_0''$, then

i) $W^u_{(w,x),2\delta}(y)$ is the graph of a C^1 function

$$g_{(w,x),y} : \bar{\mathbf{R}}^u(2\delta l(w,x)^{-1}) \to \bar{\mathbf{R}}^{c+s}(4\delta l(w,x)^{-1})$$

with $Lip(g_{(w,x),y}) < 1$;

ii) $W^u_{(w,x),2\delta}(y) \subset S^{cu}_{4\delta}(w,x)$;

2) Let $0 < \delta \leq \frac{1}{4} e^{-2(\lambda_0+\varepsilon)}$ and let $(w,x) \in \Delta_0''$. If $y \in \Phi_{(w,x)} S^{cu}_\delta(w,x)$ with $(w,y) \in \Delta_0''$ and $f_w^1 y \in \Phi_{G(w,x)} S^{cu}_\delta(G(w,x))$, then

$$H_{(w,x)} W^u_{(w,x),2\delta}(y) \cap [\bar{\mathbf{R}}^u(2\delta l(G(w,x))^{-1}) \times \bar{\mathbf{R}}^{c+s}(4\delta l(G(w,x))^{-1})] \subset W^u_{G(w,x),2\delta}(f_w^1 y);$$

3) Let $0 < \delta \leq \frac{1}{4} e^{-2(\lambda_0+\varepsilon)}$. Then for μ^*-a.e. $(w,x) \in \Delta_0''$, if $y \in \Phi_{(w,x)} S^{cu}_\delta(w,x)$ with $(w,y) \in \Delta_0''$, one has

$$S^{cu}_{2\delta}(w,x) \cap \Phi^{-1}_{(w,x)} W^u(w,y) \subset W^u_{(w,x),2\delta}(y) \subset S^{cu}_{4\delta}(w,x) \cap \Phi^{-1}_{(w,x)} W^u(w,y).$$

Proof. Let $0 < \delta \leq \frac{1}{4} e^{-(\lambda_0+\varepsilon)}$, $(w,x) \in \Delta_0''$ and $y \in \Phi_{(w,x)} S^{cu}_\delta(w,x)$ with $(w,y) \in \Delta_0''$. Denote $\zeta_y = \Phi^{-1}_{(w,x)} y$. Write

$$T_{\zeta_y} H^{-1}_{(w,x)} = \begin{bmatrix} H_{11} & H_{12} \\ H_{21} & H_{22} \end{bmatrix} : \mathbf{R}^u \times \mathbf{R}^{c+s} \to \mathbf{R}^u \times \mathbf{R}^{c+s},$$

$$r_{(w,x),y} = H_{(w,x)}^{-1} - \begin{bmatrix} H_{11} & 0 \\ 0 & H_{22} \end{bmatrix} :\{\zeta \in \mathbf{R}^{u+c+s} : \|\zeta - \zeta_y\| \leq 3\delta l(w,x)^{-1}\} \\ \to \mathbf{R}^{u+c+s}.$$

By Proposition 3.1 one can easily verify the following estimates:

$$\|H_{11}\xi\| \leq (e^{-\lambda^+ + \varepsilon} + \varepsilon\delta)\|\xi\| \qquad \text{for} \quad \xi \in \mathbf{R}^u,$$
$$\|H_{22}\eta\| \geq (e^{-\varepsilon} - \varepsilon\delta)\|\eta\| \qquad \text{for} \quad \eta \in \mathbf{R}^{c+s},$$
$$\mathrm{Lip}(r_{(w,x),y}) \leq \min\{e^{-\lambda^+ + 3\varepsilon} - (e^{-\lambda^+ + \varepsilon} + \varepsilon\delta), (e^{-\varepsilon} - \varepsilon\delta) - e^{-\lambda^+ + 3\varepsilon}\}.$$

Then by arguments analogous to those in Steps 1-6 in the proof of Theorem III.3.1 one can prove that there is a C^1 map

$$h_{(w,x),y} : \bar{\mathbf{R}}^u(3\delta l(w,x)^{-1}) \to \bar{\mathbf{R}}^{c+s}(3\delta l(w,x)^{-1}) \tag{4.1}$$

with $h_{(w,x),y}(0) = 0$ and $\mathrm{Lip}(h_{(w,x),y}) < 1$ such that

$$\Phi_{(w,x)}(\zeta_y + \mathrm{Graph}(h_{(w,x),y})) \subset W^u(w,y)$$

and moreover

$$\zeta_y + \mathrm{Graph}(h_{(w,x),y}) \\ = \{\zeta \in \bar{\mathbf{R}}(l(w,x)^{-1}) : \|H_{(w,x)}^{-n}\zeta - H_{(w,x)}^{-n}\zeta_y\| \leq 3\delta l(w,x)^{-1}e^{-(\lambda^+ - 3\varepsilon)n}, n \in \mathbf{Z}^+\}.$$

Define

$$g_{(w,x),y} : \bar{\mathbf{R}}^u(2\delta l(w,x)^{-1}) \to \bar{\mathbf{R}}^{c+s}(4\delta l(w,x)^{-1}), \\ \xi \mapsto h_{(w,x),y}(\xi - \xi_y) + \eta_y$$

where $\zeta_y = (\xi_y, \eta_y) \in \mathbf{R}^u \times \mathbf{R}^{c+s}$. Then 1) follows immediately. Let now $0 < \delta \leq \frac{1}{4}e^{-2(\lambda_0+\varepsilon)}$. We have

$$H_{(w,x)}W^u_{(w,x),2\delta}(y) \subset H_{(w,x)}(\zeta_y + \mathrm{Graph}(h_{(w,x),y})) \\ \subset \zeta_{f_w^1 y} + \mathrm{Graph}(\hat{h}_{G(w,x),f_w^1 y}) \tag{4.2}$$

where $\hat{h}_{G(w,x),f_w^1 y}$ is the function defined analogously to (4.1), corresponding to $\hat{\delta} = \frac{1}{4}e^{-(\lambda_0+\varepsilon)}$. But, if $f_w^1 y \in \Phi_{G(w,x)}S_\delta^{cu}(G(w,x))$, then

$$[\zeta_{f_w^1 y} + \mathrm{Graph}(\hat{h}_{G(w,x),f_w^1 y})] \cap [\bar{\mathbf{R}}^u(2\delta l(G(w,x))^{-1}) \times \bar{\mathbf{R}}^{c+s}(4\delta l(G(w,x))^{-1})] \\ = W^u_{G(w,x),2\delta}(f_w^1 y).$$

This proves 2). The proof of 3) is almost identical to that of Lemma 4.2. □

We remark that in general $S_\delta^{cu}(w,x)$ is a rather messy set. Among other things we think of it as containing pieces of local unstable manifolds (in view of Lemma 4.3 1) ii)). In the case when there is not zero exponent, $S_\delta^{cu}(w,x)$ is equal to $W^u_{(w,x),\delta}(x)$.

B. Some Estimates

We present here some estimates which will be used in later sections. Let $\{\Phi_{(w,x)}\}_{(w,x)\in\Delta_0''}$ be a system of (ε, l)-charts. When working in charts, we use ζ_u to denote the u-coordinate of the point $\zeta \in \bar{\mathbf{R}}(l(w,x)^{-1})$. Other notations such as ζ_s and ζ_{cu} are understood to have analogous meanings.

Lemma 4.4. *Let $0 < \delta \leq e^{-(\lambda_0+\varepsilon)}$ and let $(w,x) \in \Delta_0''$.*

1) If $\zeta, \zeta' \in \bar{\mathbf{R}}(\delta l(w,x)^{-1})$ and $\|\zeta - \zeta'\| = \|\zeta_u - \zeta_u'\|$, then

$$\begin{aligned}\|H_{(w,x)}\zeta - H_{(w,x)}\zeta'\| &= \|(H_{(w,x)}\zeta)_u - (H_{(w,x)}\zeta')_u\| \\ &\geq e^{\lambda^+ - 2\varepsilon}\|\zeta - \zeta'\|;\end{aligned}$$

2) If u in 1) is replaced by cu, then the conclusion holds with λ^+ being replaced by 0;

3) If $\zeta, \zeta' \in S_\delta^{cu}(w,x)$, then

$$\|H_{(w,x)}^{-1}\zeta - H_{(w,x)}^{-1}\zeta'\| \leq e^{2\varepsilon}\|\zeta - \zeta'\|.$$

Proof. By Proposition 3.1 2) ii) one has

$$\mathrm{Lip}((H_{(w,x)} - T_0 H_{(w,x)})|_{\bar{\mathbf{R}}(\delta l(w,x)^{-1})}) \leq \varepsilon\delta$$

which together with Proposition 3.1 2) i) yields 1) and 2) by a simple calculation. We now prove 3). First we claim that $\|\zeta - \zeta'\| = \|\zeta_{cu} - \zeta_{cu}'\|$. Indeed, if it is not this case, by applying 1) to $H_{(w,x)}^{-1}$ we have

$$\begin{aligned}\|H_{(w,x)}^{-1}\zeta - H_{(w,x)}^{-1}\zeta'\| &= \|(H_{(w,x)}^{-1}\zeta)_s - (H_{(w,x)}^{-1}\zeta')_s\| \\ &\geq e^{-\lambda^- - 2\varepsilon}\|\zeta - \zeta'\|\end{aligned}$$

which implies by induction

$$\begin{aligned}\|H_{(w,x)}^{-n}\zeta - H_{(w,x)}^{-n}\zeta'\| &= \|(H_{(w,x)}^{-n}\zeta)_s - (H_{(w,x)}^{-n}\zeta')_s\| \\ &\geq e^{-(\lambda^- + 2\varepsilon)n}\|\zeta - \zeta'\|\end{aligned}$$

for all $n \geq 0$. This contradicts the fact that $\zeta, \zeta' \in S_\delta^{cu}(w,x)$. Now this argument also applies to $H_{(w,x)}^{-1}\zeta$ and $H_{(w,x)}^{-1}\zeta'$, since they belong to $S_\delta^{cu}(G^{-1}(w,x))$. Then it follows from 2) that

$$\|\zeta - \zeta'\| \geq e^{-2\varepsilon}\|H_{(w,x)}^{-1}\zeta - H_{(w,x)}^{-1}\zeta'\|$$

which is the desired conclusion. □

Lemma 4.5. *Assume that $0 < \delta \leq \frac{1}{4}e^{-2(\lambda_0+\varepsilon)}$ and $(w,x) \in \Delta_0''$. Let $y \in \Phi_{(w,x)}S_\delta^{cu}(w,x)$ with $(w,y) \in \Delta_0''$, $\eta = (\{0\} \times \mathbf{R}^{c+s}) \cap W_{(w,x),2\delta}^u(y)$ and $\eta' =$*

$(\{0\} \times \mathbf{R}^{c+s}) \cap W^u_{G(w,x),2\hat{\delta}}(f^1_w y)$, *where* $\hat{\delta} = \frac{1}{4}e^{-(\lambda_0+\varepsilon)}$. *Then*

$$\|\eta'\| \le e^{3\varepsilon}\|\eta\|.$$

Proof. Since, by (4.2),

$$H_{(w,x)}W^u_{(w,x),2\delta}(y) \subset \zeta_{f^1_w y} + \mathrm{Graph}(\hat{h}_{G(w,x),f^1_w y}) \tag{4.3}$$

and

$$\begin{aligned}&W^u_{G(w,x),2\hat{\delta}}(f^1_w y)\\ =&[\zeta_{f^1_w y} + \mathrm{Graph}(\hat{h}_{G(w,x),f^1_w y})] \cap [\bar{\mathbf{R}}^u(2\hat{\delta}l(G(w,x))^{-1}) \times \bar{\mathbf{R}}^{c+s}(4\hat{\delta}l(G(w,x))^{-1})],\end{aligned} \tag{4.4}$$

one has

$$\|\eta'\| \le \|(H_{(w,x)}\eta)_{cs}\| + \|(H_{(w,x)}\eta)_u\|.$$

By Proposition 3.1 2), $\|(H_{(w,x)}\eta)_{cs}\| \le (e^\varepsilon + \varepsilon)\|\eta\|$ and $\|(H_{(w,x)}\eta)_u\| \le \varepsilon\|\eta\|$. Therefore, $\|\eta'\| \le (e^\varepsilon + 2\varepsilon)\|\eta\| \le e^{3\varepsilon}\|\eta\|$. □

C. Lipschitz Property of Unstable Subspaces Within Center Unstable Sets

Let $\{\Phi_{(w,x)}\}_{(w,x)\in\Delta_0''}$ be a system of (ε, l)-charts, $(w,x) \in \Delta_0''$ and $\hat{\delta} = \frac{1}{4}e^{-(\lambda_0+\varepsilon)}$. Denote by $L(\mathbf{R}^u, \mathbf{R}^{c+s})$ the space of all linear maps from $\mathbf{R}^u$ to $\mathbf{R}^{c+s}$. By Lemma 4.3 1) we know that , if $y \in \Phi_{(w,x)}S^{cu}_{\hat{\delta}}(w,x)$ with $(w,y) \in \Delta_0''$, then there exists a unique $P_y \in L(\mathbf{R}^u, \mathbf{R}^{c+s})$ with $\|P_y\| < 1$ such that

$$T_y\Phi^{-1}_{(w,x)}E^u_{(w,y)} = \mathrm{Graph}(P_y).$$

Define

$$\begin{aligned}\mathcal{L}_{(w,x)} :&\{y : y \in \Phi_{(w,x)}S^{cu}_{\hat{\delta}}(w,x) \quad \text{with} \quad (w,y) \in \Delta_0''\} \to L(\mathbf{R}^u, \mathbf{R}^{c+s}),\\ &y \mapsto P_y.\end{aligned}$$

In the sequel we show that the map $\mathcal{L}_{(w,x)}$ is Lipschitz. For this purpose we make use of the following lemma:

Lemma 4.6. *Let* $(E, |\cdot|)$ *be a Banach space,* $E = E^1 \oplus E^2$ *a decomposition of* E *into the direct sum of two subspaces* E^1 *and* E^2 *and suppose that the norm* $|\cdot|$ *satisfies* $|\xi| = \max\{|\xi_1|, |\xi_2|\}$ *for each* $\xi = \xi_1 + \xi_2 \in E^1 \oplus E^2$. *Let* $L(E,E)(L(E^1,E^2))$ *denote the space of all bounded linear operators from* $E(E^1)$ *to itself* (E^2). *Given positive numbers* λ, K *and* δ *such that* $e^\lambda < K$ *and* $0 <$

$\alpha \overset{\text{def.}}{=} (e^{\delta}+\delta)/(e^{\lambda-\delta}-\delta) < 1$, *we define*

$$X = \{\widetilde{A} = \{A_n\}_{n\in\mathbf{N}} : A_n \in L(E,E) \quad \text{is invertible with} \quad |A_n| \le K, \text{ and}$$
$$A_n = \begin{bmatrix} A_{11}^{(n)} & A_{12}^{(n)} \\ A_{21}^{(n)} & A_{22}^{(n)} \end{bmatrix} : E^1 \oplus E^2 \to E^1 \oplus E^2 \quad \text{satisfies}$$
$$|(A_{11}^{(n)})^{-1}| \le e^{-\lambda+\delta}, |A_{22}| \le e^{\delta}, |A_{12}^{(n)}| \le \delta \quad \text{and}$$
$$|A_{21}^{(n)}| \le \delta, n \in \mathbf{N}\},$$

$$Y = \{\widetilde{P} = \{P_n\}_{n\in\mathbf{N}} : P_n \in L(E^1,E^2) \quad \text{and} \quad |P_n| \le \alpha, n \in \mathbf{N}\}.$$

Let X and Y have respectively the following metrics:

$$d(\widetilde{A},\widetilde{A}') = \sum_{n=1}^{+\infty} \alpha_0^n |A_n - A_n'|, \quad \widetilde{A},\widetilde{A}' \in X,$$

$$d(\widetilde{P},\widetilde{P}') = \sum_{n=1}^{+\infty} \alpha_0^n |P_n - P_n'|, \quad \widetilde{P},\widetilde{P}' \in Y,$$

where $\alpha_0 = \frac{1}{2}(1+\alpha)$. Then for each $\widetilde{A} \in X$, there exists a unique $\widetilde{P} = \{P_n\}_{n\in\mathbf{N}} (\overset{\text{def.}}{=} \varphi(\widetilde{A})) \in Y$ such that

$$A_n^{-1}\,Graph(P_n) = Graph(P_{n+1})$$

for all $n \in \mathbf{N}$. This unique $\widetilde{P} = \varphi(\widetilde{A})$ also satisfies for each $\zeta \in Graph(P_1)$ and $n \in \mathbf{N}$

$$|\widetilde{A}^{-n}\zeta| \le \gamma^n |\zeta|,$$

where $\widetilde{A}^{-n} = A_n^{-1} \circ \cdots \circ A_1^{-1}$ and $\gamma = (e^{\lambda-\delta}-\delta)^{-1}$. Moreover, the map $\varphi : X \to Y, \widetilde{A} \mapsto \varphi(\widetilde{A})$ is Lipschitz and $Lip(\varphi) \le C$ with respect to the metrics defined above, where C is a number depending only on λ, K and δ.

Proof. Clearly, both (X,d) and (Y,d) are complete metric spaces. Define

$$\theta : X \times Y \to Y, (\widetilde{A},\widetilde{P}) \mapsto \widetilde{Q}$$

where $\widetilde{Q} = \{Q_n\}_{n\in\mathbf{N}}$ is given by

$$Q_n = (A_{21}^{(n)} + A_{22}^{(n)}P_{n+1})(A_{11}^{(n)} + A_{12}^{(n)}P_{n+1})^{-1}, \quad n \in \mathbf{N}.$$

Let us first check that this definition makes sense. Suppose that $(\widetilde{A},\widetilde{P}) \in X \times Y$. Since

$$|A_{12}^{(n)}P_{n+1}| \le \delta < e^{\lambda-\delta} \le |(A_{11}^{(n)})^{-1}|^{-1},$$

$A_{11}^{(n)} + A_{12}^{(n)}P_{n+1}$ is invertible and

$$|(A_{11}^{(n)} + A_{12}^{(n)}P_{n+1})^{-1}| \le (e^{\lambda-\delta}-\delta)^{-1}.$$

Thus $Q_n \in L(E^1, E^2)$ and

$$|Q_n| \leq (e^{\delta} + \delta)/(e^{\lambda - \delta} - \delta) = \alpha.$$

So θ is well defined.

Secondly, we verify that θ is a uniform contraction on the factor $\widetilde{P}$. Note that for each $\widetilde{A} \in X$ and $n \in \mathbf{N}$ the map

$$\begin{aligned} \Gamma_{A_n} :& \{S \in L(E^1, E^2) : |S| < 1\} \to \{S \in L(E^1, E^2) : |S| < 1\} \\ & S \mapsto (A_{21}^{(n)} + A_{22}^{(n)} S)(A_{11}^{(n)} + A_{12}^{(n)} S)^{-1} \end{aligned}$$

is a C^1 map with

$$(T_{S_0}\Gamma_{A_n})S = (A_{22}^{(n)} - \Gamma_{A_n}(S_0) A_{12}^{(n)}) S (A_{11}^{(n)} + A_{12}^{(n)} S_0)^{-1}, \quad S \in L(E^1, E^2)$$

for each $S_0 \in L(E^1, E^2)$ with $|S_0| < 1$. By a simple calculation one has $|T_{S_0}\Gamma_{A_n}| \leq \alpha$ for each $S_0 \in \{S \in L(E^1, E^2) : |S| < 1\}$. Γ_{A_n} is therefore a Lipschitz map with $\mathrm{Lip}(\Gamma_{A_n}) \leq \alpha$. From this it follows that, if $\widetilde{A} \in X$, then

$$d(\theta(\widetilde{A}, \widetilde{P}), \theta(\widetilde{A}, \widetilde{P}')) \leq \frac{\alpha}{\alpha_0} d(\widetilde{P}, \widetilde{P}')$$

for any $\widetilde{P}, \widetilde{P}' \in Y$, i.e. θ is an α/α_0-contraction on the second factor.

Let $\widetilde{A} \in X$. Then there is a unique point $\widetilde{P} = \{P_n\}_{n \in \mathbf{N}} \overset{\text{def.}}{=} \varphi(\widetilde{A}) \in Y$ such that

$$\theta(\widetilde{A}, \widetilde{P}) = \widetilde{P},$$

i.e.

$$A_n^{-1}\mathrm{Graph}(P_n) = \mathrm{Graph}(P_{n+1})$$

for all $n \in \mathbf{N}$. Also, if $n \in \mathbf{N}$ and $(\xi, P_{n+1}\xi) \in \mathrm{Graph}(P_{n+1})$, then

$$\begin{aligned} |A_n(\xi, P_{n+1}\xi)| &= |(A_{11}^{(n)} + A_{12}^{(n)} P_{n+1})\xi| \\ &\geq (e^{\lambda - \delta} - \delta)|\xi| = (e^{\lambda - \delta} - \delta)|(\xi, P_{n+1}\xi)|. \end{aligned}$$

This proves the first two conclusions of the lemma.

We now prove the last conclusion. Endow $X \times Y$ with the product metric $d(\ ,\ )$ (see Lemma III.2.1). Then by a simple calculation one can check that $\theta : (X \times Y, d) \to (Y, d)$ is a Lipschitz map and there is a number $l > 0$ depending only on λ, K and δ such that $\mathrm{Lip}(\theta) \leq l$. This, by Lemma III.2.1, asserts that $\varphi : X \to Y, \widetilde{A} \mapsto \varphi(\widetilde{A})$ is Lipschitz and $\mathrm{Lip}(\varphi) \leq (1 - \alpha/\alpha_0)^{-1} l \overset{\text{def.}}{=} C$. □

Lemma 4.6. *Let $\{\Phi_{(w,x)}\}_{(w,x) \in \Delta_0''}$ be a system of (ε, l)-charts with ε being small enough so that $e^{-\lambda^+ + 10\varepsilon} + e^{5\varepsilon} < 2$. Then for each $(w, x) \in \Delta_0''$, $\mathcal{L}_{(w,x)}$ is a Lipschitz map and*

$$Lip(\mathcal{L}_{(w,x)}) \leq D_0 l(w, x)^2$$

where $D_0 > 0$ is a number depending only on the exponents and ε.

Proof. Here we keep the notations introduced in Lemma 4.5. Put $\lambda = \lambda^+, K = e^{\lambda_0}$ and $\delta = 2\varepsilon$. Let now $(w,x) \in \Delta_0''$. For $y \in \Phi_{(w,x)} S_\delta^{cu}(w,x)$ with $(w,y) \in \Delta_0''$, set $\zeta_y = \Phi_{(w,x)}^{-1} y$ and define $\widetilde{A}_{(w,x)}(y) = \{T_{H_{(w,x)}^{-n}\zeta_y} H_{G^{-n}(w,x)}\}_{n\in\mathbf{N}}$. It is easy to see that $\widetilde{A}_{(w,x)}(y) \in X$. Write then $\varphi(\widetilde{A}_{(w,x)}(y)) = \{P_{(w,x),n}(y)\}_{n\in\mathbf{N}}$. By Lemma 4.5 we also have

$$\mathcal{L}_{(w,x)}(y) = P_{(w,x),1}(y).$$

Hence , for any $z, z' \in \{y : y \in \Phi_{(w,x)} S_\delta^{cu}(w,x) \quad \text{with} \quad (w,y) \in \Delta_0''\}$,

$$\begin{aligned}
&\|\mathcal{L}_{(w,x)}(z) - \mathcal{L}_{(w,x)}(z')\| \\
=&\|P_{(w,x),1}(z) - P_{(w,x),1}(z')\| \\
\le& C\sum_{n=1}^{+\infty} \alpha_0^n \|T_{H_{(w,x)}^{-n}\zeta_z} H_{G^{-n}(w,x)} - T_{H_{(w,x)}^{-n}\zeta_{z'}} H_{G^{-n}(w,x)}\| \qquad \text{(by Lemma 4.5)} \\
\le& C\sum_{n=1}^{+\infty} \alpha_0^n \varepsilon l(G^{-n}(w,x)) \|H_{(w,x)}^{-n}\zeta_z - H_{(w,x)}^{-n}\zeta_{z'}\| \qquad \text{(by Proposition 3.1 2))} \\
\le& C\sum_{n=1}^{+\infty} \alpha_0^n \varepsilon l(G^{-n}(w,x))(e^{2\varepsilon})^n \|\zeta_z - \zeta_{z'}\| \qquad \text{(by Lemma 4.4 3))} \\
\le& C\sum_{n=1}^{+\infty} \varepsilon(\alpha_0 e^{3\varepsilon})^n l(w,x) \|\zeta_z - \zeta_{z'}\| \\
\le& C\sum_{n=1}^{+\infty} \varepsilon(\alpha_0 e^{3\varepsilon})^n l(w,x)^2 d(z,z') \\
=& D_0 l(w,x)^2 d(z,z')
\end{aligned}$$

where $D_0 = C\sum_{n=1}^{+\infty} \varepsilon(\alpha_0 e^{3\varepsilon})^n < +\infty$ since $\alpha_0 e^{3\varepsilon} < 1$. The proof is completed. □

§5 Related Measurable Partitions

A. Partitions Adapted to Lyapunov Charts

In order to make use of the geometry of Lyapunov charts in the calculation of entropy, it is convenient to have partitions whose elements lie in charts. If $\mathcal{P}$ is a measurable partition of $\Omega^{\mathbf{Z}} \times M$, write $\mathcal{P}^+ = \bigvee_{n=0}^{+\infty} G^n \mathcal{P}$ and denote by $\mathcal{P}_w$ the partition $\mathcal{P}|_{\{w\}\times M}$ for $w \in \Omega^{\mathbf{Z}}$. In what follows we shall identify $\{w\} \times M$ with M for any $w \in \Omega^{\mathbf{Z}}$. Let $\{\Phi_{(w,x)}\}_{(w,x)\in\Delta_0''}$ be a system of (ε, l)-charts and $0 < \delta \le 1$.

Definition 5.1. *A measurable partition $\mathcal{P}$ of $\Omega^{\mathbf{Z}} \times M$ is said to be adapted to $(\{\Phi_{(w,x)}\}_{(w,x)\in\Delta_0''}, \delta)$ if $\mathcal{P}_w^+(x) \subset \Phi_{(w,x)} S_\delta^{cu}(w,x)$ for μ^*-a.e. (w,x).*

Our purpose in this subsection is to show that there always exists such a partition $\mathcal{P}$ with $H_{\mu^*}(\mathcal{P}|\sigma) < +\infty$. We shall use the following two preliminary lemmas whose ideas are given in [Man]$_2$ in the deterministic case.

Lemma 5.1. *If $x_n \in [0,1]$ for $n \geq 0$, then*

$$-\sum_{n=0}^{+\infty} x_n \log x_n \leq \sum_{n=1}^{+\infty} n x_n + c_0 \tag{5.1}$$

where we admit $0 \log 0 = 0$ and where $c_0 = 2[e(1-e^{-\frac{1}{2}})]^{-1}$.

Proof. If $\sum_{n=1}^{+\infty} n x_n = +\infty$, (5.1) is trivial. We now assume that $\sum_{n=1}^{+\infty} n x_n < +\infty$. Let S be the set of integers n such that $x_n > 0$ and $-\log x_n < n$. Then

$$\begin{aligned} -\sum_{n=0}^{+\infty} x_n \log x_n &= -\sum_{n\in S} x_n \log x_n - \sum_{n\notin S} x_n \log x_n \\ &\leq \sum_{n\in S} n x_n - \sum_{n\notin S} x_n \log x_n . \end{aligned}$$

Note that $n \notin S$ implies $x_n \leq e^{-n}$. On the other hand,

$$-\sqrt{t}\log t \leq \frac{2}{e}$$

for all $t \in [0,1]$. Hence, we obtain

$$-\sum_{n\notin S} x_n \log x_n \leq \sum_{n\notin S} \frac{2}{e}\sqrt{x_n} \leq \frac{2}{e}\sum_{n=0}^{+\infty} e^{-\frac{1}{2}n}.$$

This completes the proof. □

Lemma 5.2. *Let $\rho : \Omega^{\mathbf{Z}} \times M \to (0,1]$ be a measurable function with $\int \log \rho d\mu^* > -\infty$. Then there exists a measurable partition $\mathcal{P}$ of $\Omega^{\mathbf{Z}} \times M$ such that:*

1) $\mathrm{diam}\mathcal{P}_w(x) < \rho(w,x)$ for each $(w,x) \in \Omega^{\mathbf{Z}} \times M$;

2) $H_{\mu^*}(\mathcal{P}|\sigma) < +\infty$.

Proof. Take numbers $C > 0$ and $r_0 > 0$ such that if $0 < r \leq r_0$ there exists a measurable partition ξ_r of M which satisfies

$$\mathrm{diam}\xi_r(x) \leq r$$

for all $x \in M$ and

$$|\xi_r| \leq C(\frac{1}{r})^{m_0}$$

where $|\xi_r|$ denotes the number of elements in ξ_r. For each $n \geq 0$, put $U_n = \{(w,x) \in \Omega^{\mathbf{Z}} \times M : e^{-(n+1)} < \rho(w,x) \leq e^{-n}\}$. The integrability of $\log \rho$ implies that for every $N \geq 1$,

$$\sum_{n=1}^{N} n\mu^*(U_n) \leq \sum_{n=1}^{N} -\int_{U_n} \log \rho d\mu^* \leq -\int_{\Omega^{\mathbf{Z}} \times M} \log \rho d\mu^*,$$

so that

$$\sum_{n=1}^{+\infty} n\mu^*(U_n) < +\infty.$$

Define a partition $\mathcal{P}$ of $\Omega^{\mathbf{Z}} \times M$ by demanding that $\mathcal{P} \geq \{U_n : n \geq 0\}$ and $\mathcal{P}|_{U_n} = \{\Omega^{\mathbf{Z}} \times A : A \in \xi_{r_n}\}|_{U_n}$, where $n \geq 0$ and $r_n = e^{-(n+1)}$. Then $\mathcal{P}$ is clearly a measurable partition satisfying 1). We now verify that $H_{\mu^*}(\mathcal{P}|\sigma) < +\infty$.

In fact, we first have

$$H_{\mu^*}(\mathcal{P}|\sigma) = \int H_{\mu_w}(\mathcal{P}_w) d\nu^{\mathbf{Z}}(w) \tag{5.2}$$

and for each $w \in \Omega^{\mathbf{Z}}$

$$H_{\mu_w}(\mathcal{P}_w) = \sum_{n=0}^{+\infty} \Big(- \sum_{\substack{P \in \mathcal{P}_w \\ P \subset (U_n)_w}} \mu_w(P) \log \mu_w(P)\Big).$$

For each $w \in \Omega^{\mathbf{Z}}$ and $n \geq 0$,

$$\begin{aligned}
&- \sum_{\substack{P \in \mathcal{P}_w \\ P \subset (U_n)_w}} \mu_w(P) \log \mu_w(P) \\
\leq &\mu_w(U_n)(\log |\xi_{r_n}| - \log \mu_w(U_n)) \\
\leq &\mu_w(U_n)(\log C + m_0 \log(1/r_n) - \log \mu_w(U_n)) \\
\leq &\mu_w(U_n) \log C + m_0(n+1)\mu_w(U_n) - \mu_w(U_n) \log \mu_w(U_n).
\end{aligned}$$

Summing over $\{n : n \geq 0\}$, we obtain

$$H_{\mu_w}(\mathcal{P}_w) \leq \log C + m_0 \sum_{n=0}^{+\infty} (n+1)\mu_w(U_n) - \sum_{n=0}^{+\infty} \mu_w(U_n) \log \mu_w(U_n)$$

which, by Lemma 5.1, implies

$$H_{\mu_w}(\mathcal{P}_w) \leq \log C + m_0 + c_0 + (m_0 + 1) \sum_{n=1}^{+\infty} n\mu_w(U_n). \tag{5.3}$$

Since for every $N \geq 1$

$$\int \sum_{n=1}^{N} n\mu_w(U_n)dv^{\mathbf{Z}}(w) = \sum_{n=1}^{N} n\mu^*(U_n)$$
$$\leq \sum_{n=1}^{+\infty} n\mu^*(U_n) < +\infty,$$

by the monotone convergence theorem, we have

$$\int \sum_{n=1}^{+\infty} n\mu_w(U_n)dv^{\mathbf{Z}}(w) \leq \sum_{n=1}^{+\infty} n\mu^*(U_n) < +\infty. \tag{5.4}$$

That $H_{\mu^*}(\mathcal{P}|\sigma) < +\infty$ follows then from (5.2)-(5.4). □

Remark 5.1. As can be seen from the proof above, Lemma 5.2 also holds true if $\mathcal{X}(M, v, \mu)$ is not ergodic.

Proposition 5.1. *Let $\{\Phi_{(w,x)}\}_{(w,x)\in\Delta_0''}$ be a system of (ε, l)-charts and $0 < \delta \leq 1$. Then there exists a measurable partition $\mathcal{P}$ of $\Omega^{\mathbf{Z}} \times M$ such that:*

1) $\mathcal{P}$ is adapted to $(\{\Phi_{(w,x)}\}_{(w,x)\in\Delta_0''}, \delta)$;

2) $H_{\mu^*}(\mathcal{P}|\sigma) < +\infty$.

Proof. Fix $l_0 > 1$ such that $\Lambda = \{(w, x) \in \Delta_0'' : l(w, x) \leq l_0\}$ has positive μ^* measure. The ergodicness of $G : (\Omega^{\mathbf{Z}} \times M, \mu^*) \hookleftarrow$ implies that there is a Borel set $\Delta \subset \Omega^{\mathbf{Z}} \times M$ with $\mu^*(\Delta) = 1$ such that each $(w, x) \in \Delta$ satisfies $G^k(w, x) \in \Lambda$ for some $k > 0$. If $(w, x) \in \Delta$, put $r(w, x) = \min\{k : k > 0, G^k(w, x) \in \Lambda\}$. Let $\Lambda' = \Lambda \cap \Delta$. Define $\varphi : \Omega^{\mathbf{Z}} \times M \to (0, 1]$ by

$$\varphi(w, x) = \begin{cases} \delta & \text{if } (w, x) \notin \Lambda', \\ \delta l_0^{-2} e^{-(\lambda_0+\varepsilon)r(w,x)} & \text{if } (w, x) \in \Lambda'. \end{cases}$$

$\log\varphi$ is μ^* -integrable since $\int_{\Lambda'} r(w, x)d\mu^* = 1$. By Lemma 5.2, there is a measurable partition $\mathcal{P}$ of $\Omega^{\mathbf{Z}} \times M$ such that $H_{\mu^*}(\mathcal{P}|\sigma) < +\infty$ and $\mathcal{P}_w(x) \subset B(x, \varphi(w, x))$ for each $(w, x) \in \Omega^{\mathbf{Z}} \times M$. We now prove that this $\mathcal{P}$ is adapted to $(\{\Phi_{(w,x)}\}_{(w,x)\in\Delta_0''}, \delta)$. In fact, it suffices to show $\mathcal{P}_w^+(x) \subset \Phi_{(w,x)}\bar{\mathbf{R}}(\delta l(w, x)^{-1})$ for each $(w, x) \in [\cup_{n=0}^{+\infty} G^n\Lambda'] \setminus [\cup_{n=0}^{+\infty} G^n(\Lambda\setminus\Lambda')]$.

First consider $(w, x) \in \Lambda'$. By the choice of $\mathcal{P}$, we have $\mathcal{P}_w^+(x) \subset \mathcal{P}_w(x) \subset B(x, \varphi(w, x))$ which is contained in $\Phi_{(w,x)}\bar{\mathbf{R}}(\delta l(w, x)^{-1})$ because $\varphi(w, x) \leq \delta l(w, x)^{-2}$. Suppose now that $(w, x) \notin \Lambda'$ but $(w, x) \in [\cup_{n=0}^{+\infty} G^n\Lambda'] \setminus [\cup_{n=0}^{+\infty} G^n(\Lambda\setminus\Lambda')]$. Let $n > 0$ be the smallest positive integer such that $G^{-n}(w, x) \in \Lambda'$. Then $f_w^{-n}\mathcal{P}_w^+(x) \subset \mathcal{P}_{\tau^{-n}w}^+(f_w^{-n}x) \subset B(f_w^{-n}x, \varphi(G^{-n}(w, x)))$. Now

$$
\begin{aligned}
&f_w^n B(f_w^{-n}x, \varphi(G^{-n}(w,x))) \\
&\subset \Phi_{(w,x)} H^n_{G^{-n}(w,x)} \bar{\mathbf{R}}(\delta l(G^{-n}(w,x))^{-1} e^{-(\lambda_0+\varepsilon) r(G^{-n}(w,x))}) \\
&\subset \Phi_{(w,x)} \bar{\mathbf{R}}(\delta l(w,x)^{-1} e^{n\varepsilon} e^{-(\lambda_0+\varepsilon) r(G^{-n}(w,x))} e^{\lambda_0 n}) \\
&\subset \Phi_{(w,x)} \bar{\mathbf{R}}(\delta l(w,x)^{-1})
\end{aligned}
$$

since $n \le r(G^{-n}(w,x))$. Note that the computation above makes sense because for every $1 \le k < n, H^k_{G^{-n}(w,x)} \bar{\mathbf{R}}(\delta l(G^{-n}(w,x))^{-1} e^{-(\lambda_0+\varepsilon) r(G^{-n}(w,x))}) \subset \bar{\mathbf{R}}(l(G^{-n+k}(w,x))^{-1}\, e^{-(\lambda_0+\varepsilon)})$. The proof is completed. □

B. Increasing Partitions Subordinate to W^u-manifolds

A measurable partition ξ of $\Omega^{\mathbf{Z}} \times M$ is said to be *increasing* if $\xi \ge G\xi$. In order to prove Theorem 1.1 2)⇒ 1), we require a family of increasing partitions of $\Omega^{\mathbf{Z}} \times M$ that are not only subordinate to W^u-manifolds of $\mathcal{X}(M, v, \mu)$ but also of some additional properties. This family of partitions are described in the proof of the following proposition.

Proposition 5.2. *There exist measurable partitions of $\Omega^{\mathbf{Z}} \times M$ each of which, written ξ, has the following properties:*

1) ξ is increasing and subordinate to W^u-manifolds of $\mathcal{X}(M, v, \mu)$;

2) For every $B \in \mathcal{B}(\Omega^{\mathbf{Z}} \times M)$, the function

$$P_B(w,x) = \lambda^u_{(w,x)}(\xi_w(x) \cap B_w)$$

(well defined for μ^*-a.e. (w,x)) is measurable and μ^* almost everywhere finite;

3) $\vee_{n=0}^{+\infty} G^{-n}\xi$ is equal to the partition into single points;

4) $\mathcal{B}(\wedge_{n=0}^{+\infty} G^n \xi) = \mathcal{B}^u(\mathcal{X}(M, v, \mu))$, μ^*-mod 0 where $\mathcal{B}(\wedge_{n=0}^{+\infty} G^n \xi)$ is the σ-algebra consisting of all measurable $\wedge_{n=0}^{+\infty} G^n \xi$-sets.

Proof. Let $\{\Phi_{(w,x)}\}_{(w,x)\in\Delta_0''}$ be a system of (ε, l)-charts. Fix $l_0 > 0$ such that the set $\{(w,x) \in \Delta_0'' : l(w,x) \le l_0\}$ has positive μ^* measure. Proposition 1.3 asserts that there exist a compact set $\Lambda \subset \{(w,x) \in \Delta_0'' : l(w,x) \le l_0\}$ with $\mu^*(\Lambda) > 0$ and a continuous family of C^1 embedded u-dimensional disks $\{W^u_{\text{loc}}(w,x)\}_{(w,x)\in\Lambda}$ such that the following (i)-(v) hold true:

(i) $W^u_{\text{loc}}(w,x) \subset \Phi_{(w,x)} W^u_{(w,x),\hat{\delta}}(x)$ for all $(w,x) \in \Lambda$, where $\hat{\delta} = \frac{1}{4} e^{-(\lambda_0+\varepsilon)}$;

(ii) There exist $\hat{\lambda} > 0$ and $\hat{\gamma} > 0$ such that for each $(w,x) \in \Lambda$, if $y, z \in W^u_{\text{loc}}(w,x)$, then for all $l \ge 0$

$$d^u(f_w^{-l} y, f_w^{-l} z) \le \hat{\gamma} e^{-\hat{\lambda} l} d^u(y,z);$$

(iii) There exist numbers $\hat{r}, \hat{\varepsilon}$ and $\hat{d}$ with $0 < \hat{r} < \rho_0/4, 0 < \hat{\varepsilon} < 1$ and $\hat{d} > 2\hat{r}$ such that for any $r \in (0, \hat{r}]$ and $(w,x) \in \Lambda$, if $(w', x') \in B_\Lambda((w,x), \hat{\varepsilon} r) \stackrel{\text{def.}}{=} \{(w', x')$

$\in \Lambda : \max\{d(w,w'), d(x,x')\} < \hat{\varepsilon} r\}$, then $W^u_{\mathrm{loc}}(w',x') \cap B(x,r)$ is connected, its d^u-diameter is less than $\hat{d}$ and the map

$$(w',x') \mapsto W^u_{\mathrm{loc}}(w',x') \cap B(x,r)$$

is a continuous map from $B_\Lambda((w,x), \hat{\varepsilon} r)$ to the space of subsets of $B(x,r)$ (endowed with the Hausdorff topology);

(iv) Let $r \in (0, \hat{r}]$ and $(w,x) \in \Lambda$. If $(w',x'), (w',x'') \in B_\Lambda((w,x), \hat{\varepsilon} r)$, then either

$$W^u_{\mathrm{loc}}(w',x') \cap B(x,r) = W^u_{\mathrm{loc}}(w',x'') \cap B(x,r)$$

or otherwise the two terms in the above equation are disjoint. In the latter case, if it is assumed moreover that $x'' \in W^u(w',x')$, then

$$d^u(y,z) > \hat{d} > 2\hat{r}$$

for any $y \in W^u_{\mathrm{loc}}(w',x') \cap B(x,r)$ and $z \in W^u_{\mathrm{loc}}(w',x'') \cap B(x,r)$;

(v) There exists $\hat{R} > 0$ such that for each $(w,x) \in \Lambda$, if $(w',x') \in B_\Lambda((w,x), \hat{\varepsilon}\hat{r})$ and $y \in W^u_{\mathrm{loc}}(w',x') \cap B(x,\hat{r})$, then $W^u_{\mathrm{loc}}(w',x')$ contains the closed ball of center y and d^u radius $\hat{R}$ in $W^u(w',x')$.

We now choose $(w_0,x_0) \in \Lambda$ such that $B_\Lambda((w_0,x_0), \hat{\varepsilon}\hat{r}/2)$ has positive μ^* measure. For each $r \in [\hat{r}/2, \hat{r}]$, put

$$S_r = \cup\{\{w\} \times [W^u_{\mathrm{loc}}(w,x) \cap B(x_0,r)] : (w,x) \in B_\Lambda((w_0,x_0), \hat{\varepsilon} r)\}$$

and let ξ_r denote the partition of $\Omega^{\mathbf{Z}} \times M$ into all the sets $\{w\} \times [W^u_{\mathrm{loc}}(w,x) \cap B(x_0,r)], (w, x) \in B_\Lambda((w_0,x_0), \hat{\varepsilon} r)$ and the set $\Omega^{\mathbf{Z}} \times M \backslash S_r$. We now define a measurable function $\beta_r : S_r \to \mathbf{R}^+$ by

$$\beta_r(w,y) = \inf_{n\geq 0} \left\{ \hat{\mathbf{R}}, \frac{1}{2\hat{\gamma}} d(f_w^{-n} y, \partial B(x_0,r)) e^{n\hat{\lambda}}, \frac{r}{\hat{\gamma}} \right\}.$$

By arguments completely analogous to those in the proof of Proposition IV.2.1 concerning the existence of a partition satisfying 1) and 2) there, we know that there exists $r' \in [\hat{r}/2, \hat{r}]$ such that $\beta_{r'} > 0$ μ^* almost everywhere on $S_{r'}$, this implying that $\xi \stackrel{\mathrm{def.}}{=} \xi_{r'}^+ = \vee_{n=0}^{+\infty} G^n \xi_{r'}$ satisfies 1) of this proposition since $\mu^*(\cup_{n=0}^{+\infty} G^n S_{r'}) = 1$. 2) is verified similarly to Proposition IV.2.1 3). We now verify that ξ also satisfies 3) and 4).

Put $\xi^- = \vee_{n=0}^{+\infty} G^{-n}\xi$. Since $G : (\Omega^{\mathbf{Z}} \times M, \mu^*) \hookleftarrow$ is ergodic, for μ^*-a.e. $(w,y) \in \Omega^{\mathbf{Z}} \times M$ there exist infinitely many positive integers $\{n_i : i = 1, 2, \cdots\}$ such that $G^{n_i}(w,y) \in S_{r'}$ for all $i \geq 1$. Then from property (ii) just above it follows that the d^u diameter of $\xi_w^-(y)$ is less than $\hat{\gamma}\hat{d}e^{-\hat{\lambda} n_i}$ for all $i \geq 1$ and hence is equal to 0. This proves that ξ^- is equal to the partition into single points.

In order to prove that $\mathcal{B}(\wedge_{n=0}^{+\infty} G^n \xi) \subset \mathcal{B}^u(\mathcal{X}(M,v,\mu))$, μ^* -mod 0, it suffices to ensure that for μ^*-a.e. $(w,y) \in \Delta_0''$, if $z \in W^u(w,y)$, then there exists $k > 0$ such that $G^{-k}(w,z) \subset \xi(G^{-k}(w,y))$. In fact, if $(w,y) \in \Delta_0''$ and $z \in W^u(w,y)$, we first have

$$\limsup_{n\to+\infty} \frac{1}{n} \log d^u(f_w^{-n} y, f_w^{-n} z) \leq -\lambda^+ < 0.$$

Define now $\beta'_{r'} : \Omega^{\mathbf{Z}} \times M \to \mathbf{R}^+$ by

$$\beta'_{r'}(w,y) = \begin{cases} \beta_{r'}(w,y) & \text{if } \ (w,y) \in S_{r'}, \\ 0 & \text{otherwise.} \end{cases}$$

According to Birkhoff ergodic theorem, one has for μ^*-a.e. $(w,y) \in \Delta''_0$

$$\lim_{n\to+\infty} \frac{1}{n} \sum_{k=0}^{n-1} \beta'_{r'}(G^{-k}(w,y)) = \int \beta'_{r'} d\mu^* > 0$$

and hence, if $z \in W^u(w,y)$, there will be some $k > 0$ such that

$$\beta'_{r'}(G^{-k}(w,y)) > d^u(f_w^{-n}y, f_w^{-n}z)$$

which implies that $G^{-k}(w,y) \in S_{r'}$ and $G^{-k}(w,z) \in \xi(G^{-k}(w,y))$. On the other hand, it is clear that $\mathcal{B}^u(\mathcal{X}(M,v,\mu)) \subset \mathcal{B}(G^n\xi)(\mu^*\text{-mod } 0)$ for all $n \geq 0$ and hence $\mathcal{B}^u(\mathcal{X}(M,v,\mu)) \subset \cap_{n=0}^{+\infty}\mathcal{B}(G^n\xi) = \mathcal{B}(\wedge_{n=0}^{+\infty}G^n\xi)(\mu^*\text{-mod } 0)$. The proof is completed. □

Remark 5.2. Each partition ξ we just constructed has the following additional characterization: Let $\hat{S}$ denote the set $S_{r'}$ introduced in the construction of ξ. If $\hat{\eta} = \vee_{n=0}^{+\infty} G^n\{\hat{S}, \Omega^{\mathbf{Z}} \times M\backslash\hat{S}\}$, then for every $(w,x) \in \Omega^{\mathbf{Z}} \times M, (w,y) \in \xi(w,x)$ if and only if $(w,y) \in \hat{\eta}(w,x)$ and $d^u(f_w^{-n}x, f_w^{-n}y) \leq \hat{d}$ whenever $G^{-n}(w,x) \in \hat{S}$ (we define $d^u(f_w^{-n}x, f_w^{-n}y) = +\infty$ if $f_w^{-n}y \notin W^u(G^{-n}(w,x))$).

Lemma 5.3. *Let ξ_1 and ξ_2 be partitions constructed in the proof of Proposition 5.2. Then*

$$h^\sigma_{\mu^*}(G^{-1},\xi_1) = h^\sigma_{\mu^*}(G^{-1},\xi_2).$$

Proof. It suffices to prove $h^\sigma_{\mu^*}(G^{-1},\xi_1 \vee \xi_2) = h^\sigma_{\mu^*}(G^{-1},\xi_1)$. Using the fact that $h^\sigma_{\mu^*}(G^{-1}) = h^\sigma_{\mu^*}(G) \leq \sum_{\lambda^{(i)}>0} \lambda^{(i)} m_i < +\infty$, by Theorem 0.5.2 we have for every $n \geq 1$,

$$\begin{aligned} h^\sigma_{\mu^*}(G^{-1},\xi_1 \vee \xi_2) &= h^\sigma_{\mu^*}(G^{-1}, G^n(\xi_1 \vee \xi_2)) \\ &= h^\sigma_{\mu^*}(G^{-1},\xi_1 \vee G^n\xi_2) \\ &= H_{\mu^*}(\xi_1 \vee G^n\xi_2 | G\xi_1 \vee G^{n+1}\xi_2 \vee \sigma) \\ &= H_{\mu^*}(\xi_1 | G\xi_1 \vee G^{n+1}\xi_2) + H_{\mu^*}(\xi_2 | G\xi_2 \vee G^{-n}\xi_1). \end{aligned}$$

By Proposition 5.2 4), one has as $n \to +\infty$

$$G\xi_1 \vee G^{n+1}\xi_2 \searrow G\xi_1 \vee (\wedge_{l=0}^{+\infty} G^l\xi_2) = G\xi_1.$$

Hence

$$H_{\mu^*}(\xi_1 | G\xi_1 \vee G^{n+1}\xi_2) \to H_{\mu^*}(\xi_1 | G\xi_1)$$

as $n \to +\infty$. Also, by Proposition 5.2 3), $G\xi_2 \vee G^{-n}\xi_1$ tends increasingly to the partition of $\Omega^{\mathbf{Z}} \times M$ into single points. Thus

$$H_{\mu^*}(\xi_2|G\xi_2 \vee G^{-n}\xi_1) \to 0$$

as $n \to +\infty$. This completes the proof. □

C. Two Useful Partitions

Let $\{\Phi_{(w,x)}\}_{(w,x)\in\Delta_0''}$ be a system of (ε, l)-charts with ε being small enough so that $e^{-\lambda^+ + 10\varepsilon} + e^{5\varepsilon} < 2$. Let ξ be a partition of the type constructed in the proof of Proposition 5.2, with $l_0, \hat{S}$ and $\hat{d}$ having the same meaning as in Subsection 5.B. Fix $0 < \delta \leq \min\{\frac{1}{16}e^{-2(\lambda_0+\varepsilon)}, \hat{d}/2K_0\}$ and let $\mathcal{P}$ be a partition adapted to $(\{\Phi_{(w,x)}\}_{(w,x)\in\Delta_0''}, \delta)$ with $H_{\mu^*}(\mathcal{P}|\sigma) < +\infty$. We require that $\mathcal{P}$ refines $\{\hat{S}, \Omega^{\mathbf{Z}} \times M \backslash \hat{S}\} \vee \sigma$ and $\{\hat{E}, \Omega^{\mathbf{Z}} \times M \backslash \hat{E}\} \vee \sigma$, where $\hat{E}$ is a Borel set of positive μ^* measure which will be specified in Subsection 5.E. Define

$$\eta_1 = \xi \vee \mathcal{P}^+,$$
$$\eta_2 = \mathcal{P}^+.$$

These two partitions will play central roles in Section 7. Some of their properties are described in the following lemma.

Lemma 5.4. *1)* $G\eta_1 \leq \eta_1, G\eta_2 \leq \eta_2$;

2) $\eta_1 \geq \eta_2$;

3) $\eta_{2w}(x) \subset \Phi_{(w,x)}S_\delta^{cu}(w,x)$ *and* $\eta_{1w}(x) \subset \Phi_{(w,x)}W^u_{(w,x),\delta}(x)$ *for* μ^**-a.e.* (w,x);

4) $h^\sigma_{\mu^*}(G^{-1}, \eta_2) = h^\sigma_{\mu^*}(G^{-1}, \mathcal{P}), h^\sigma_{\mu^*}(G^{-1}, \eta_1) = h^\sigma_{\mu^*}(G^{-1}, \xi)$.

Proof. Properties 1) and 2) and the first half of 3) follow from the definitions of η_1 and η_2. The second half of 3) is a consequence of Lemma 4.2. The first half of 4) is straightforward. We now prove the last assertion.

In view of the fact that $h^\sigma_{\mu^*}(G^{-1}) < +\infty$ and $H_{\mu^*}(\mathcal{P}|\sigma) < +\infty$, by Theorem 0.5.2 we have for every $n \geq 1$

$$\begin{aligned} h^\sigma_{\mu^*}(G^{-1}, \eta_1) &= h^\sigma_{\mu^*}(G^{-1}, G^n\xi \vee G^n\mathcal{P}^+) \\ &= h^\sigma_{\mu^*}(G^{-1}, \xi \vee G^n\mathcal{P}^+) \\ &= H_{\mu^*}(\xi \vee G^n\mathcal{P}^+ | G\xi \vee G^{n+1}\mathcal{P}^+ \vee \sigma) \\ &\leq H_{\mu^*}(\xi|G\xi) + H_{\mu^*}(\mathcal{P}^+|G^{-n}\xi \vee G\mathcal{P}^+). \end{aligned}$$

$H_{\mu^*}(\mathcal{P}^+|G^{-n}\xi \vee G\mathcal{P}^+) \to 0$ as $n \to +\infty$ since $G^{-n}\xi$ increasingly tends to the partition into single points and $H_{\mu^*}(\mathcal{P}^+|\xi \vee G\mathcal{P}^+) \leq H_{\mu^*}(\mathcal{P}|\sigma) < +\infty$. Hence

$$h^\sigma_{\mu^*}(G^{-1}, \eta_1) \leq H_{\mu^*}(\xi|G\xi) = h^\sigma_{\mu^*}(G^{-1}, \xi).$$

On the other hand, also by Theorem 0.5.2, we have

$$h^\sigma_{\mu^*}(G^{-1}, \eta_1) = h^\sigma_{\mu^*}(G^{-1}, \xi \vee \mathcal{P}) \geq h^\sigma_{\mu^*}(G^{-1}, \xi)$$

since $H_{\mu^*}(\eta_1 | \vee_{n=1}^{+\infty} G^n(\xi \vee \mathcal{P}) \vee \sigma) < +\infty$ and $H_{\mu^*}(\xi \vee \mathcal{P} | G\xi \vee \sigma) < +\infty$. This completes the proof. □

D. Quotient Structure of η_2/η_1

Since $\eta_1 \geq \eta_2$, for each $w \in \Omega^{\mathbf{Z}}$ and $x \in M$ we can view η_{1w} restricted to $\eta_{2w}(x)$, written $\eta_{1w}|_{\eta_{2w}(x)}$, as a subpartition of $\eta_{2w}(x)$. Let $(w,x) \in \Delta_0''$ such that $\eta_{2w}(x) \subset \Phi_{(w,x)} S_\delta^{cu}(w,x)$. Recall that for every $y \in \Phi_{(w,x)} S_\delta^{cu}(w,x)$ with $(w,y) \in \Delta_0''$, $W_{(w,x),2\delta}^u(y)$ is the graph of a function from $\bar{\mathbf{R}}^u(2\delta l(w,x)^{-1})$ to $\bar{\mathbf{R}}^{c+s}(4\delta l(w,x)^{-1})$. The restriction of these graphs to $\Phi_{(w,x)}^{-1}\eta_{2w}(x)$ gives, roughly speaking, a natural partition of $\Phi_{(w,x)}^{-1}\eta_{2w}(x)$. The next lemma says that this corresponds to $\eta_{1w}|_{\eta_{2w}(x)}$.

Lemma 5.5. *For μ^*-a.e. $(w,x) \in \Delta_0''$, if $y \in \eta_{2w}(x)$ with $(w,y) \in \Delta_0''$ and $\xi_w(y) \subset W^u(w,y)$, then*

$$\Phi_{(w,x)} W_{(w,x),2\delta}^u(y) \cap \eta_{2w}(x) = \eta_{1w}(y).$$

Proof. First consider $z \in \Phi_{(w,x)} W_{(w,x),2\delta}^u(y) \cap \eta_{2w}(x)$. We shall prove that $z \in \xi_w(y)$. Since $\mathcal{P}$ refines $\{\hat{S}, \Omega^{\mathbf{Z}} \times M \backslash \hat{S}\}$ and $z \in \mathcal{P}_w^+(y)$, in view of Remark 5.2, it suffices to show that $d^u(f_w^{-n}y, f_w^{-n}z) \leq \hat{d}$ whenever $G^{-n}(w,y) \in \hat{S}$. This is in fact true for all $n \geq 0$, since by Lemma 4.3 3) and Lemma 4.4 1) one has for all $n \geq 0$

$$\begin{aligned} \|H_{(w,x)}^{-n}\Phi_{(w,x)}^{-1}y - H_{(w,x)}^{-n}\Phi_{(w,x)}^{-1}z\| &\leq e^{-(\lambda^+ - 2\varepsilon)n}\|\Phi_{(w,x)}^{-1}y - \Phi_{(w,x)}^{-1}z\| \\ &\leq \|\Phi_{(w,x)}^{-1}y - \Phi_{(w,x)}^{-1}z\| \\ &\leq 2\delta l(w,x)^{-1} \end{aligned}$$

which implies $d^u(f_w^{-n}y, f_w^{-n}z) \leq K_0 2\delta l(w,x)^{-1} \leq \hat{d}$. Thus $\Phi_{(w,x)} W_{(w,x),2\delta}^u(y) \cap \eta_{2w}(x) \subset \eta_{1w}(y)$. The reverse containment follows from $\xi_w(y) \subset W^u(w,y)$ and Lemma 4.3 3). Noting that the above argument holds true for μ^*-a.e. (w,x), we complete the proof. □

This lemma allows us to regard the factor-space $\eta_{2w}(x)/(\eta_{1w}|_{\eta_{2w}(x)})$, or written simply $\eta_{2w}(x)/\eta_{1w}$, as a subset of $\mathbf{R}^{c+s}$ via the correspondence $\eta_{1w}(y) \leftrightarrow W_{(w,x),2\delta}^u(y) \cap (\{0\} \times \mathbf{R}^{c+s})$. If we identify $\eta_1(w,x)$ and $\eta_2(w,x)$ with $\eta_{1w}(x)$ and $\eta_{2w}(x)$ respectively, the next lemma tells then that $G|_{G^{-1}(\eta_2(w,x))}$: $G^{-1}(\eta_2(w,x)) \longrightarrow \eta_2(w,x)$ acts like a skew product with respect to the above quotient structure.

Lemma 5.6. *For μ^*-a.e. $(w,x) \in \Delta_0''$, if $(w,y) \in \eta_2(w,x)$ with $(w,y) \in \Delta_0''$, $\xi_w(y) \subset W^u(w,y)$ and $\xi_{\tau^{-1}w}(f_w^{-1}y) \subset W^u(G^{-1}(w,y))$, then we have*

$$G^{-1}(\eta_1(w,y)) = \eta_1(G^{-1}(w,y)) \cap G^{-1}(\eta_2(w,x)).$$

Proof. From the definitions of η_1 and η_2 it follows clearly that $G^{-1}(\eta_1(w,y)) \subset \eta_1(G^{-1}(w, y)) \cap G^{-1}(\eta_2(w,x))$ for every $(w,x) \in \Omega^{\mathbf{Z}} \times M$ and any $(w,y) \in \eta_2(w,x)$. On the other hand, if both (w,x) and $G^{-1}(w,x)$ meet the requirement of Lemma 5.5 and (w,y) is a point in $\eta_2(w,x)$ such that $(w,y) \in \Delta_0''$, $\xi_w(y) \subset W^u(w,y)$ and $\xi_{\tau^{-1}w}(f_w^{-1}y) \subset W^u(G^{-1}(w,y))$, then the reverse containment follows from Lemma 5.5 and Lemma 4.3 2). □

E. Transverse Metrics

As we have said at the beginning of Section 3, the first main point for the proof of Theorem 1.1 2)⇒ 1) is to prove that the entropy $h_\mu(\mathcal{X}(M,v))$ is determined by actions of $f_w^n, w \in \Omega^{\mathbf{Z}}, n \in \mathbf{Z}$ on the W^u-manifolds of $\mathcal{X}(M,v,\mu)$, or more precisely, to prove that $h_{\mu^*}^\sigma(G) = H_{\mu^*}(\xi|G\xi)$ where ξ is a certain increasing partition subordinate to W^u-manifolds of $\mathcal{X}(M,v,\mu)$. In order to use the fact that all the expansion of $\mathcal{X}(M,v,\mu)$ occurs along the W^u-manifolds to prove this assertion, we need to show that the action induced by G on $(G^{-1}(\eta_2(w,x)))/\eta_1 \to \eta_2(w,x)/\eta_1$ does not expand distances. For this purpose we define in this subsection a metric on the factor-space $\eta_2(w,x)/\eta_1$ for μ^* -a.e.(w,x). This will be referred to as a *transverse metric.*

We shall actually deal with η_1 and η_2 restricted to a certain measurable set of full μ^* measure. Now we choose a measurable set $\Delta_0''' \subset \Delta_0''$ with $\mu^*(\Delta_0''') = 1$ and $G\Delta_0''' = \Delta_0'''$ such that for each $(w,x) \in \Delta_0''', \xi_w(x) \subset W^u(w,x), \eta_{2w}(x) \subset \Phi_{(w,x)}S_\delta^{cu}(w,x)$ and the requirements of Lemmas 5.5 and 5.6 are satisfied. We then put

$$\eta_1' = \eta_1|_{\Delta_0'''},$$

$$\eta_2' = \eta_2|_{\Delta_0'''}.$$

In what follows we define a transverse metric on $\eta_2'(w,x)/\eta_1'$ for μ^*-a.e. $(w,x) \in \Delta_0'''$.

First we give a point-dependent definition. Let $(w,x) \in \Delta_0'''$. From Lemma 4.3 we know that for every $y \in \eta_{2w}'(x), W^u_{(w,x),2\delta}(y)$ intersects $\{0\} \times \mathbf{R}^{c+s}$ at exactly one point. We denote this point by ζ_y. For $(w,y),(w,y') \in \eta_2'(w,x)$, define

$$d_{(w,x)}((w,y),(w,y')) = \|\zeta_y - \zeta_{y'}\|.$$

By Lemma 5.5, $d_{(w,x)}(\ ,\ )$ induces a metric on $\eta_2'(w,x)/\eta_1'$, but in general, $d_{(w,x')}(\ ,\ ) \neq d_{(w,x)}(\ ,\ )$ for $(w,x') \in \eta_2'(w,x)$ with $(w,x') \neq (w,x)$.

Now we need to rectify this situation to give a point-independent definition. To this end we shall first specify $\hat{E}$ (see Subsection 5.C) and then choose a reference plane T and standardize all measurements with respect to T. By Proposition 1.3, there exists a compact set $\Delta \subset \{(w,x) \in \Delta_0'' : l(w,x) \leq l_0\}$ with $\mu^*(\Delta) > 0$ and meeting the following two requirements:

(i) $E^u_{(w,x)}$ and $E^{c+s}_{(w,x)} \overset{\text{def.}}{=} E^c_{(w,x)} \oplus E^s_{(w,x)}$ depend continuously on $(w,x) \in \Delta$;

(ii) There exists a number $0 < \alpha_0 < \rho_0/8$ and for each $(w,x) \in \Delta$ there exists a C^1 map

$$h_{(w,x)} : \{\xi \in E^u_{(w,x)} : |\xi| < \alpha_0\} \to E^{c+s}_{(w,x)}$$

such that $h_{(w,x)}(0) = 0$, $\mathrm{Lip}(h_{(w,x)}) \leq 1/3, D_{(w,x)} \stackrel{\text{def.}}{=} \exp_x \mathrm{Graph}(h_{(w,x)}) \subset W^u(w,x)$ and $\{D_{(w,x)}\}_{(w,x)\in\Delta}$ is a continuous family of C^1 embedded disks of dimension u.

For $(w,x) \in \Delta$ and $\rho > 0$, we put $E_{(w,x)}(\rho) = E^u_{(w,x)}(\rho) \times E^{c+s}_{(w,x)}(\rho)$ where $E^u_{(w,x)}(\rho) = \{\xi \in E^u_{(w,x)} : |\xi| < \rho\}$ and $E^{c+s}_{(w,x)}(\rho) = \{\eta \in E^{c+s}_{(w,x)} : |\eta| < \rho\}$. From the compactness of Δ it follows that there are positive numbers t_0 and s_0 with $t_0 \leq \alpha_0/2$ such that for each $(w,x) \in \Delta$ and any $(w',x') \in B_\Delta((w,x), s_0)$ the following (a) and (b) hold true:

(a) $\exp_x E_{(w,x)}(t_0) \subset \Phi_{(w,x)}\bar{\mathbf{R}}(\delta l(w,x)^{-1})$;

(b) The map $I_{(w',x'),(w,x)} \stackrel{\text{def.}}{=} \exp_{x'}^{-1} \circ \exp_x : E_{(w,x)}(\alpha_0) \to T_{x'}M$ is well defined with $\mathrm{Lip}(I_{(w',x'),(w,x)}) \leq 2$. $\exp_{x'}^{-1} x \in E_{(w',x')}(t_0)$ and $\exp_{x'}^{-1} D_{(w,x)}$ intersects $\{0\} \times E^{c+s}_{(w',x')}$ at exactly one point. Moreover,

$$\exp_{x'}^{-1} D_{(w,x)} \cap E_{(w',x')}(t_0) = \mathrm{Graph}(h_{(w',x'),(w,x)})$$

where

$$h_{(w',x'),(w,x)} : E^u_{(w',x')}(t_0) \to E^{c+s}_{(w',x')}(t_0)$$

is a C^1 map with $\mathrm{Lip}(h_{(w',x'),(w,x)}) \leq 1/2$, and

$$I_{(w',x'),(w,x)}(\{0\} \times E^{c+s}_{(w,x)}(2t_0)) \cap E_{(w',x')}(t_0) = \mathrm{Graph}(g_{(w',x'),(w,x)})$$

where $g_{(w',x'),(w,x)} : E^{c+s}_{(w',x')}(t_0) \to E^u_{(w',x')}(t_0)$ is a C^1 map with $\mathrm{Lip}(g_{(w',x'),(w,x)}) \leq 1/100$.

Choose now $(w_0,x_0) \in \Delta$ such that $B_\Delta((w_0,x_0), s_0/2)$ has positive μ^* measure. Then we define

$$\hat{E} = B_\Delta((w_0,x_0), s_0/2),$$
$$T = \exp_{x_0}(\{0\} \times E^{c+s}_{(w_0,x_0)}(2t_0)).$$

With $\hat{E}$ and T thus specified, noting that $\mathcal{P}$ is required to refine $\{\hat{E}, \Omega^{\mathbf{Z}} \times M \backslash \hat{E}\}$ (see Subsection 5.C), we define now a metric on $\eta_2'(w,x)/\eta_1'$ for every $(w,x) \in \cup_{n=0}^{+\infty} G^n \hat{E}'$ where

$$\hat{E}' = \hat{E} \cap \Delta_0'''.$$

First take an isomorphism $I_{(w_0,x_0)} : E^{c+s}_{(w_0,x_0)} \to \mathbf{R}^{c+s}$ and define a function $\pi : \cup_{n=0}^{+\infty} G^n \hat{E}' \to \mathbf{R}^{c+s}$ as follows: For $(w,x) \in \hat{E}'$, let

$$\pi(w,x) = (I_{(w_0,x_0)} \circ \exp_{x_0}^{-1})\{T \cap D_{(w,x)}\}$$

and in general, let

$$\pi(w,x) = \pi(G^{-n(w,x)}(w,x))$$

where $n(w,x)$ is the smallest nonnegative integer such that $G^{-n(w,x)}(w,x) \in \hat{E}'$. Then define for each $(w,x) \in \cup_{n=0}^{+\infty} G^n \hat{E}'$

$$d^T_{(w,x)}((w,y),(w,y')) = \|\pi(w,y) - \pi(w,y')\|$$

if $(w,y),(w,y') \in \eta_2'(w,x)$, where $\|\cdot\|$ denotes the usual Euclidean distance. We now explain why $d^T_{(w,x)}(\ ,\ )$ induces a metric on $\eta_2'(w,x)/\eta_1'$.

Let $(w,x) \in \cup_{n=0}^{+\infty} G^n \hat{E}'$. Since $\mathcal{P} \geq \{\hat{E}, \Omega^{\mathbf{Z}} \times M \backslash \hat{E}\} \vee \sigma$ and $\eta_2' = \mathcal{P}^+|_{\Delta_0'''}$, for every $n \geq 0$ either $G^{-n}(\eta_2'(w,x)) \subset \hat{E}'$ or $G^{-n}(\eta_2'(w,x)) \cap \hat{E}' = \emptyset$. Moreover, when $G^{-n}(w,x) \in \hat{E}'$, for each $(w,y) \in \eta_2'(w,x)$ one can inductively prove by using Lemma 5.6

$$G^{-n}(\eta_1'(w,y)) = \eta_1'(G^{-n}(w,y)) \cap G^{-n}(\eta_2'(w,x))$$

and, using Lemma 5.5, one can easily obtain

$$\eta_1'(G^{-n}(w,y)) = (\{\tau^{-n}w\} \times D_{G^{-n}(w,y)}) \cap \eta_2'(G^{-n}(w,x)),$$

hence

$$G^{-n}(\eta_1'(w,y)) = (\{\tau^{-n}w\} \times D_{G^{-n}(w,y)}) \cap G^{-n}(\eta_2'(w,x)).$$

This guarantees that $d^T_{(w,x)}(\ ,\ )$ induces a genuine metric on $\eta_2'(w,x)/\eta_1'$ and that for any $(w,x') \in \eta_2'(w,x)$, $d^T_{(w,x)}(\ ,\ ) = d^T_{(w,x')}(\ ,\ )$.

Lemma 5.7. *Let $\hat{E}'$ and T be as introduced above. Then there is a number $N = N(l_0)$ such that for all $(w,x) \in \hat{E}'$,*

$$\frac{1}{N} d_{(w,x)}(\ ,\ ) \leq d^T_{(w,x)}(\ ,\ ) \leq N d_{(w,x)}(\ ,\ ).$$

Proof. Let $(w,x) \in \hat{E}'$. We define the Poincaré map

$$\theta : \{(\{0\} \times E^{c+s}_{(w,x)}) \cap \mathrm{Graph}(h_{(w,x),(w,x')}) : x' \in \eta_{2w}'(x)\} \to \exp_x^{-1} T$$

by sliding along $\mathrm{Graph}(h_{(w,x),(w,x')})$. Lemma 4.6 tells us that there is a number $D' = D'(\lambda_0, K_0, \lambda^+, \varepsilon)$ such that

$$\max\{\mathrm{Lip}(\theta), \mathrm{Lip}(\theta^{-1})\} \leq D' l(w,x)^2$$

where θ^{-1} is understood to be defined on the image of θ. Thus, if $(w,y),(w,y') \in \eta_2'(w,x)$ and $\zeta_{(w,y)}$ and $\zeta_{(w,y')}$ are respectively the points of intersection of $\mathrm{Graph}(h_{(w,x),(w,y)})$ and $\mathrm{Graph}(h_{(w,x),(w,y')})$ with $\exp_x^{-1} T$, then

$$d_{\exp_x^{-1} T}(\zeta_{(w,y)}, \zeta_{(w,y')}) \leq K_0 b(\rho_0/2) D' l_0^2 d_{(w,x)}((w,y),(w,y'))$$

where $d_{\exp_x^{-1}T}(\quad,\quad)$ is the distance along the submanifold $\exp_x^{-1} T$. Therefore we have

$$\begin{aligned} d^T_{(w,x)}((w,y),(w,y')) &= \|\pi(w,y) - \pi(w,y')\| \\ &= |I_{(w_0,x_0),(w,x)}\zeta_{(w,y)} - I_{(w_0,x_0),(w,x)}\zeta_{(w,y')}| \\ &\le 2K_0 b(\rho_0/2) D' l_0^2 d_{(w,x)}((w,y),(w,y')). \end{aligned}$$

The other inequality is proved similarly. □

As is evident from the proof, the number N depends only on the charts and on l_0. It is independent of η_1 and η_2, or the choice of $\hat{E}$ and T (provided of course that everything is as described before).

Finally, what we have done in the last two subsections is, roughly speaking, to present $(\Omega^{\mathbf{Z}} \times M)/\eta_1$ as a subset of $(\Omega^{\mathbf{Z}} \times M)/\eta_2 \times \mathbf{R}^{c+s}$, and to define transverse metrics on $\eta_2(w,x)/\eta_1$ that correspond to the Euclidean distance on $\mathbf{R}^{c+s}$. This Euclidean space geometry plays a role in some of the averaging arguments in the next section.

§6 Some Consequences of Besicovitch's Covering Theorem

For $x \in \mathbf{R}^n$, let $B(x,r)$ denote the ball of radius r centered at x and let $\bar{B}(x,r)$ denote the associated closed ball. All distances are the usual Euclidean ones in this section. The covering theorem of Besicovitch (see [Guz]) that follows is a valuable tool in the theory of differentiation and in many other fields of analysis.

Theorem. (Besicovitch's Covering Theorem (BCT)) *Let A be a bounded subset of $\mathbf{R}^n$. For each $x \in A$ a closed ball $\bar{B}(x, r(x))$ with center x and radius $r(x)$ is given. Put $\mathcal{A} = \{\bar{B}(x,r(x))\}_{x\in A}$. Then there exists a subset $\mathcal{A}'$ of $\mathcal{A}$ such that $\mathcal{A}'$ covers A and no point in $\mathbf{R}^n$ lies in more than $c(n)$ elements of $\mathcal{A}'$, $c(n)$ depending only on n.*

Remark 6.1. If in BCT A is not bounded but

$$\sup\{r(x) : x \in A\} = R < +\infty,$$

the above covering theorem is still valid with the constant $c(n)$ changing conveniently. To show this , it is sufficient to partition $\mathbf{R}^n$ into disjoint sets $A_i \overset{\text{def}}{=} \{x \in \mathbf{R}^n : 3iR \le \|x\| < 3(i+1)R\}$, $i \in \mathbf{Z}^+$ and apply BCT to the intersection of A with each one of these sets A_i.

In what follows we derive some useful lemmas from BCT. Now let m be a Borel probability measure on $\mathbf{R}^n$. The next two lemmas are standard when m is Lebesgue. When working with arbitrary finite Borel measures, we use BCT instead of Vitali's covering theorem. Let $g \in L^1(\mathbf{R}^n, m)$ and define for each $\delta > 0$

$$g_\delta(x) = \frac{1}{m(B(x,\delta))} \int_{B(x,\delta)} g dm,$$

$x \in \mathbf{R}^n$ (we admit here that $0/0 = 1$). If $g \geq 0$ m almost everywhere, we further define

$$g^* = \sup_{\delta > 0} g_\delta$$

and

$$g_* = \inf_{\delta > 0} g_\delta.$$

g^* and g_* are Borel measurable functions since for each $t \in \mathbf{R}$ the sets $\{x : g^*(x) > t\}$ and $\{x : g_*(x) < t\}$ are open.

Lemma 6.1. *1) For any $t \geq 0$,*

$$m(\{g^* > t\}) \leq \frac{c(n)}{t} \int g dm;$$

2) Let ν be defined by $d\nu = g dm$. Then for any $t \geq 0$,

$$\nu(\{g_* < t\}) \leq c(n) t.$$

Proof. Let $A = \{g^* > t\} \cap C$ where C is an arbitrarily fixed bounded Borel set. For each $x \in A$ we can choose $\delta(x)$ such that $\int_{\bar{B}(x,\delta(x))} g dm > t m(\bar{B}(x,\delta(x)))$. Letting $\mathcal{A} = \{\bar{B}(x,\delta(x))\}_{x \in A}$ and choosing $\mathcal{A}'$ as in BCT, we have

$$\begin{aligned} m(A) &\leq \sum_{B \in \mathcal{A}'} m(B) \\ &\leq \sum_{B \in \mathcal{A}'} \frac{1}{t} \int_B g dm \leq \frac{c(n)}{t} \int_{\mathbf{R}^n} g dm \end{aligned}$$

which together with the arbitrariness of C proves 1). Part 2) is proved similarly. □

Lemma 6.2. *Let $g \in L^1(\mathbf{R}^n, m)$. Then $g_\delta \to g$ m almost everywhere as $\delta \to 0$.*

Proof. It is sufficient to verify that for each $t > 0$, the set

$$A_t = \left\{ x \in \mathbf{R}^n : \limsup_{\delta \to 0} \left| \frac{1}{m(B(x,\delta))} \int_{B(x,\delta)} g dm - g(x) \right| > t \right\}$$

is of m measure zero. We prove $m(A_t) = 0$ in the following way. Given $\varepsilon > 0$, we take a continuous function f such that $h \overset{\text{def.}}{=} g - f$ satisfies $||h||_1 \leq \varepsilon$ where $||h||_1$

is the L^1 -norm of h in $L^1(\mathbf{R}^n, m)$. For f we have obviously at m-a.e. $x \in \mathbf{R}^n$

$$\lim_{\delta \to 0} \frac{1}{m(B(x,\delta))} \int_{B(x,\delta)} f dm = f(x),$$

and so $m(A_t) = m(B_t)$ where

$$B_t = \left\{ x \in \mathbf{R}^n : \limsup_{\delta \to 0} \left| \frac{1}{m(B(x,\delta))} \int_{B(x,\delta)} h dm - h(x) \right| > t \right\}.$$

Notice that

$$B_t \subset \left\{ x \in \mathbf{R}^n : \limsup_{\delta \to 0} \left| \frac{1}{m(B(x,\delta))} \int_{B(x,\delta)} h dm \right| > \frac{t}{2} \right\} \cup \left\{ x \in \mathbf{R}^n : |h(x)| > \frac{t}{2} \right\}$$

$$\stackrel{\text{def.}}{=} B_t^1 \cup B_t^2.$$

Since for each $x \in \mathbf{R}^n$

$$\limsup_{\delta \to 0} \left| \frac{1}{m(B(x,\delta))} \int_{B(x,\delta)} h dm \right| \le |h|^*(x)$$

where $|h|^*$ is defined analogously to g^*, by Lemma 6.1 1) we have

$$m(B_t^1) \le m(\{|h|^* > \frac{t}{2}\}) \le \frac{2c(n)}{t} ||h||_1 \le \frac{2c(n)}{t} \varepsilon.$$

Also

$$m(B_t^2) \le \frac{2}{t} ||h||_1 \le \frac{2}{t} \varepsilon.$$

Since $\varepsilon > 0$ is arbitrary, we obtain $m(A_t) = 0$. □

The next lemma (see also [Led]$_2$) is usually stated in a slightly different way in the literature. For geometric reasons we average over balls instead of taking conditional expectations with respect to fixed partitions.

Lemma 6.3. *Suppose that $(X, \mathcal{B}, m)$ is a Lebesgue space and $\pi : X \to \mathbf{R}^n$ is a measurable map. Let $\{m_\xi\}_{\xi \in \mathbf{R}^n}$ be a canonical system of conditional measures of m associated with the partition $\{\pi^{-1}\{\xi\}\}_{\xi \in \mathbf{R}^n}$. Let α be a measurable partition of X with $H_m(\alpha) < +\infty$. For $\xi \in \mathbf{R}^n$ and $A \in \alpha$, define*

$$g^A(\xi) = m_\xi(A).$$

Let g_δ^A and g_*^A be functions on $\mathbf{R}^n$ defined as above. Define g, g_δ and $g_* : X \to \mathbf{R}$ by

$$g(x) = \sum_{A\in\alpha} \chi_A(x) g^A(\pi x),$$
$$g_\delta(x) = \sum_{A\in\alpha} \chi_A(x) g_\delta^A(\pi x),$$
$$g_*(x) = \sum_{A\in\alpha} \chi_A(x) g_*^A(\pi x).$$

Then $g_\delta \to g$ m almost everywhere on X and

$$\int -\log g_* dm \leq H_m(\alpha) + \log c(n) + 1$$

where $c(n)$ is as in BCT.

Proof. First by Lemma 6.2 we have $g_\delta^A \to g^A$ πm -a.e. on $\mathbf{R}^n$ as $\delta \to 0$ for each $A \in \alpha$ and hence $g_\delta \to g$ m-a.e. on X since $H_m(\alpha) < +\infty$. Note also that the function $h : \mathbf{R}^+ \to \mathbf{R}^+, s \mapsto m(\{-\log g_* > s\})$ is continuous almost everywhere (in the sense of Lebesgue) and hence is Riemann integrable on any interval $[0, b], b > 0$. From this it follows that

$$\begin{aligned}\int -\log g_* dm &= \int_0^{+\infty} m(\{-\log g_* > s\}) ds \\ &= \int_0^{+\infty} \sum_{A\in\alpha} m(A \cap \{g_*^A \circ \pi < e^{-s}\}) ds.\end{aligned}$$

Now for each $A \in \alpha$

$$m(A \cap \{g_*^A \circ \pi < e^{-s}\}) \leq m(A)$$

and

$$\begin{aligned}m(A \cap \{g_*^A \circ \pi < e^{-s}\}) &= \int \chi_A \chi_{\{g_*^A\circ\pi < e^{-s}\}} dm \\ &= \int E_m(\chi_A \chi_{\{g_*^A\circ\pi<e^{-s}\}} | \mathcal{B}(\{\pi^{-1}\{\xi\}\}_{\xi\in\mathbf{R}^n})) dm \\ &= \int (g^A \circ \pi) \chi_{\{g_*^A\circ\pi<e^{-s}\}} dm \\ &\leq c(n) e^{-s},\end{aligned}$$

the last estimate follows from Lemma 6.1 2). Thus

$$\begin{aligned}\int -\log g_* dm &\leq \sum_{A\in\alpha} \int_0^{+\infty} \min\{m(A), c(n)e^{-s}\} ds \\ &\leq H_m(\alpha) + \log c(n) + 1\end{aligned}$$

by a simple calculation. □

Another consequence of BCT is the following result.

Lemma 6.4. *Let m be a finite Borel measure on $\mathbf{R}^n$. Then*

$$\inf_{0<\varepsilon\leq 1} \frac{m(B(x,\varepsilon))}{\varepsilon^n} > 0$$

for m-a.e. $x \in \mathbf{R}^n$. In particular,

$$\limsup_{\varepsilon\to 0} \frac{\log m(B(x,\varepsilon))}{\log \varepsilon} \leq n$$

for m-a.e. $x \in \mathbf{R}^n$.

Proof. Let N be a positive integer. Put

$$A_N = \left\{ x \in \mathbf{R}^n : \|x\| \leq N \quad \text{and} \quad \inf_{0<\varepsilon\leq 1} \frac{m(B(x,\varepsilon))}{\varepsilon^n} = 0 \right\}.$$

It is sufficient to prove that $m(A_N) = 0$. Let now $\delta > 0$ be given arbitrarily. For each $x \in A_N$ there exists a number $0 < \varepsilon(x) \leq 1$ such that $m(\bar{B}(x,\varepsilon(x))) \leq \delta\varepsilon(x)^n$. Letting $\mathcal{A} = \{\bar{B}(x,\varepsilon(x))\}_{x\in A_N}$ and choosing $\mathcal{A}' \subset \mathcal{A}$ as in BCT, we have

$$\begin{aligned} m(A_N) &\leq \sum_{B\in\mathcal{A}'} m(B) \\ &\leq \sum_{B\in\mathcal{A}'} \delta n^n \lambda(B) \leq \delta n^n c(n)\lambda(\bar{B}(0,N+1)) \end{aligned}$$

where λ is the Lebesgue measure on $\mathbf{R}^n$. Since δ is arbitrary, we obtain $m(A_N) = 0$. □

§7 The Main Proposition

Using the machinary developed in Sections 3-6 we can now complete the first step of the proof of Theorem 1.1 2) $\Rightarrow$ 1), i.e. we can now prove that $h^{\sigma}_{\mu^*}(G^{-1})$ is equal to the σ-conditional entropies of G^{-1} with respect to certain partitions subordinate to W^u-manifolds of $\mathcal{X}(M,v,\mu)$.

Proposition 7.1. *Let $\mathcal{X}(M,v,\mu)$ be given with μ being ergodic. Then for any $\beta > 0$, there exists a measurable partition ξ_β of $\Omega^{\mathbf{Z}} \times M$ and of the type as constructed in the proof of Proposition 5.2 such that*

$$\beta(c+s) \geq (1-\beta)[h^{\sigma}_{\mu^*}(G^{-1}) - h^{\sigma}_{\mu^*}(G^{-1},\xi_\beta) - \beta].$$

Proof. The strategy is to construct ξ_β as in Subsection 5.B and to use it to construct η_1 and η_2 as in Subsection 5.C with $h^\sigma_{\mu^*}(G^{-1}, \eta_2) \geq h^\sigma_{\mu^*}(G^{-1}) - \beta/3$. Let $\{\mu^*_{\eta_1(w,x)}\}$ and $\{\mu^*_{\eta_2(w,x)}\}$ be respectively (μ^*-mod 0 unique) canonical systems of conditional measures of μ^* associated with η_1 and η_2 and denote them respectively by $\{\mu^1_{(w,x)}\}$ and $\{\mu^2_{(w,x)}\}$ for simplicity of notations. We shall prove that, if $B^T((w,x),\rho) = \{(w,y) \in \eta_2(w,x) : d^T_{(w,x)}((w,x),(w,y)) < \rho\}$, then

$$\beta \cdot \limsup_{\rho \to 0} \frac{\log \mu^2_{(w,x)} B^T((w,x),\rho)}{\log \rho} \geq (1-\beta)[h^\sigma_{\mu^*}(G^{-1},\eta_2) - h^\sigma_{\mu^*}(G^{-1},\eta_1) - 2\beta/3]$$

for μ^* -a.e. (w,x). The desired conclusion then follows immediately from this and Lemmas 5.4 and 6.4.

We divide the proof into five parts.

(A) We start by enumerating the specifications on ξ_β, η_1 and η_2. First fix $\varepsilon > 0$ such that $\varepsilon < \min\{\beta/3, \lambda^+/100m_0, -\lambda^-/100m_0\}$ and $e^{-\lambda^+ + 10\varepsilon} + e^{5\varepsilon} < 2$(see Lemma 4.6). Let $\{\Phi_{(w,x)}\}_{(w,x)\in\Delta''_0}$ be a system of (ε, l) -charts as described in Section 3. Using these charts, we construct an increasing measurable partition ξ_β as in the proof of Proposition 5.2 with $l_0, \hat{S}$ and $\hat{d}$ having the same meaning as in that proof. Let $\delta_0 = \min\{(\frac{1}{4}e^{-(\lambda_0+\varepsilon)})^2, \hat{d}/2K_0\}$. Choose $\hat{E}$ and T as in Subsection 5.E. We assume that $e^{-\beta\varepsilon} N^{4\mu^*(\hat{E})} < 1$ where $N = N(l_0)$ is the number introduced in Lemma 5.7. Now we take a measurable partition $\mathcal{P}$ of $\Omega^{\mathbf{Z}} \times M$ adapted to $(\{\Phi_{(w,x)}\}_{(w,x)\in\Delta''_0}, \delta_0)$ such that $H_{\mu^*}(\mathcal{P}|\sigma) < +\infty, \mathcal{P} \geq \{\hat{S}, \Omega^{\mathbf{Z}} \times M\backslash\hat{S}\} \vee \{\hat{E}, \Omega^{\mathbf{Z}} \times M\backslash\hat{E}\} \vee \sigma$ and $h^\sigma_{\mu^*}(G^{-1},\mathcal{P}) \geq h^\sigma_{\mu^*}(G^{-1}) - \varepsilon$. Then we set $\eta_1 = \xi_\beta \vee \mathcal{P}^+$ and $\eta_2 = \mathcal{P}^+$. Let Δ'''_0 be a set as chosen in Subsection 5.E. Recalling that $\mu^*(\Delta'''_0) = 1$ and $G\Delta'''_0 = \Delta'''_0$, for the sake of presentation we may assume that $\Delta'''_0 = \Omega^{\mathbf{Z}} \times M$ since otherwise the discussions below also apply to the system $G : (\Delta'''_0, \mu^*) \hookleftarrow$ and lead to the same conclusion. With η_1 and η_2 so constructed, $\eta_2(w,x)/\eta_1$ has then a nice quotient structure endowed with a transverse metric $d^T_{(w,x)}(\ ,\ )$ for μ^*-a.e. (w,x).

(B) Before proceeding with the main argument, we record some estimates derived from the results of Section 6. For $\delta > 0$, define g, g_δ and $g_* : \Omega^{\mathbf{Z}} \times M \to \mathbf{R}$ by

$$g(w,y) = \mu^1_{(w,y)}(G^{-1}\eta_2)(w,y),$$

$$g_\delta(w,y) = \frac{1}{\mu^2_{(w,y)}B^T((w,y),\delta)} \int_{B^T((w,y),\delta)} \mu^1_{(w,z)}(G^{-1}\eta_2)(w,y)d\mu^2_{(w,y)}(w,z)$$

$$g_*(w,y) = \inf_{\delta\in Q} g_\delta(w,y)$$

where $Q = \{e^{-\beta l}N^{2j} : l,j \in \mathbf{Z}^+\}$. By Lemma 5.6 we know that $g(w,y)$ is also equal to $\mu^1_{(w,y)}(G^{-1}\eta_1)(w,y)$ for μ^*-a.e. (w,y). For each $\delta > 0$, one can check that the functions $(w,y) \mapsto \mu^2_{(w,y)}B^T((w,y),\delta)$ and $(w,y) \mapsto \mu^*_{(G^{-1}\eta_2)(w,y)}B^T($ $(w,y),\delta)$ are measurable and $\mu^2_{(w,y)}B^T((w,y),\delta) > 0$ for μ^*-a.e. (w,y). Since

$H_{\mu^*}(G^{-1}\eta_2|\eta_2) < +\infty$, for μ^*-a.e. (w, y) one has $\mu^2_{(w,y)}(G^{-1}\eta_2)(w, y) > 0$ and hence

$$g_\delta(w, y) = \frac{\mu^2_{(w,y)}(B^T((w, y), \delta) \cap (G^{-1}\eta_2)(w, y))}{\mu^2_{(w,y)}B^T((w, y), \delta)}$$
$$= \frac{\mu^*_{(G^{-1}\eta_2)(w,y)}B^T((w, y), \delta)}{\mu^2_{(w,y)}B^T((w, y), \delta)} \cdot \mu^2_{(w,y)}(G^{-1}\eta_2)(w, y).$$

g_δ is therefore measurable for each fixed $\delta > 0$. The measurability of g_* is obvious.

We claim that $g_\delta \to g$ μ^* -a.e. on $\Omega^{\mathbf{Z}} \times M$ when $\delta \in Q$ and $\delta \to 0$ and that $\int -\log g_* d\mu^* < +\infty$. To see this, first consider one element of η_2 at a time. Fix (w, x). Substitute $(\eta_2(w, x), \mu^2_{(w,x)})$ for (X, m) in Lemma 6.3, let $\pi : \eta_2(w, x) \to \mathbf{R}^{c+s}$ be the π defined in Subsection 5.E and let $\alpha = (G^{-1}\eta_2)|_{\eta_2(w,x)}$. Then we can conclude that $g_\delta \to g$ $\mu^2_{(w,x)}$ -a.e. as $\delta \in Q$ and $\delta \to 0$ and that $\int -\log g_* d\mu^2_{(w,x)} \leq \int -\log(\inf_{\delta>0} g_\delta(w, y))d\mu^2_{(w,x)}(y) \leq H_{\mu^2_{(w,x)}}(G^{-1}\eta_2)+\log c(n)+1$. Integrating over $\Omega^{\mathbf{Z}} \times M$, this gives $\int -\log g_* d\mu^* \leq H_{\mu^*}(G^{-1}\eta_2|\eta_2)+\log c(n)+1 < +\infty$.

(C) The purpose of this step is to study the induced action of G on $G^{-1}(\eta_2(w, x))/\eta_1 \to \eta_2(w, x)/\eta_1$ with respect to the metrics $d^T_{G^{-1}(w,x)}(\ ,\ )$ and $d^T_{(w,x)}(\ ,\ )$. Consider $(w, x) \in \Omega^{\mathbf{Z}} \times M$. The point (w, x) will be subjected to a finite number of a.e. assumptions. Let $r_0 < r_1 < r_2 < \cdots$ be the successive times t when $G^t(w, x) \in \hat{E}$ with $r_0 \leq 0 < r_1$. Note that r_0 is constant on $\eta_2(w, x)$. For large n and $0 \leq k < n$, define $a((w, x), k)$ as follows: If $r_j \leq k < r_{j+1}$, then

$$a((w, x), k) = B^T(G^k(w, x), e^{-\beta(n-r_j)}N^{2j}).$$

We now claim that

$$a((w, x), k) \cap (G^{-1}\eta_2)(G^k(w, x)) \subset G^{-1}a((w, x), k + 1). \tag{7.1}$$

In fact, if $k \neq r_j - 1$ for any j, then we have $Ga((w, x), k) \cap \eta_2(G^{k+1}(w, x)) = a((w, x), k + 1)$ automatically since $d^T_{G^k(w,x)}(\ ,\ )$ and $d^T_{G^{k+1}(w,x)}(\ ,\ )$ are defined by pulling back to $\hat{E}$. The case when $k = r_j - 1$ for some j reduces to the following consideration : Let $(w, y) \in \hat{E}$ and let $r > 0$ be the smallest integer such that $G^r(w, y) \in \hat{E}$. Let $(w, z) \in (G^{-r}\eta_2)(w, y)$. It suffices to show that

$$d^T_{G^r(w,y)}(G^r(w, y), G^r(w, z)) \leq N^2 e^{r\beta} d^T_{(w,y)}((w, y), (w, z)).$$

First we have $d_{(w,y)}((w, y), (w, z)) \leq N d^T_{(w,y)}((w, y), (w, z))$ (for the definition of
$d_{(w,y)}(\ ,\ )$ see Subsection 5.E). Then for $i = 1, 2, \cdots, r$, Lemma 4.5 tells us that $d_{G^i(w,y)}(G^i(w, y), G^i(w, z)) \leq e^{\beta i} d_{(w,y)}((w, y), (w, z))$. We pick up another factor of N when converting back to the d^T-metric at $G^r(w, y)$. What we claimed above is thus proved.

(D) It is easy to see that there exists a Borel set $\Delta \subset \Omega^{\mathbf{Z}} \times M$ with $\mu^*(\Delta) = 1$ and $G\Delta = \Delta$ such that, if $(w,x) \in \Delta$, then $\mu^2_{(w,x)} B^T((w,x),\delta) > 0$ for all $\delta \in Q$. We now estimate $\mu^2_{(w,x)} B^T((w,x), e^{-\beta(n-r_0(w,x))}) = \mu^2_{(w,x)} a((w,x),0)$ for $(w,x) \in \Delta$ which will be subjected to a finite number of a.e. assumptions. Write

$$\begin{aligned}
&\mu^2_{(w,x)} a((w,x),0) \\
&= \prod_{k=0}^{p-1} \frac{\mu^2_{G^k(w,x)} a((w,x),k)}{\mu^2_{G^{k+1}(w,x)} a((w,x),k+1)} \cdot \mu^2_{G^p(w,x)} a((w,x),p)
\end{aligned}$$

where $p = [n(1-\varepsilon)]$. First note that the last term ≤ 1. For each $0 \leq k < p$,

$$\frac{\mu^2_{G^k(w,x)} a((w,x),k)}{\mu^2_{G^{k+1}(w,x)} a((w,x),k+1)} = \mu^2_{G^k(w,x)} a((w,x),k) \cdot \frac{\mu^2_{G^k(w,x)} G^{-1}(\eta_2(G^{k+1}(w,x)))}{\mu^2_{G^k(w,x)} G^{-1}(a((w,x),k+1))}$$

by the G-invariance of μ^* and by uniqueness of conditional measures. This is

$$\leq \frac{\mu^2_{G^k(w,x)} a((w,x),k)}{\mu^2_{G^k(w,x)}((G^{-1}\eta_2)(G^k(w,x)) \cap a((w,x),k))} \cdot \mu^2_{G^k(w,x)}(G^{-1}\eta_2)(G^k(w,x)) \tag{7.2}$$

by (7.1). If g_δ is defined as in (B), the first quotient in (7.2) is equal to

$$[g_{\delta((w,x),n,k)}(G^k(w,x))]^{-1}$$

where

$$\delta((w,x),n,k) = e^{-\beta(n-r_j(w,x))} N^{2j}$$

and

$$j = \#\{0 < i \leq k : G^i(w,x) \in \hat{E}\}.$$

Write $I(w,x) = -\log \mu^2_{(w,x)}(G^{-1}\eta_2)(w,x)$. Then the second term in (7.2) is equal to $e^{-I(G^k(w,x))}$. Thus

$$\begin{aligned}
&\log \mu^2_{(w,x)} B^T((w,x), e^{-\beta(n-r_0(w,x))}) \\
&\leq -\sum_{k=0}^{p-1} \log g_{\delta((w,x),n,k)}(G^k(w,x)) - \sum_{k=0}^{p-1} I(G^k(w,x)).
\end{aligned}$$

Multiplying by $-1/n$ and taking liminf on both sides of this inequality, we have

$$\begin{aligned}
&\beta \cdot \limsup_{\rho \to 0} \frac{\log \mu^2_{(w,x)} B^T((w,x),\rho)}{\log \rho} \\
&\geq \beta \cdot \liminf_{n \to +\infty} \frac{\log \mu^2_{(w,x)} B^T((w,x), e^{-\beta(n-r_0(w,x))})}{\log e^{-\beta n}} \\
&\geq \liminf_{n \to +\infty} \frac{1}{n} \sum_{n=0}^{[n(1-\varepsilon)]} \log g_{\delta((w,x),n,k)}(G^k(w,x)) + \lim_{n \to +\infty} \frac{1}{n} \sum_{n=0}^{[n(1-\varepsilon)]} I(G^k(w,x)).
\end{aligned}$$

The last limit$= (1-\varepsilon)H_{\mu^*}(G^{-1}\eta_2|\eta_2) \geq (1-\varepsilon)(h^\sigma_{\mu^*}(G^{-1})-\varepsilon)$. Thus Proposition 7.1 is proved if we show that

$$\limsup_{n\to+\infty} \frac{1}{n} \sum_{n=0}^{[n(1-\varepsilon)]} \log g_{\delta((w,x),n,k)}(G^k(w,x)) \leq (1-\varepsilon)(h^\sigma_{\mu^*}(G^{-1},\eta_1)+2\varepsilon). \quad (7.3)$$

(E) We now prove this last assertion (7.3). It follows from (B) that there is a measurable function $\delta : \Omega^{\mathbf{Z}} \times M \to \mathbf{R}^+$ such that for μ^*-a.e.(w,x), if $\delta \in Q$ and $\delta \leq \delta(w,x)$, then $-\log g_\delta(w,x) \leq -\log g(w,x) + \varepsilon$. Also, since $\int -\log g_* d\mu^* < +\infty$, there is a number δ_1 such that if $A = \{(w,x) : \delta(w,x) > \delta_1\}$ then $\int_{\Omega^{\mathbf{Z}}\times M\setminus A} -\log g_* d\mu^* \leq \varepsilon$.

We claim that for μ^*-a.e. (w,x), if n is sufficiently large, then $\delta((w,x),n,k) \leq \delta_1$ for all $k \leq n(1-\varepsilon)$. First, by Birkhoff ergodic theorem, there is a positive integer $N(w,x)$ such that for $n \geq N(w,x)$, $\#\{i : 0 \leq i < n, G^i(w,x) \in \hat{E}\} \leq 2n\mu^*(\hat{E})$. If $n \geq N(w,x)$, then for each $k \leq n(1-\varepsilon)$

$$\begin{aligned} \delta((w,x),n,k) &= e^{-\beta(n-r_j(w,x))}N^{2j} \\ &\leq e^{-\beta\varepsilon n}N^{2\cdot 2n\mu^*(\hat{E})}. \end{aligned}$$

Since $e^{-\beta\varepsilon}N^{4\mu^*(\hat{E})} < 1$, $\delta((w,x),n,k)$ is less than δ_1 for sufficiently large n. Thus

$$\begin{aligned} &\sum_{n=0}^{[n(1-\varepsilon)]} -\log g_{\delta((w,x),n,k)}(G^k(w,x)) \\ &\leq \sum_{\substack{k=0 \\ G^k(w,x)\in A}}^{[n(1-\varepsilon)]} (-\log g(G^k(w,x)) + \varepsilon) + \sum_{\substack{k=0 \\ G^k(w,x)\notin A}}^{[n(1-\varepsilon)]} -\log g_*(G^k(w,x)) \end{aligned}$$

and the limsup we wish to estimate in (7.3) is bounded above by

$$(1-\varepsilon)\left[\int -\log g d\mu^* + \varepsilon + \int_{\Omega^{\mathbf{Z}}\times M\setminus A} -\log g_* d\mu^*\right].$$

Recalling that $g(w,x) = \mu^1_{(w,x)}(G^{-1}\eta_1)(w,x)$ for μ^*-a.e.(w,x), we have $\int -\log g \, d\mu^* = h^\sigma_{\mu^*}(G^{-1},\eta_1)$ which is equal to $h^\sigma_{\mu^*}(G^{-1},\xi_\beta)$ by Lemma 5.4. This completes the proof. □

Corollary 7.1. *Let $\mathcal{X}(M,v,\mu)$ be given with μ being ergodic. Then for any partition ξ of the type as constructed in the proof of Proposition 5.2, we have*

$$h^\sigma_{\mu^*}(G^{-1},\xi) = h^\sigma_{\mu^*}(G^{-1}).$$

Proof. For any $\beta > 0$, by Lemma 5.3 we have $h^\sigma_{\mu^*}(G^{-1},\xi) = h^\sigma_{\mu^*}(G^{-1},\xi_\beta)$ where ξ_β is as in Proposition 7.1. Letting $\beta \to 0$, we obtain the desired conclusion. □

§8 SBR Sample Measures: Necessity For Entropy Formula

In this section we complete the proof of Theorem 1.1 2) $\Rightarrow$ 1).

A. Proof of Theorem 1.1 2) $\Rightarrow$ 1): The Ergodic Case

We have indicated that Theorem 1.1 2) $\Rightarrow$ 1) is completely trivial if $u = 0$. We now assume that $u > 0$. Let ξ be a partition of $\Omega^{\mathbf{Z}} \times M$ subordinate to W^u-manifolds of $\mathcal{X}(M, v, \mu)$, as constructed in the proof of Proposition 5.2 with the associated ε satisfying $e^{-\lambda^+ + 10\varepsilon} + e^{5\varepsilon} < 2$. By Corollary 7.1, $h^\sigma_{\mu^*}(G^{-1}, \xi) = H_{\mu^*}(\xi|G\xi) = h^\sigma_{\mu^*}(G^{-1}) = h^\sigma_{\mu^*}(G)$. Let $\{\mu^*_{\xi(w,x)}\}$ be a (essentially unique) canonical system of conditional measures of μ^* associated with ξ and let $\lambda^u_{(w,x)}$ be the Lebesgue measure on $W^u(w,x)$. We shall prove that

$$H_{\mu^*}(\xi|G\xi) = \sum_{\lambda^{(i)}>0} \lambda^{(i)} m_i \Rightarrow \mu^*_{\xi(w,x)} << \lambda^u_{(w,x)} \quad \text{for } \mu^* - \text{a.e.}(w,x)$$

where $\xi(w,x)$ is identified with $\xi_w(x)$. The idea of the proof is as follows. Put $J^u(w,x) = |\det(T_x f_0(w)|_{E^u_{(w,x)}})|$ for μ^*-a.e.$(w,x) \in \Omega^{\mathbf{Z}} \times M$. Then by Oseledec multiplicative ergodic theorem, $\int \log J^u d\mu^* = \sum_{\lambda^{(i)}>0} \lambda^{(i)} m_i$. Suppose we know that $\mu^*_{\xi(w,x)} << \lambda^u_{(w,x)}$ for μ^*-a.e.(w,x). Then $d\mu^*_{\xi(w,x)} = \rho d\lambda^u_{(w,x)}$, μ^*-a.e.(w,x) for some function $\rho : \Omega^{\mathbf{Z}} \times M \to \mathbf{R}^+$. This function must satisfy for μ^*-a.e.(w,x)

$$\int_{\xi_w(x)} \rho(w,y) d\lambda^u_{(w,x)}(y) = 1$$

and on $\xi_w(x)$

$$\rho(w,y) = \frac{1}{\mu^*_{\xi(w,x)}((G^{-1}\xi)(w,x))} \cdot \frac{\rho(G^{-1}(w,y))}{J^u(G^{-1}(w,y))}, \ \lambda^u_{(w,x)}\text{-a.e.} y$$

by the formula for change of variables (see the proof of Claim 2.1). From this one can guess that for μ^*-a.e.(w,x), if $(w,y) \in \xi(w,x)$,

$$\Delta((w,x),(w,y)) \overset{\text{def.}}{=} \frac{\rho(w,y)}{\rho(w,x)} = \prod_{k=1}^{+\infty} \frac{J^u(G^{-k}(w,x))}{J^u(G^{-k}(w,y))}.$$

A candidate for ρ is then

$$\rho(w,y) = \frac{\Delta((w,x),(w,y))}{L(w,x)}$$

if $(w,y) \in \xi(w,x)$, where

$$L(w,x) = \int_{\xi_w(x)} \Delta((w,x),(w,y)) d\lambda^u_{(w,x)}(y).$$

In the sequel we prove rigorously that all this makes sense.

Lemma 8.1. *For μ^*-a.e. (w,x), $y \mapsto \log \Delta((w,x),(w,y))$ is a well-defined Lipschitz function on $\Phi_{(w,x)} W^u_{(w,x),\hat{\delta}}(x)$ where $\hat{\delta} = \frac{1}{4}e^{-(\lambda_0+\varepsilon)}$. It follows from this that for μ^*-a.e. (w,x), $y \mapsto \Delta((w,x),(w,y))$ is a well-defined function on $\xi_w(x)$ and is uniformly bounded away from 0 and $+\infty$ on $\xi_w(x)$.*

Proof. First, by Condition (1.1) there is a Borel set $\Gamma' \subset \Omega^{\mathbf{Z}}$ with $v^{\mathbf{Z}}(\Gamma') = 1$ and $\tau\Gamma' = \Gamma'$ and there is a Borel function $l' : \Gamma' \to [1, +\infty)$ such that for each $w \in \Gamma'$ and each $n \geq 0$,

$$\max\{|f_0(w)^{-1}|_{C^1}, |f_0(w)|_{C^2}\} \leq l'(w)$$

and

$$l'(\tau^{-n}w) \leq l'(w)e^{\varepsilon n}.$$

Secondly, by Lemma 4.6, there is a number $C_0 > 0$ such that for each $(w,x) \in \Delta_0''$ (see Section 3), if $y, z \in \Phi_{(w,x)} W^u_{(w,x),\hat{\delta}}(x)$, then

$$|J^u(w,y) - J^u(w,z)| \leq C_0 l(w,x)^8 |f_0(w)|_{C^2} d(y,z)$$

and also

$$\max\{J^u(w,y)^{-1}, J^u(w,z)^{-1}\} \leq |f_0(w)^{-1}|^{m_0}_{C^1}.$$

Let now $(w,x) \in \Delta_0''$ with $w \in \Gamma'$. For any $y, z \in \Phi_{(w,x)} W^u_{(w,x),\hat{\delta}}(x)$, by Lemma 4.1 and Lemma 4.4 1) we have

$$\begin{aligned}
&\sum_{k=1}^{+\infty} \left| \log \frac{J^u(G^{-k}(w,y))}{J^u(G^{-k}(w,z))} \right| \\
&= \sum_{k=1}^{+\infty} |\log J^u(G^{-k}(w,y)) - \log J^u(G^{-k}(w,z))| \\
&\leq \sum_{k=1}^{+\infty} |f_0(\tau^{-k}w)|^{m_0}_{C^1} |J^u(G^{-k}(w,y)) - J^u(G^{-k}(w,z))| \\
&\leq \sum_{k=1}^{+\infty} |f_0(\tau^{-k}w)|^{m_0}_{C^1} C_0 l(\tau^{-k}(w,x))^8 |f_0(\tau^{-k}w)|_{C^2} d(f_w^{-k}y, f_w^{-k}z) \\
&\leq \sum_{k=1}^{+\infty} C_0 K_0 l'(w)^{m_0+1} l(w,x)^9 e^{[-\lambda^+ + (m_0+11)\varepsilon]k} d(y,z).
\end{aligned}$$

From this the first part of the lemma follows clearly. Since for μ^*-a.e. (w,x) there is $n > 0$ such that $G^{-n}(\xi(w,x)) \subset \{w'\} \times \Phi_{(w',x')} W^u_{(w',x'),\hat{\delta}}(x')$ for some $(w',x') \in \Delta_0'' \cap (\Gamma' \times M)$, the second part follows immediately. $\square$

Lemma 8.2. *There exists a measurable function $\rho : \Omega^{\mathbf{Z}} \times M \to \mathbf{R}^+$ such that for μ^*-a.e. (w,x), $\rho(w,y) = \Delta((w,x),(w,y))/L(w,x)$ for each $(w,y) \in \xi(w,x)$.*

Proof. We define a sequence of functions $\rho_n : \Omega^{\mathbf{Z}} \times M \to \mathbf{R}^+, n \geq 1$ in the following way: Let $(w,x) \in \Omega^{\mathbf{Z}} \times M$. If $G^{-l}(\xi(w,x)) \subset \{w'\} \times \Phi_{(w',x')} W^u_{(w',x'),\hat{\delta}}(x')$ for some $l \geq 0$ and some $(w',x') \in \Delta''_0 \cap (\Gamma' \times M)$, then define for each $n \geq 1$ and $(w,y) \in \xi(w,x)$

$$\rho_n(w,y) = \frac{\prod_{k=1}^{n} J^u(G^{-k}(w,y))^{-1}}{\int_{\xi_w(x)} \prod_{k=1}^{n} J^u(G^{-k}(w,y))^{-1} d\lambda^u_{(w,x)}(y)}$$

$$= \frac{\prod_{k=1}^{n} \frac{J^u(G^{-k}(w,x))}{J^u(G^{-k}(w,y))}}{\int_{\xi_w(x)} \prod_{k=1}^{n} \frac{J^u(G^{-k}(w,x))}{J^u(G^{-k}(w,y))} d\lambda^u_{(w,x)}(y)}.$$

Otherwise, we define

$$\rho_n(w,y) = 1$$

for each $n \geq 1$ and $(w,y) \in \xi(w,x)$. From the construction of ξ it is easy to see that ρ_n is measurable on $\Omega^{\mathbf{Z}} \times M$ for all $n \geq 1$ and, by Lemma 8.1, for each (w,x) the limit

$$\lim_{n \to +\infty} \rho_n(w,y) \stackrel{\text{def.}}{=} \rho(w,y) \tag{8.1}$$

exists for all $(w,y) \in \xi(w,x)$. Let $\rho : \Omega^{\mathbf{Z}} \times M \to \mathbf{R}^+$ be defined by (8.1). Then it satisfies clearly the requirement of this lemma. □

Suppose that $\rho : \Omega^{\mathbf{Z}} \times M \to \mathbf{R}^+$ is as defined above. We now define a measure ν on $\Omega^{\mathbf{Z}} \times M$ by

$$\nu(A) = \int \left[\int_{\xi_w(x)} \chi_A(w,y)\rho(w,y) d\lambda^u_{(w,x)}(y) \right] d\mu^*(w,x), A \in \mathcal{B}(\Omega^{\mathbf{Z}} \times M). \tag{8.2}$$

By Proposition 5.2 2), using standard arguments from measure theory one can easily verify that ν is indeed a well-defined Borel probability measure. Also, from the definition of ν it follows clearly that, if $\{\nu_{\xi(w,x)}\}$ is a canonical system of conditional measures of ν associated with ξ, then $d\nu_{\xi(w,x)} = \rho d\lambda^u_{(w,x)}$ for μ^*-a.e. (w,x) and that ν coincides with μ^* on $\mathcal{B}(\xi)$ (the σ-algebra consisting of all measurable ξ-sets).

Lemma 8.3. $\int -\log \nu_{\xi(w,x)}((G^{-1}\xi)(w,x))d\mu^* = \int \log J^u d\mu^*$.

Proof. Define $q(w,x) = \nu_{\xi(w,x)}((G^{-1}\xi)(w,x))$. Then for μ^*-a.e. (w,x)

$$q(w,x) = \frac{\int_{(G^{-1}\xi)_w(x)} \Delta((w,x),(w,y)) d\lambda^u_{(w,x)}(y)}{L(w,x)}$$

$$= \frac{L(G(w,x))}{L(w,x)} \cdot \frac{1}{J^u(w,x)} \tag{8.3}$$

Since, by Lemma 8.1,

$$L(w,x) = \lim_{n\to+\infty} \int_{\xi_w(x)} \prod_{k=1}^{n} \frac{J^u(G^{-k}(w,x))}{J^u(G^{-k}(w,y))} d\lambda^u_{(w,x)}(y)$$

for μ^*-a.e.(w,x), it follows that L is a positive finite-valued measurable function on $\Omega^{\mathbf{Z}} \times M$ with

$$\int \log^+ \frac{L \circ G}{L} d\mu^* \leq \int \log^+ J^u d\mu^* < +\infty.$$

Thus, by Lemma I.3.1, $\int \log \frac{L\circ G}{L} d\mu^* = 0$. The lemma follows then from (8.3). □

We have indicated that $\nu = \mu^*$ on $\mathcal{B}(\xi)$. The next lemma and an induction show that $\nu = \mu^*$ on $\mathcal{B}(G^{-n}\xi)$ for all $n \geq 0$ and hence they are equal on $\vee_{n=0}^{+\infty} \mathcal{B}(G^{-n}\xi) = \mathcal{B}(\vee_{n=0}^{+\infty} G^{-n}\xi) = \mathcal{B}_{\mu^*}(\Omega^{\mathbf{Z}} \times M)$.

Lemma 8.4. *$\int \log J^u d\mu^* = H_{\mu^*}(\xi|G\xi)$ implies $\nu = \mu^*$ on $\mathcal{B}(G^{-1}\xi)$.*

Proof. For $(w,y) \in \Omega^{\mathbf{Z}} \times M$, define

$$P(w,y) = \frac{\nu_{\xi(w,y)}((G^{-1}\xi)(w,y))}{\mu^*_{\xi(w,y)}((G^{-1}\xi)(w,y))}.$$

P is well defined μ^* almost everywhere since $H_{\mu^*}(G^{-1}\xi|\xi) < +\infty$. Noting that for μ^*-a.e. $(w,x), (G^{-1}\xi)|_{\xi(w,x)}$ is ($\mu^*_{\xi(w,x)}$-mod 0) a countable partition, by the convexity of the function $\log x$ we have

$$\int \log P d\mu^* \leq \log \int P d\mu^* \leq 0$$

(we admit here $\log 0 = -\infty$) with $\int \log P d\mu^* = 0$ if and only if $P = 1$ μ^* almost everywhere. But we know that $\int \log P d\mu^* = 0$, since Lemma 8.3 says that

$$\begin{aligned} -\int \log \nu_{\xi(w,x)}((G^{-1}\xi)(w,x)) d\mu^* &= \int \log J^u d\mu^* \\ &= H_{\mu^*}(G^{-1}\xi|\xi) \\ &= -\int \log \mu^*_{\xi(w,x)}((G^{-1}\xi)(w,x)) d\mu^*. \end{aligned}$$

Thus $\nu = \mu^*$ on $\mathcal{B}(G^{-1}\xi)$. □

Now let η be an arbitrary measurable partition of $\Omega^{\mathbf{Z}} \times M$ subordinate to W^u-manifolds of $\mathcal{X}(M,v,\mu)$ and let $\{\mu^*_{\eta(w,x)}\}$ be a canonical system of conditional measures of μ^* associated with η. In order to prove that $\mu^*_{\eta(w,x)} << \lambda^u_{(w,x)}$ for μ^* -a.e. (w,x), we take a partition ξ as dealt with above. Suppose that $\{\mu^*_{(\xi\vee\eta)(w,x)}\}$ is a canonical system of conditional measures of μ^* associated with

$\xi \vee \eta$. Noting that $(\xi \vee \eta)|_{\xi(w,x)}$ and $(\xi \vee \eta)|_{\eta(w,x)}$ are countable partitions for μ^*-a.e.(w,x), we have $\mu^*_{\xi(w,x)}((\xi \vee \eta)(w,x)) > 0, \mu^*_{\eta(w,x)}((\xi \vee \eta)(w,x)) > 0$ and

$$\mu^*_{(\xi\vee\eta)(w,x)}(\cdot) = \frac{\mu^*_{\xi(w,x)}(\cdot)}{\mu^*_{\xi(w,x)}((\xi \vee \eta)(w,x))} = \frac{\mu^*_{\eta(w,x)}(\cdot)}{\mu^*_{\eta(w,x)}((\xi \vee \eta)(w,x))}$$

for μ^*-a.e. (w,x). From this it follows clearly that $\mu^*_{\eta(w,x)} << \lambda^u_{(w,x)}, \mu^*$-a.e.$(w,x)$. The proof of Theorem 1.1 2) $\Rightarrow$ 1) of the ergodic case is completed. □

Corollary 8.1. *Let $\mathcal{X}(M,v,\mu)$ be given ergodic such that Pesin's entropy formula holds true. Let η be a partition of $\Omega^{\mathbf{Z}} \times M$ subordinate to W^u-manifolds of $\mathcal{X}(M,v,\mu)$ and let ρ be the density of $\mu^*_{\eta(w,x)}$ with respect to $\lambda^u_{(w,x)}$. Then for μ^*-a.e.(w,x), there exist a countable number of disjoint open subsets $U_n(w,x)$, $n \in \mathbf{N}$ of $W^u(w,x)$ such that $\cup_{n\in\mathbf{N}} U_n(w,x) \subset \eta_w(x)$, $\lambda^u_{(w,x)}(\eta_w(x) \setminus \cup_{n\in\mathbf{N}} U_n(w,x)) = 0$ and on each $U_n(w,x)$ ρ is a strictly positive function satisfying*

$$\frac{\rho(w,y)}{\rho(w,z)} = \prod_{k=1}^{+\infty} \frac{J^u(G^{-k}(w,z))}{J^u(G^{-k}(w,y))}, \qquad y,z \in U_n(w,x),$$

in particular, $\log\rho$ restricted to $U_n(w,x)$ is Lipschitz along $W^u(w,x)$.

B. Proof of Theorem 1.1 2)$\Rightarrow$ 1): The General Case

We reduce the general case to its ergodic one. Let $\mathcal{X}(M,v,\mu)$ be given. Let $\hat{\eta}_0$ be the measurable partition of $\Omega^{\mathbf{Z}} \times M$ into disjoint sets $\{\hat{C}_\alpha\}_{\alpha\in\mathcal{A}}$ and $\Omega^{\mathbf{Z}} \times M \setminus \cup_{\alpha\in\mathcal{A}} \hat{C}_\alpha$, as introduced in Subsection 1.B for $\mathcal{X}(M,v,\mu)$.

Suppose that

$$h^\sigma_{\mu^*}(G) = \int \sum_i \lambda^{(i)}(w,x)^+ m_i(w,x) d\mu^*. \tag{8.4}$$

By Theorem I.2.6 and (8.4) we have

$$\begin{aligned} h^\sigma_{\mu^*}(G) &= \int h^\sigma_{\rho^*_\alpha}(G) d\mu^*_{\hat{\eta}_0}(\alpha) \\ &= \int\int \sum_i \lambda^{(i)}(w,x)^+ m_i(w,x) d\rho^*_\alpha(w,x) d\mu^*_{\hat{\eta}_0}(\alpha) \end{aligned}$$

where $\mu^*_{\hat{\eta}_0}$ is the measure induced by μ^* on the factor-space $\Omega^{\mathbf{Z}} \times M/\hat{\eta}_0$. But, by Theorem II.0.1 together with Remark 1.1 and (1.2),

$$h^\sigma_{\rho^*_\alpha}(G) \le \int \sum_i \lambda^{(i)}(w,x)^+ m_i(w,x) d\rho^*_\alpha$$

for each $\alpha \in \mathcal{A}$. Hence, there exists a measurable set $\mathcal{A}_1 \subset \mathcal{A}$ with $\mu^*_{\tilde{\eta}_0}(\mathcal{A}_1) = 1$ such that for each $\alpha \in \mathcal{A}_1$

$$h^{\sigma}_{\rho^*_\alpha}(G) = \int \sum_i \lambda^{(i)}(w,x)^+ m_i(w,x) d\rho^*_\alpha.$$

Using a variant of Theorem III.3.1 for $\mathcal{X}(M, v, \mu)$ and a procedure analogous to part of the proof of Proposition IV.2.1, one can construct a measurable partition ξ satisfying the requirements of Proposition 2.1. Moreover, the partition ξ can be constructed to have the additional property that there is a measurable function $\rho : \Omega^{\mathbf{Z}} \times M \to \mathbf{R}^+$ such that for μ^* -a.e. (w, x),

$$\rho(w,y) = \frac{\Delta((w,x),(w,y))}{L(w,x)}, \quad \forall (w,y) \in \xi(w,x) \tag{8.5}$$

if $\lambda^{(i)}(w,x) > 0$ for some i and

$$\rho(w,y) = 1, \ \forall (w,y) \in \xi(w,x) \tag{8.6}$$

if $\lambda^{(i)}(w,x) \leq 0$ for all $1 \leq i \leq r(w,x)$. Then we can define a measure $\hat{\nu}$ on $\Omega^{\mathbf{Z}} \times M$ in a way completely analogous to (8.2), i.e.

$$\hat{\nu}(A) = \int \left[\int_{\xi_w(x)} \chi_A(w,y)\rho(w,y) d\lambda^u_{(w,x)}(y) \right] d\mu^*(w,x) \tag{8.7}$$

for each $A \in \mathcal{B}(\Omega^{\mathbf{Z}} \times M)$. (8.7) can be also written as

$$\hat{\nu}(A) = \int \left\{ \int \left[\int_{\xi_w(x)} \chi_A(w,y)\rho(w,y) d\lambda^u_{(w,x)}(y) \right] d\rho^*_\alpha(w,x) \right\} d\mu^*_{\tilde{\eta}_0}(\alpha) \tag{8.8}$$

for each $A \in \mathcal{B}(\Omega^{\mathbf{Z}} \times M)$. Clearly, if $\{\hat{\nu}_{\xi(w,x)}\}$ is a canonical system of conditional measures of $\hat{\nu}$ associated with ξ, then $d\hat{\nu}_{\xi(w,x)} = \rho d\lambda^u_{(w,x)}$ for μ^*-a.e.(w,x).

Let ξ be as given above. Then there is a measurable set $\mathcal{A}_2 \subset \mathcal{A}_1$ with $\mu^*_{\tilde{\eta}_0}(\mathcal{A}_2) = 1$ such that for every $\alpha \in \mathcal{A}_2$, $\xi_w(x) \subset W^u(w,x)$ and contains an open neighbourhood of x in $W^u(w,x)$ for ρ^*_α-a.e.(w,x),i.e. ξ is indeed a measurable partition subordinate to W^u-manifolds of $\mathcal{X}(M, v, \rho_\alpha)$. Let now $\Lambda \subset \Omega^{\mathbf{Z}} \times M$ be a measurable set of full μ^* measure such that (8.5) and (8.6) hold true for each $(w,x) \in \Lambda$. Put $\mathcal{A}_3 = \{\alpha \in \mathcal{A}_2 : \rho^*_\alpha(\Lambda) = 1\}$. Then for each $\alpha \in \mathcal{A}_3$, we can appeal to the proof of the ergodic case and conclude that

$$\rho^*_\alpha(A) = \int \left[\int_{\xi_w(x)} \chi_A(w,y)\rho(w,y) d\lambda^u_{(w,x)}(y) \right] d\rho^*_\alpha(w,x), \quad \forall A \in \mathcal{B}(\Omega^{\mathbf{Z}} \times M).$$

This together with (8.8) yields

$$\mu^* = \hat{\nu}.$$

Therefore, if $\{\mu^*_{\xi(w,x)}\}$ is a canonical system of conditional measures of μ^* associated with ξ, we have $\mu^*_{\xi(w,x)} << \lambda^u_{(w,x)}$ and $d\mu^*_{\xi(w,x)}/d\lambda^u_{(w,x)} = \rho$ for μ^*-a.e. (w,x).

Let now η be an arbitrary measurable partition of $\Omega^{\mathbf{Z}} \times M$ subordinate to W^u-manifolds of $\mathcal{X}(M, v, \mu)$. By arguments analogous to those in the paragraph before Corollary 8.1 one easily proves that $\mu^*_{\eta(w,x)} << \lambda^u_{(w,x)}$ for μ^*-a.e. (w, x). This completes the proof of Theorem 1.1 2) $\Rightarrow$ 1) in the general situation. □

As can be seen from the discussion just above, we have the following

Corollary 8.2. *Corollary 8.1 holds true in the nonergodic case.*

Chapter VII Random Perturbations of Hyperbolic Attractors

While in the preceding chapters we intended to make the presentation self-contained as much as we could, in this chapter, it is not the case. It is actually devoted to the derivation of a set of new results, based on what we have already obtained. We shall study invariant measures for random perturbations of hyperbolic attractors. Leaving precise statements for later, we first give a description of the main result of this chapter. Let f be a twice differentiable diffeomorphism on a Riemannian manifold N and let Λ be a hyperbolic attractor of f with basin of attraction U. As we have indicated at the beginning of Chapter VI, there is a unique f-invariant measure ρ with support in Λ that is characterized by each of the following properties: (a) ρ has absolutely continuous conditional measures on unstable manifolds; (b) Pesin's entropy formula holds true for the system (N, f, ρ); (c) For Leb. -a.e. $x \in U$ one has $\lim_{n\to+\infty} \frac{1}{n}\sum_{k=0}^{n-1} \delta_{f^k x} = \rho$. The measure ρ is called the SBR measure on the attractor Λ. Our main purpose here is to show that the same kind of result holds true as well if $f : U \to U$ is subjected to certain random perturbations. This is described in the following paragraph.

Now we consider the case when the system $f : U \to U$ is subjected to certain random perturbation which can be viewed as random compositions of maps from U into itself nearby f. Let v be a Borel probability measure on $\Omega = C^2(U, U)$ concentrated on those maps which are sufficiently nearly f. Let $G : \Omega^{\mathbf{Z}} \times U \hookleftarrow$ be as defined in Chapter VI and let $P_1 : \Omega^{\mathbf{Z}} \times U \to \Omega^{\mathbf{Z}}$ be the projection on the first factor. Then our main result of this chapter says that there is a unique G-invariant Borel probability measure $\tilde{\rho}$ with $P_1\tilde{\rho} = v^{\mathbf{Z}}$ and with support in $\tilde{\Lambda} = \bigcap_{n\geq 0} G^n(\Omega^{\mathbf{Z}} \times U)$ such that it is characterized by each of the following properties: (1) $\tilde{\rho}$ has absolutely continuous conditional measures on unstable manifolds; (2) An entropy formula of Pesin's type holds true for the system $(\Omega^{\mathbf{Z}} \times U, G, \tilde{\rho})$, i.e. $h_{\tilde{\rho}}^{\sigma}(G) = \int \Sigma_i \lambda^{(i)}(w, x)^+ m_i(w, x) d\tilde{\rho}$; (3) For $v^{\mathbf{Z}} \times$ Leb. -a.e. $(w, x) \in \Omega^{\mathbf{Z}} \times U$ it holds that $\lim_{n\to+\infty} \frac{1}{n}\sum_{k=0}^{n-1} \delta_{G^k(w,x)} = \tilde{\rho}$. In addition, if μ is an absolutely continuous (with respect to the Lebesgue measure on U) invariant measure of the system $\mathcal{X}(U, v)$ (see Section 1 of this chapter) then $\tilde{\rho}$ is just the measure μ^* introduced in a way analogous to Proposition I. 1.2.

The existence and the characterizations (1) and (3) of such a measure $\tilde{\rho}$ are due to [You] and the characterization (2) of the measure is due to $[\text{Liu}]_5$. One of the purposes of this chapter is to apply our general argument to the measure $\tilde{\rho}$ given by [You].

Finally we remark that in this chapter we study random perturbations of the hyperbolic attractor Λ as random compositions of maps nearby $f : U \hookleftarrow$ and the main result described above is then actually a consequence of the persistence of the hyperbolic structure on Λ for the randomly composed maps. Y. Kifer also studied random perturbations of hyperbolic attractors from the point of view of Markov processes. For further information we refer the reader to Kifer's book [Kif]$_2$.

§1 Definitions and Statements of Results

In this chapter we assume that M is a C^∞ (maybe not compact) connected Riemannian manifold without boundary. Let U_0 be an open subset of M with compact closure $\overline{U}_0$, and let $f : U_0 \to fU_0$ be a C^2 diffeomorphism. A set $\Lambda \subset U_0$ is said to be f-invariant if $f\Lambda = \Lambda$.

Definition 1.1. *A compact f-invariant set $\Lambda \subset U_0$ is said to be uniformly hyperbolic or simply hyperbolic if there is a continuous splitting of the tangent bundle $T_\Lambda M$ over Λ into a direct sum of two subbundles $E^u \oplus E^s$ and there are also numbers $0 < \lambda_0 < 1$ and $C_1, C_2 > 0$ such that for all $x \in \Lambda$ and $n \geq 0$*

$$T_x f E_x^u = E_{f(x)}^u, \quad T_x f E_x^s = E_{f(x)}^s$$

and

$$|T_x f^n \xi| \geq C_1 \lambda_0^{-n} |\xi|, \quad \xi \in E_x^u,$$

$$|T_x f^n \eta| \leq C_2 \lambda_0^n |\eta|, \quad \eta \in E_x^s.$$

Remark 1.1. Via a change of Riemannian metric we may–and will–assume that $C_1 = C_2 = 1$. This means that there always exists an equivalent (to $\langle\ ,\ \rangle$) Riemannian metric $<<\ ,\ >>$ on M, called an *adapted Riemannian metric*, such that for all $x \in \Lambda$ and $n \geq 0$

$$\|T_x f^n \xi\| \geq \alpha^{-n} \|\xi\|, \quad \xi \in E_x^u,$$

$$\|T_x f^n \eta\| \leq \alpha^n \|\eta\|, \quad \eta \in E_x^s,$$

where $0 < \alpha < 1$ is a constant and $\|\cdot\|$ is the norm on TM induced by $<<\ ,\ >>$. This can be shown in the following way. Take a C^∞ function $\theta : M \to \mathbf{R}^+$ such that $\theta(x) = 1$ for all $x \in \Lambda$, $\theta(x) = 0$ for all $x \in M \backslash U_0$ and $0 \leq \theta(x) \leq 1$ for all $x \in M$. Then we define for $x \in M$ and $\xi, \eta \in T_x M$

$$\ll \xi, \eta \gg = \begin{cases} \langle \xi, \eta \rangle + \sum_{l=1}^{q-1} \langle \theta(x) T_x f^l \xi, \theta(x) T_x f^l \eta \rangle & \text{if } x \in U_0 \\ \langle \xi, \eta \rangle & \text{if } x \notin U_0 \end{cases}$$

where q is a positive integer such that $C_1 \lambda_0^{-q} > 1$ and $C_2 \lambda_0^q < 1$. Let $\|\cdot\|$ be the norm on TM induced by $\ll\ ,\ \gg$. Then one can easily check that for some number $0 < \alpha < 1$

$$\|Tf\xi\| \geq \alpha^{-1} \|\xi\| \text{ for } \xi \in E^u, \quad \|Tf\eta\| \leq \alpha \|\eta\| \text{ for } \eta \in E^s. \tag{1.1}$$

Note that the metric $\ll \;,\; \gg$ thus defined may not be smooth. However, one can approximate $\ll \;,\; \gg$ by a smooth Riemannian metric, denoted by the same notation $\ll \;,\; \gg$, such that (1.1) will remain true but with possibly a little bigger $\alpha < 1$.

Definition 1.2. *An f-invariant set $\Lambda \subset U_0$ is called a hyperbolic attractor with basin of attraction U if the following hold true:*

1) *U is an open set such that $\overline{U} \subset U_0, f\overline{U} \subset U$ and*

$$\bigcap_{n\geq 0} f^n U = \Lambda;$$

2) *Λ is a uniformly hyperbolic set;*
3) *$f|_\Lambda$ has a dense orbit.*

In the sequel we shall always assume that Λ is a hyperbolic attractor of f with basin of attraction U. We now review briefly some relevant results concerning the SBR measure on Λ. To this end we first present some related concepts.

For $x \in \Lambda$, the *unstable manifold* $W^u(f,x)$ of f at x is defined as

$$W^u(f,x) = \{y \in U_0 : d(f^{-n}x, f^{-n}y) \to 0 \text{ as } n \to +\infty\}.$$

Then for each $x \in \Lambda, W^u(f,x)$ is an immersed C^1 (actually C^2) submanifold of U_0 and $W^u(f,x) \subset \Lambda$ ([Hir]$_2$). Let μ be a Borel probability measure concentrated on Λ.

Definition 1.3. *A measurable partition η of Λ is said to be subordinate to W^u-manifolds with respect to μ if for μ-a.e. $x \in \Lambda, \eta(x) \subset W^u(f,x)$ and it contains an open neighbourhood of x in $W^u(f,x)$, this neighbourhood being taken in the submanifold topology of $W^u(f,x)$.*

Definition 1.4. *We say that μ has absolutely continuous conditional measures on W^u-manifolds if for any measurable partition η of Λ subordinate to W^u-manifolds with respect to μ one has*

$$\mu_x^\eta \ll \lambda_{(f,x)}^u \quad \textit{for } \mu - a.e.x \in \Lambda,$$

where $\{\mu_x^\eta\}_{x\in\Lambda}$ is a canonical system of conditional measures of μ associated with η and $\lambda_{(f,x)}^u$ is the Lebesgue measure on $W^u(f,x)$ induced by its inherited Riemannian metric as a submanifold of M.

Proposition 1.1. *Let Λ be a hyperbolic attractor of f with basin of attraction U. Then there exists a unique f-invariant Borel probability measure ρ with support in Λ such that it is characterized by each of the following properties:*

1) *ρ has absolutely continuous conditional measures on W^u-manifolds;*
2) *Pesin's entropy formula holds true for the system $f : (U, \rho) \hookleftarrow$, i.e.*

$$h_\rho(f) = \int_\Lambda \sum_i \lambda^{(i)}(x)^+ m_i(x) d\rho$$

where $\lambda^{(1)}(x) < \cdots < \lambda^{(r(x))}(x)$ *denote the Lyapunov exponents of* f *at* x, $\{m_i(x)\}_{i=1}^{r(x)}$ *their multiplicities respectively, and* $h_\rho(f)$ *denotes the usual measure-theoretic entropy of the system* $f : (U, \rho) \hookleftarrow$;

3) *For Leb.-a.e.* $x \in U$, $\frac{1}{n}\sum_{k=0}^{n-1}\delta_{f^k x}$ *weakly converges to* ρ *as* $n \to +\infty$.

This above result is due to Sinai [Sin], Bowen and Ruelle [Rue]$_3$, [Bow]$_2$. The measure ρ is thus called the *SBR measure* on the attractor Λ. As a particular case, a proof of the proposition will also be included in Section 3, where we shall prove a more general random version of the result.

Now we begin to consider the case when $f : U \hookleftarrow$ is subjected to certain random perturbation which can be viewed as random compositions of maps from U into itself nearby f. Denote by $C^r(U,U)(r \geq 1)$ the space of all C^r maps from U into itself, equipped with the C^r compact-open topology (see [Hir]$_1$). If $g \in C^2(U,U)$ and $g : U \to gU$ is a C^2 diffeomorphism, in a way analogous to (1.1) in Chapter II we define

$$||g||_{C^2,U} = \sup\{T_\xi Tg : \xi \in T_U M, |\xi| \leq 1\}$$

and

$$||g||_{C^2,U}^{-} = \sup\{T_\xi Tg^{-1} : \xi \in T_{gU} M, |\xi| \leq 1\}.$$

An equivalent definition of the C^2-norms $||\cdot||_{C^2,U}$ and $||\cdot||_{C^2,U}^{-}$ can be given by using local charts as in the definition of $|\cdot|_{C^2}$ in Section I. 1.

Choose a neighbourhood $\mathcal{U}_1(f)$ of f in $C^1(U,U)$ such that if $g \in \mathcal{U}_1(f)$ then $g : U \to gU$ is a C^1 diffeomorphism. Write now

$$B_{f,U} = \max\{||f||_{C^2,U}, ||f||_{C^2,U}^{-}\}.$$

Given a neighbourhood $\mathcal{U}(f)$ of f in $C^1(U,U)$ with $\mathcal{U}(f) \subset \mathcal{U}_1(f)$ and a number $B \geq B_{f,U}$, we put

$$\Omega_{\mathcal{U}(f),B} = \{g \in \mathcal{U}(f) \cap C^2(U,U) : ||g||_{C^2,U} \leq B, ||g||_{C^2,U}^{-} \leq B\}$$

and let $\Omega_{\mathcal{U}(f),B}$ have the inherited C^1 topology as a subset of $C^1(U,U)$. By Arzela-Ascoli theorem, it is easy to see that $\Omega_{\mathcal{U}(f),B}$ is a compact subset of $C^1(U,U)$. Also, if $g \in \Omega_{\mathcal{U}(f),B}$ then g is a uniformly continuous map from U into itself and so it can be uniquely extended to a continuous map defined on $\overline{U}$. In the sequel, for a given neighbourhood $\mathcal{U}(f)$ of f in $C^1(U,U)$ with $\mathcal{U}(f) \subset \mathcal{U}_1(f)$ and a given number $B \geq B_{f,U}$ we shall write

$$\Omega = \Omega_{\mathcal{U}(f),B}$$

for simplicity of notation. This will not cause any confusion.

Given $\Omega = \Omega_{\mathcal{U}(f),B}$, let $\Omega^{\mathbf{Z}}$ be the bi-infinite product of copies of Ω and let $\Omega^{\mathbf{Z}}$ have the product σ-algebra $\mathcal{B}(\Omega)^{\mathbf{Z}}$ and the product topology. For $w =$

$(\cdots, g_{-1}(w), g_0(w), g_1(w), \cdots) \in \Omega^{\mathbf{Z}}$ and $n > 0$ we write

$$g_w^0 = id,$$

$$g_w^n = g_{n-1}(w) \circ \cdots \circ g_0(w),$$

$$g_w^{-n} = g_{-n}(w)^{-1} \circ \cdots \circ g_{-1}(\omega)^{-1}$$

defined wherever they make sense. If v is a Borel probability measure on Ω, we shall denote by $\mathcal{X}(U, v)$ the random dynamical system generated by actions on U of $g_w^n, n \in \mathbf{Z}$ with w being chosen according to law $v^{\mathbf{Z}}$, where $v^{\mathbf{Z}}$ is the bi-infinite product of copies of v.

Let $\mathcal{X}(U, v)$ be given. We define

$$G : \Omega^{\mathbf{Z}} \times U \to \Omega^{\mathbf{Z}} \times U, \quad (w, x) \longmapsto (\tau w, g_0(w)x),$$

where τ is the shift operator on $\Omega^{\mathbf{Z}}$ (see Section I. 1), and projections

$$P_1 : \Omega^{\mathbf{Z}} \times U \to \Omega^{\mathbf{Z}}, \quad (w, x) \longmapsto w,$$

$$P_2 : \Omega^{\mathbf{Z}} \times U \to U, \quad (w, x) \longmapsto x.$$

Now suppose that $\tilde{\mu}$ is a Borel probability measure on $\Omega^{\mathbf{Z}} \times U$ such that

$$G\tilde{\mu} = \tilde{\mu}, \quad P_1\tilde{\mu} = v^{\mathbf{Z}}. \tag{1.2}$$

By the definition of Ω there holds clearly the following integrability condition:

$$\int_{\Omega^{\mathbf{Z}} \times U} \log^+ |T_x g_w^1| d\tilde{\mu} < +\infty.$$

From this and Oseledec multiplicative ergodic theorem we obtain the following

Proposition 1.2. *Let $G : (\Omega^{\mathbf{Z}} \times U, \tilde{\mu}) \hookleftarrow$ be as given above. Then for $\tilde{\mu}$-a.e. $(w, x) \in \Omega^{\mathbf{Z}} \times U$ there exist measurable (in (w, x)) numbers $r(w, x)$ and*

$$\lambda^{(1)}(w, x) < \cdots < \lambda^{(r(w,x))}(w, x),$$

and also an associated measurable (in (ω, x)) filtration by linear subspaces of $T_x M$

$$\{0\} = V_{(w,x)}^{(0)} \subset V_{(w,x)}^{(1)} \subset \cdots \subset V_{(w,x)}^{(r(w,x))} = T_x M$$

such that

$$\lim_{n \to +\infty} \frac{1}{n} \log |T_x g_w^n \xi| = \lambda^{(i)}(w, x)$$

for each $\xi \in V_{(w,x)}^{(i)} \backslash V_{(w,x)}^{(i-1)}, 1 \leq i \leq r(w, x)$.

As usual, $\lambda^{(i)}(w,x), 1 \le i \le r(w,x)$ are called the *Lyapunov exponents* of $G : (\Omega^{\mathbf{Z}} \times U, \tilde{\mu}) \hookleftarrow$ at point (w,x), and $m_i(w,x) \stackrel{\text{def}}{=} \dim V^{(i)}_{(w,x)} - \dim V^{(i-1)}_{(w,x)}$ is called the *multiplicity* of $\lambda^{(i)}(w,x)$.

We next turn to an (measure-theoretic) entropy characteristic of the system $G : (\Omega^{\mathbf{Z}} \times U, \tilde{\mu}) \hookleftarrow$. Let $\{\mu_w\}_{w\in\Omega^{\mathbf{Z}}}$ be a ($v^{\mathbf{Z}}$-mod 0 unique) canonical system of conditional measures of $\tilde{\mu}$ associated with the partition $\{\{w\} \times U : w \in \Omega^{\mathbf{Z}}\}$ of $\Omega^{\mathbf{Z}} \times U$. Identifying $\{w\} \times U$ with U, we regard μ_w as a Borel probability measure on U and call $\{\mu_w\}_{w\in\Omega^{\mathbf{Z}}}$ the *family of sample measures* of $\tilde{\mu}$. From the G-invariance of $\tilde{\mu}$ it follows clearly that for each $k \in \mathbf{Z}^+$

$$g^k_w \mu_w = \mu_{\tau^k w}, \quad v^{\mathbf{Z}} - \text{a.e.}\, w. \tag{1.3}$$

Proposition 1.3. *Let $G : (\Omega^{\mathbf{Z}} \times U, \tilde{\mu}) \hookleftarrow$ be as given above.*

1) Let ξ be a finite measurable partition of U. Then the limit

$$h_{\tilde{\mu}}(\mathcal{X}(U,v),\xi) \stackrel{\text{def}}{=} \lim_{n\to+\infty} \frac{1}{n} H_{\mu_w}\left(\bigvee_{k=0}^{n-1} (g^k_w)^{-1}\xi\right)$$

exists and is constant for $v^{\mathbf{Z}}$-a.e. $w \in \Omega^{\mathbf{Z}}$;

2) Define

$$h_{\tilde{\mu}}(\mathcal{X}(U,v)) = \sup\{h_{\tilde{\mu}}(\mathcal{X}(U,v),\xi) : \xi \text{ is a finite measurable partition of } U\}. \tag{1.4}$$

Then

$$h_{\tilde{\mu}}(\mathcal{X}(U,v)) = h^{\sigma}_{\tilde{\mu}}(G) \tag{1.5}$$

where $\sigma = \{\Gamma \times U : \Gamma \in \mathcal{B}(\Omega)^{\mathbf{Z}}\}$.

Remark 1.2. Bogenschütz introduced in [Bog] the notion of measure-theoretic entropy for a general random dynamical system. The definition of $h_{\tilde{\mu}}(\mathcal{X}(U,v))$ in (1.4) fits into that context.

Proof. 1) Put

$$a_n(w,\xi) = H_{\mu_w}\left(\bigvee_{k=0}^{n-1} (g^k_w)^{-1}\xi\right).$$

Then

$$\begin{aligned}
a_{n+m}(w,\xi) = &\ H_{\mu_w}\left(\bigvee_{k=0}^{n+m-1}(g_w^k)^{-1}\xi\right) \\
\le &\ H_{\mu_w}\left(\bigvee_{k=0}^{n-1}(g_w^k)^{-1}\xi\right) + H_{\mu_w}\left((g_w^n)^{-1}\bigvee_{k=0}^{m-1}(g_{\tau^n w}^k)^{-1}\xi\right) \\
= &\ a_n(w,\xi) + H_{g_w^n\mu_w}\left(\bigvee_{k=0}^{m-1}(g_{\tau^n w}^k)^{-1}\xi\right) \\
= &\ a_n(w,\xi) + a_m(\tau^n w,\xi)
\end{aligned}$$

for $v^{\mathbf{Z}}$-a.e. w. By the subadditive ergodic theorem (Theorem I. 3.1), the limit function $a(w,\xi) = \lim_{n\to+\infty}\frac{1}{n}a_n(w,\xi)$ exists and is τ-invariant and thus constant $v^{\mathbf{Z}}$ almost everywhere on $\Omega^{\mathbf{Z}}$.

2) We first prove that, if ξ is a finite measurable partition of U and ζ is a countable measurable partition of $\Omega^{\mathbf{Z}}$, then

$$h_{\tilde{\mu}}(\mathcal{X}(U,v),\xi) = h_{\tilde{\mu}}^{\sigma}(G,\zeta\times\xi), \tag{1.6}$$

where $\zeta\times\xi = \{\Gamma\times A : \Gamma\in\zeta, A\in\xi\}$. In fact,

$$\begin{aligned}
&h_{\tilde{\mu}}^{\sigma}(G,\zeta\times\xi) \\
= &\lim_{n\to+\infty}\frac{1}{n}H_{\tilde{\mu}}\left(\bigvee_{k=0}^{n-1}G^{-k}(\zeta\times\xi)\,\Big|\,\sigma\right) \\
= &\lim_{n\to+\infty}\frac{1}{n}\int H_{\mu_w}\left(\left(\bigvee_{k=0}^{n-1}G^{-k}(\zeta\times\xi)\right)\Big|_{\{w\}\times U}\right)dv^{\mathbf{Z}}(w) \\
= &\lim_{n\to+\infty}\frac{1}{n}\int H_{\mu_w}\left(\bigvee_{k=0}^{n-1}(g_w^k)^{-1}\xi\right)dv^{\mathbf{Z}}(w) \\
= &\ h_{\tilde{\mu}}(\mathcal{X}(U,v),\xi).
\end{aligned}$$

This proves (1.6).

Now from (1.6) it follows clearly that

$$h_{\tilde{\mu}}(\mathcal{X}(U,v)) \le h_{\tilde{\mu}}^{\sigma}(G).$$

Then what remains is to prove that

$$h_{\tilde{\mu}}(\mathcal{X}(U,v)) \ge h_{\tilde{\mu}}^{\sigma}(G).$$

By (1.6) it suffices to show that for every finite measurable partition α of $\Omega^{\mathbf{Z}}\times U$ and every $\varepsilon>0$ there exists a measurable partition β of $\Omega^{\mathbf{Z}}\times U$ of the type $\zeta\times\xi$ as explained above such that

$$h_{\tilde{\mu}}^{\sigma}(G,\alpha) \le h_{\tilde{\mu}}^{\sigma}(G,\beta)+\varepsilon.$$

Since $\overline{U}$ is compact, one can easily find an increasing sequence of finite measurable partitions $\{\xi_n\}_{n=1}^{+\infty}$ of U such that $\bigvee_{n=1}^{+\infty} \xi_n$ is the partition of U into single points. Define $\beta_n = \{\Omega^{\mathbf{Z}}\} \times \xi_n, n \geq 1$. By (3.8) of Chapter 0 one has

$$H_{\tilde{\mu}}(\alpha \mid \beta_n \vee \sigma) \to 0$$

as $n \to +\infty$. This together with 4) of Theorem 0.4.2 yields that

$$\begin{aligned} h^{\sigma}_{\tilde{\mu}}(G,\alpha) &\leq h^{\sigma}_{\tilde{\mu}}(G,\beta_n) + H_{\tilde{\mu}}(\alpha \mid \beta_n \vee \sigma) \\ &\leq h^{\sigma}_{\tilde{\mu}}(G,\beta_n) + \varepsilon \end{aligned}$$

for sufficiently large n. The proof is completed. □

Given $\Omega = \Omega_{\mathcal{U}(f),B}$, we put

$$\tilde{\Lambda}_{\mathcal{U}(f),B} = \bigcap_{n \geq 0} G^n(\Omega^{\mathbf{Z}} \times U)$$

and also write

$$\tilde{\Lambda} = \tilde{\Lambda}_{\mathcal{U}(f),B}$$

for simplicity of notation. For $(w,x) \in \tilde{\Lambda}$, the *unstable manifold* $W^u(w,x)$ of G at (w,x) is defined as

$$W^u(w,x) = \{y \in U : d(g_w^{-n}x, g_w^{-n}y) \to 0 \text{ as } n \to +\infty\}.$$

If $\mathcal{U}(f)$ is given sufficiently small, then for any given $B \geq B_{f,U}, \tilde{\Lambda} = \tilde{\Lambda}_{\mathcal{U}(f),B}$ is a compact subset of $\Omega^{\mathbf{Z}} \times U, G\tilde{\Lambda} = \tilde{\Lambda}$, and $W^u(w,x)$ is a C^1 immersed submanifold of U and $W^u(w,x) \subset \tilde{\Lambda}$ for each $(w,x) \in \tilde{\Lambda}$. Proofs of these results will be given in the next section.

Now let $\mathcal{U}(f)$ be such a sufficiently small neighbourhood of f in $C^1(U,U)$ and let $B \geq B_{f,U}$. Let $\tilde{\mu}$ be a G-invariant Borel probability measure on $\Omega^{\mathbf{Z}} \times U$ (where $\Omega = \Omega_{\mathcal{U}(f),B}$). Clearly, $\tilde{\mu}$ is concentrated on $\tilde{\Lambda} = \tilde{\Lambda}_{\mathcal{U}(f),B}$. By $\sigma_{\tilde{\Lambda}}$ we denote the measurable partition $\{\{w\} \times U : w \in \Omega^{\mathbf{Z}}\} |_{\tilde{\Lambda}}$.

Definition 1.5. *A measurable partition η of $\tilde{\Lambda}$ is said to be subordinate to W^u-manifolds with respect to $\tilde{\mu}$ if $\eta \geq \sigma_{\tilde{\Lambda}}$ and for $\tilde{\mu}$-a.e. $(w,x) \in \tilde{\Lambda}, \eta_w(x) \stackrel{\text{def}}{=} \{y : (w,y) \in \eta(w,x)\} \subset W^u(w,x)$ and it contains an open neighbourhood of x in $W^u(w,x)$, this neighbourhood being taken in the submanifold topology of $W^u(w,x)$.*

Definition 1.6. *We say that $\tilde{\mu}$ has absolutely continuous conditional measures on W^u-manifolds if for any measurable partition η of $\tilde{\Lambda}$ subordinate to W^u-manifolds with respect to $\tilde{\mu}$ one has*

$$\tilde{\mu}^{\eta}_{(w,x)} \ll \lambda^u_{(w,x)}, \quad \tilde{\mu} - a.e.(w,x) \in \tilde{\Lambda},$$

where $\{\tilde{\mu}^{\eta}_{(w,x)}\}_{(w,x)\in\tilde{\Lambda}}$ *is a canonical system of conditional measures of* $\tilde{\mu}$ *associated with* η, $\tilde{\mu}^{\eta}_{(w,x)}$ *is treated as a measure on* $\eta_w(x)$ *by identifying* $\{w\}\times\eta_w(x)$ *with* $\eta_w(x)$, *and* $\lambda^u_{(w,x)}$ *is the Lebesgue measure on* $W^u(w,x)$ *induced by its inherited Riemannian metric as a submanifold of* M.

Clearly, this property of $\tilde{\mu}$ can also be characterized by the corresponding property of its sample measures.

We now state the main result of this chapter in the following theorem:

Theorem 1.1. *Let* Λ *be a hyperbolic attractor of* f *with basin of attraction* U. *Given* $B \geq B_{f,U}$, *if* $\mathcal{U}(f)$ *is a sufficiently small neighbourhood of* f *in* $C^1(U,U)$ *and* v *is a Borel probability measure on* $C^1(U,U)$ *which is concentrated on* $\Omega = \Omega_{\mathcal{U}(f),B}$, *then there exists a unique* G*-invariant Borel probability measure* $\tilde{\rho}$ *with support in* $\tilde{\Lambda} = \tilde{\Lambda}_{\mathcal{U}(f),B}$ *and with* $P_1\tilde{\rho} = v^{\mathbf{Z}}$ *such that it is characterized by each of the following properties:*

1) $\tilde{\rho}$ *has absolutely continuous conditional measures on* W^u*-manifolds;*

2) $h_{\tilde{\rho}}(\mathcal{X}(U,v)) = \int \sum_i \lambda^{(i)}(w,x)^+ m_i(w,x)d\tilde{\rho}$;

3) *For* $v^{\mathbf{Z}}\times$ *Leb.-a.e.* $(w,x)\in\Omega^{\mathbf{Z}}\times U$ *one has as* $n\to+\infty$

$$\frac{1}{n}\sum_{k=0}^{n-1}\delta_{G^k(w,x)}\to\tilde{\rho}. \tag{1.7}$$

In addition, $G:(\tilde{\Lambda},\tilde{\rho})\hookleftarrow$ *is ergodic.*

The proof of this theorem will be given in the next two sections.

Given $\Omega = \Omega_{\mathcal{U}(f),B}$ and a Borel probability measure v on Ω, the *transition probabilities* $P(x,\cdot), x\in U$ of $\mathcal{X}(U,v)$ are defined by

$$P(x,A) = v(\{g\in\Omega : gx\in A\})$$

for $x\in U$ and $A\in\mathcal{B}(U)$. We say that the transition probabilities of $\mathcal{X}(U,v)$ have a density if there is a Borel function $p: U\times U\to\mathbf{R}^+$ such that for every $x\in U$ one has

$$P(x,A) = \int_A p(x,y)d\lambda(y)$$

for all $A\in\mathcal{B}(U)$, where λ denotes the Lebesgue measure on U. A Borel probability measure μ on U is said to be $\mathcal{X}(U,v)$*-invariant* if

$$\int g\mu dv(g) = \mu$$

or equivalently

$$\int P(x,A)d\mu(x) = \mu(A)$$

for all $A \in \mathcal{B}(U)$. It is easy to see that if $\mathcal{U}(f)$ is given sufficiently small such that $gU \subset W$ for all $g \in \mathcal{U}(f)$ and for some compact neighbourhood W of Λ in U then there exists at least one $\mathcal{X}(U,v)$-invariant measure. As we have shown in Section IV. 1, if the transition probabilities of $\mathcal{X}(U,v)$ have a density then any $\mathcal{X}(U,v)$-invariant measure is absolutely continuous with respect to the Lebesgue measure on U.

For a reason analogous to Proposition I. 1.2, if μ is an $\mathcal{X}(U,v)$-invariant measure then $G^n(v^{\mathbf{Z}} \times \mu)$ weakly converges as $n \to +\infty$ to a G-invariant measure μ^* on $\Omega^{\mathbf{Z}} \times U$ which satisfies $P_1\mu^* = v^{\mathbf{Z}}$ and $P_2\mu^* = \mu$. As a consequence of Theorem 1.1 we have

Corollary 1.1. *In the circumstances of Theorem 1.1, if we assume moreover that the transition probabilities of $\mathcal{X}(U,v)$ have a density and μ is an $\mathcal{X}(U,v)$-invariant measure, then*

$$\tilde{\rho} = \mu^*$$

where $\tilde{\rho}$ is the measure defined in Theorem 1.1 and μ^* is as introduced just above.

Proof. Since the transition probabilities of $\mathcal{X}(U,v)$ have a density, we know that $\mu \ll$ Leb.. By 3) of Theorem 1.1, this implies that $v^{\mathbf{Z}} \times \mu$-a.e. $(w,x) \in \Omega^{\mathbf{Z}} \times U$ satisfies (1.7). Let $\varphi : \Omega^{\mathbf{Z}} \times U \to \mathbf{R}$ be a bounded continuous function. Then

$$\begin{aligned} \int \varphi d\tilde{\rho} &= \int \lim_{n\to+\infty} \frac{1}{n} \sum_{k=0}^{n-1} \varphi \circ G^k(w,x) dv^{\mathbf{Z}} \times \mu \\ &= \lim_{n\to+\infty} \int \varphi d\left(\frac{1}{n} \sum_{k=0}^{n-1} G^k(v^{\mathbf{Z}} \times \mu) \right) \\ &= \int \varphi d\mu^* \end{aligned}$$

and thus $\tilde{\rho} = \mu^*$. □

Remark 1.3. Given $B \geq B_{f,U}$, let $\mathcal{U}(f)$ be a sufficiently small neighbourhood of f in $C^1(U,U)$, and let $v_\varepsilon, \varepsilon > 0$ be a family of Borel probability measures on $C^1(U,U)$ with support in $\Omega = \Omega_{\mathcal{U}(f),B}$ and with $v_\varepsilon \to \delta_f$ as $\varepsilon \to 0$. Let $\tilde{\rho}_\varepsilon$ and $\tilde{\rho}_f$ be the measures on $\Omega^{\mathbf{Z}} \times U$ given by Theorem 1.1 corresponding to $v = v_\varepsilon$ and $v = \delta_f$ respectively. Then it can be shown that

$$\tilde{\rho}_\varepsilon \to \tilde{\rho}_f$$

as $\varepsilon \to 0$ ([You]). Thus, if for each $\varepsilon > 0$ the transition probabilities of $\mathcal{X}(U,v_\varepsilon)$ have a density and μ_ε is an $\mathcal{X}(U,v_\varepsilon)$-invariant measure, then, by Corollary 1.1,

$$\mu_\varepsilon^* \to \rho^*$$

as $\varepsilon \to 0$, where ρ is the SBR measure on Λ. From this it follows that

$$P_2\mu_\varepsilon^* \to P_2\rho^*$$

as $\varepsilon \to 0$, i.e.

$$\mu_\varepsilon \to \rho$$

as $\varepsilon \to 0$. This result can be interpreted as a statement of stochastic stability of SBR measures on hyperbolic attractors ([You]).

§2 Technical Preparations for the Proof of the Main Rusult

Here we present some technical preparations for the proof of Theorem 1.1. We shall only outline main arguments or principle ideas of the proofs of the results presented in this section, leaving details to the reader. Throughout this section we shall always assume that $f : U_0 \to M$ is as given in Section 1, Λ is a hyperbolic attractor of f with basin of attraction U, and $0 < \lambda_0 < 1$ is the number introduced in Definition 1.1 corresponding to Λ.

By using standard machinery of fibre bundle theory, one can extend $T_\Lambda M = E^u \oplus E^s$ to a continuous splitting $T_{U'}M = E^1 \oplus E^2$ of the tangent bundle $T_{U'}M$ over an open neighbourhood U' of Λ (see [Hir]$_3$). We now fix an open neighbourhood V of Λ such that $\overline{V} \subset U \cap U'$. On $T_V M = E^1 \oplus E^2$ we introduce a new norm $\|\cdot\|$ by

$$\|\xi\| = \max\{|\xi_1|, |\xi_2|\}, \quad \xi = \xi_1 + \xi_2 \in E^1 \oplus E^2.$$

The norms $|\cdot|$ and $\|\cdot\|$ on $T_V M$ are clearly equivalent.

Given $\Omega = \Omega_{\mathcal{U}(f),B}$, let $\tilde{\Lambda} = \tilde{\Lambda}_{\mathcal{U}(f),B}$ be as defined in Section 1. It is easy to see that there exists a neighbourhood $\mathcal{U}_2(f)$ of f in $C^1(U,U)$ with $\mathcal{U}_2(f) \subset \mathcal{U}_1(f)$ (see Section 1) such that if $\mathcal{U}(f) \subset \mathcal{U}_2(f)$ and $B \geq B_{f,U}$ then

$$\tilde{\Lambda} \subset \Omega^{\mathbf{Z}} \times V. \tag{2.1}$$

When (2.1) holds true, we denote by $E_{\tilde{\Lambda}}$ the pull-back of $T_V M$ by means of the projection $P_2 : \tilde{\Lambda} \to V, (w,x) \longmapsto x$.

Proposition 2.1. *There exists a neighbourhood $\mathcal{U}_3(f)$ of f in $C^1(U,U)$ with $\mathcal{U}_3(f) \subset \mathcal{U}_2(f)$ such that, if $\mathcal{U}(f) \subset \mathcal{U}_3(f)$ and $B \geq B_{f,U}$, then there is a continuous splitting of $E_{\tilde{\Lambda}}$ into $E^u_{\tilde{\Lambda}} \oplus E^s_{\tilde{\Lambda}}$ and there exists a number $0 < \lambda < 1$ (depending only on $\mathcal{U}_3(f)$) such that the following hold true:*

1) $T_x g_0(w) E^u_{(w,x)} = E^u_{G(w,x)}$ and $T_x g_0(w) E^s_{(w,x)} = E^s_{G(w,x)}$ for each $(w,x) \in \tilde{\Lambda}$;

2) For each $(w,x) \in \tilde{\Lambda}$,

$$\|T_x g_0(w)\xi\| \geq \lambda^{-1}\|\xi\|, \quad \xi \in E^u_{(w,x)},$$

$$\|T_x g_0(w)\eta\| \leq \lambda\|\eta\|, \quad \eta \in E^s_{(w,x)}.$$

Proof. Let $E^1_{\tilde\Lambda}$ and $E^2_{\tilde\Lambda}$ be respectively the pull-backs of E^1 and E^2 by means of the map $P_2 : \tilde\Lambda \to V, (w,x) \longmapsto x$. As usual, by $L(E^1_{\tilde\Lambda}, E^2_{\tilde\Lambda})$ we denote the fibre bundle over $\tilde\Lambda$ whose fibre at point (w,x) is the space of all linear maps from $E^1_{(w,x)}$ to $E^2_{(w,x)}$, and by a continuous section σ of $L(E^1_{\tilde\Lambda}, E^2_{\tilde\Lambda})$ we mean a continuous map $\sigma : \tilde\Lambda \to L(E^1_{\tilde\Lambda}, E^2_{\tilde\Lambda})$ satisfying $\pi \circ \sigma = id$ where $\pi : L(E^1_{\tilde\Lambda}, E^2_{\tilde\Lambda}) \to \tilde\Lambda$ is the natural projection. Let now S^c denote the space of all continuous sections of $L(E^1_{\tilde\Lambda}, E^2_{\tilde\Lambda})$, equipped with the norm $\|\cdot\|$ defined by

$$\|\sigma\| \stackrel{\text{def}}{=} \sup_{(w,x)\in\tilde\Lambda} \|\sigma(w,x)\| < +\infty.$$

It is easy to see that $(S^c, \|\cdot\|)$ is a Banach space with respect to the natural operations of addition and scalar multiplication.

For each $(w,x) \in \tilde\Lambda$, write

$$T_x g^1_w = \begin{bmatrix} G_{11}(w,x) & G_{12}(w,x) \\ G_{21}(w,x) & G_{22}(w,x) \end{bmatrix} : E^1_{(w,x)} \oplus E^2_{(w,x)} \to E^1_{G(w,x)} \oplus E^2_{G(w,x)},$$

$$T_x g_w^{-1} = \begin{bmatrix} \hat G_{11}(w,x) & \hat G_{12}(w,x) \\ \hat G_{21}(w,x) & \hat G_{22}(w,x) \end{bmatrix} : E^1_{(w,x)} \oplus E^2_{(w,x)} \to E^1_{G^{-1}(w,x)} \oplus E^2_{G^{-1}(w,x)}.$$

Choose a number $0 < \varepsilon < 1$ satisfying

$$\lambda_0 + 2\varepsilon < 1, \quad (\lambda_0 + \varepsilon)^{-1} - \varepsilon > 1$$

and write

$$\mu = \frac{\lambda_0 + 2\varepsilon}{(\lambda_0 + \varepsilon)^{-1} - \varepsilon} < 1,$$

$$\lambda = [(\lambda_0 + \varepsilon)^{-1} - \varepsilon]^{-1} < 1.$$

Let now $\mathcal{U}_3(f)$ be a neighbourhood of f in $C^1(U,U)$ with $\mathcal{U}_3(f) \subset \mathcal{U}_2(f)$ such that if $\mathcal{U}(f) \subset \mathcal{U}_3(f)$ and $B \geq B_{f,U}$ then for each $(w,x) \in \tilde\Lambda$

$$\max\{\|G_{11}(w,x)^{-1}\|, \|G_{22}(w,x)\|, \quad \|\hat G_{11}(w,x)\|, \quad \|\hat G_{22}(w,x)^{-1}\|\} < \lambda_0 + \varepsilon,$$

$$\max\{\|G_{12}(w,x)\|, \|G_{21}(w,x)\|, \quad \|\hat G_{12}(w,x)\|, \quad \|\hat G_{21}(w,x)\|\} < \varepsilon.$$

Put $S^c_1 = \{\sigma \in S^c : \|\sigma\| \leq 1\}$ and define a map $\Gamma : S^c_1 \to S^c_1$ by

$$(\Gamma\sigma)(w,x) = \quad [G_{21}(G^{-1}(w,x)) + G_{22}(G^{-1}(w,x)) \circ \sigma(G^{-1}(w,x))]$$

$$\circ [G_{11}(G^{-1}(w,x)) + G_{12}(G^{-1}(w,x)) \circ \sigma(G^{-1}(w,x))]^{-1}$$

for $\sigma \in S^c_1$ and $(w,x) \in \tilde\Lambda$. In a way analogous to the proof of Lemma VI. 4.6 one can check that Γ is a μ-contraction and the unique fixed point, written σ^u, has the following properties:

(i) $\|\sigma^u\| \leq \mu$;

(ii) The vector bundle $E^u_{\tilde\Lambda}$ defined by

$$E^u_{(w,x)} = (id, \sigma^u(w,x))E^1_{(w,x)}, \quad (w,x) \in \tilde\Lambda$$

satisfies

$$T_x g_0(w) E^u_{(w,x)} = E^u_{G(w,x)}, \qquad (w,x) \in \tilde\Lambda;$$

(iii) If $(w,x) \in \tilde\Lambda$ and $\xi \in E^u_{(w,x)}$, then

$$\|T_x g_0(w)\xi\| \geq \lambda^{-1}\|\xi\|.$$

By a completely analogous argument one can also prove that there exists a continuous section σ^s of $L(E^2_{\tilde\Lambda}, E^1_{\tilde\Lambda})$ such that:

(i)′ $\|\sigma^s\| \leq \mu$;

(ii)′ The vector bundle $E^s_{\tilde\Lambda}$ defined by

$$E^s_{(w,x)} = (\sigma^s(w,x), id)E^2_{(w,x)}, \quad (w,x) \in \tilde\Lambda$$

satisfies

$$T_x g_0(w) E^s_{(w,x)} = E^s_{G(w,x)}, \quad (w,x) \in \tilde\Lambda;$$

(iii)′ If $(w,x) \in \tilde\Lambda$ and $\eta \in E^s_{(w,x)}$, then

$$\|T_x g_0(w)\eta\| \leq \lambda\|\eta\|.$$

Thus we obtain a continuous splitting $E_{\tilde\Lambda} = E^u_{\tilde\Lambda} \oplus E^s_{\tilde\Lambda}$ which obviously satisfies the requirements of the proposition. □

Proposition 2.2. *There exists a neighbourhood $\mathcal{U}_4(f)$ of f in $C^1(U,U)$ with $\mathcal{U}_4(f) \subset \mathcal{U}_3(f)$ such that if $\mathcal{U}(f) \subset \mathcal{U}_4(f)$ and $B \geq B_{f,U}$ then there holds the following conclusion: There are numbers (depending only on $\mathcal{U}_4(f)$ and B) $\alpha_u, \beta_u, \gamma_u > 0$ and $0 < \theta_u < 1$, and for each $0 < \delta \leq \alpha_u$ there exists a continuous family of C^1 embedded k_u-dimensional ($k_u = \dim E^u_{(w,x)}, (w,x) \in \tilde\Lambda$) discs $\{W^u_\delta(w,x)\}_{(w,x)\in\tilde\Lambda}$ in U such that the following hold true for each $(w,x) \in \tilde\Lambda$:*

1) $W^u_\delta(w,x) = \exp_x$ Graph $(h_{(w,x)}|_{\{\xi \in E^u_{(w,x)}: \|\xi\| < \delta\}})$ where

$$h_{(w,x)} : \{\xi \in E^u_{(w,x)} : \|\xi\| < \alpha_u\} \to E^s_{(w,x)}$$

is a $C^{1,1}$ map satisfying

(i) $h_{(w,x)}(0) = 0, T_0 h_{(w,x)} = 0$;

(ii) $\mathrm{Lip}(h_{(w,x)}) \leq \beta_u, \mathrm{Lip}(T.h_{(w,x)}) \leq \beta_u$, where Lip$(\cdot)$ is taken with respect to the norm $|\cdot|$;

2) $g_w^{-1} W^u_\delta(w,x) \subset W^u_\delta(G^{-1}(w,x))$;

3) $d^u(g_w^{-n}y, g_w^{-n}z) \leq \gamma_u \theta_u^n d^u(y,z), y, z \in W^u_\delta(w,x), n \in \mathbf{Z}^+$, where $d^u(\ ,\ )$ denotes the distances along $W^u_\delta(G^{-n}(w,x)), n \in \mathbf{Z}^+$.

The proof of this proposition is actually a simplification of that of Theorem III. 3.1. The details are omitted here. We shall call $W_\delta^u(w,x)$ a *local unstable manifold* of $G:\Omega^{\mathbf{Z}}\times U \hookleftarrow$ at (w,x).

Let $\mathcal{U}(f)\subset\mathcal{U}_4(f)$ and $B\geq B_{f,U}$. From 2) of Proposition 2.2 it follows clearly that for each $(w,x)\in\tilde{\Lambda}$ and $0<\delta\leq\alpha_u$,

$$\{w\}\times W_\delta^u(w,x)\subset\tilde{\Lambda}. \tag{2.2}$$

Suppose that $(w,x)\in\tilde{\Lambda}$ and $W^u(w,x)$ is defined as in Section 1. Then, in a way analogous to the proof of Theorem III. 3.2, one has

$$W^u(w,x)=\bigcup_{n\geq 0} g^n_{\tau^{-n}w}W_\delta^u(G^{-n}(w,x))$$

for each $0<\delta\leq\alpha_u$. Thus $W^u(w,x)$ is the image of $E^u_{(w,x)}$ under an injective immersion of class $C^{1,1}$ and is tangent to $E^u_{(w,x)}$ at point x. Moreover, by (2.2) we have

$$\{w\}\times W^u(w,x)\subset\tilde{\Lambda}.$$

The next result is an analogue of Proposition 2.2 for local stable manifolds of $G:\Omega^{\mathbf{Z}}\times U\hookleftarrow$ and the proof is also completely analogous.

Proposition 2.3. *There exists a neighbourhood $\mathcal{U}_5(f)$ of f in $C^1(U,U)$ with $\mathcal{U}_5(f)\subset\mathcal{U}_4(f)$ such that if $\mathcal{U}(f)\subset\mathcal{U}_5(f)$ and $B\geq B_{f,U}$ then there holds the following conclusion: There are numbers (depending only on $\mathcal{U}_5(f)$ and B) $\alpha_s,\beta_s,\gamma_s>0$ and $0<\theta_s<1$, and for each $0<\delta\leq\alpha_s$ there exists a continuous family of C^1 embedded k_s-dimensional ($k_s=\dim E^s_{(w,x)}$, $(w,x)\in\tilde{\Lambda}$) discs $\{W_\delta^s(w,x)\}_{(w,x)\in\tilde{\Lambda}}$ in U such that the following hold true for each $(w,x)\in\tilde{\Lambda}$:*

1) $W_\delta^s(w,x)=\exp_x Graph(l_{(w,x)}|_{\{\eta\in E^s_{(w,x)}:\|\eta\|<\delta\}})$ *where*

$$l_{(w,x)}:\{\eta\in E^s_{(w,x)}:\|\eta\|<\alpha_s\}\to E^u_{(w,x)}$$

is a $C^{1,1}$ map satisfying

(i) $l_{(w,x)}(0)=0, T_0l_{(w,x)}=0$;

(ii) $\mathrm{Lip}(l_{(w,x)})\leq\beta_s$, $\mathrm{Lip}(T.l_{(w,x)})\leq\beta_s$, *where $Lip(\cdot)$ is taken with respect to the norm $|\cdot|$;*

2) $g_w^1W_\delta^s(w,x)\subset W_\delta^s(G(w,x))$;

3) $d^s(g_w^ny,g_w^nz)\leq\gamma_s\theta_s^nd^s(y,z), y,z\in W_\delta^s(w,x), n\in\mathbf{Z}^+$ *where $d^s(\ ,\ )$ denotes the distances along $W_\delta^s(G^n(w,x)), n\in\mathbf{Z}^+$.*

Let $\mathcal{U}(f)\subset\mathcal{U}_5(f)$ and $B\geq B_{f,U}$. For $(w,x)\in\tilde{\Lambda}$ and $y\in W^s_{\alpha_s}(w,x)$, we denote by $E^s_{(w,y)}$ the subspace of T_yM tangent to $W^s_{\alpha_s}(w,x)$ at point y.

Proposition 2.4. *There is a neighbourhood $\mathcal{U}_6(f)$ of f in $C^1(U,U)$ with $\mathcal{U}_6(f)\subset\mathcal{U}_5(f)$, for which there holds the following conclusion: If $\mathcal{U}(f)\subset\mathcal{U}_6(f)$ and $B\geq B_{f,U}$, then there exist numbers (depending only on $\mathcal{U}_6(f)$ and B)*

$0 < \delta_0 \leq \min\{\alpha_u, \alpha_s\}, \alpha_0 > 0$ and $L_0 > 0$ such that for each $(w, x) \in \tilde{\Lambda}$ and $0 < \delta \leq \delta_0$ the family of spaces $\{E^s_{(w,y)} : y \in \cup_{z \in W^u_\delta(w,x)} W^s_\delta(w,z)\}$ is Hölder continuous in y on $\cup_{z \in W^u_\delta(w,x)} W^s_\delta(w,z)$ with exponent α_0 and constant L_0 (for the definition see Definition III. 4.2).

Proof. Let V be as given at the beginning of this section. By an argument analogous to the proof of Lemma III. 4.2 we know that for any given $\Omega = \Omega_{\mathcal{U}(f),B}$ there is a constant $C > 0$ (depending only on B) such that

$$\prod_{i=0}^{n-1} |T(g_i(w)|_V)|_{H^1} \leq C^n \tag{2.3}$$

for all $w \in \Omega^{\mathbf{Z}}$ and $n \geq 1$.

Now fix another open neighbourhood W of Λ which satisfies $\overline{W} \subset V$. Clearly, there exists a neighbourhood $\mathcal{U}_6(f)$ of f in $C^1(U, U)$ with $\mathcal{U}_6(f) \subset \mathcal{U}_5(f)$ such that if $\mathcal{U}(f) = \mathcal{U}_6(f)$ and $B \geq B_{f,U}$ then for some $0 < \hat{\delta} \leq \min\{\alpha_u, \alpha_s\}$ one has $W^s_{\hat{\delta}}(w, x) \subset W$ for all $(w, x) \in \tilde{\Lambda}$. In this case, note that, if $(w, x) \in \tilde{\Lambda}$ and $\langle\ ,\ \rangle_{(w,x)}$ is an inner product on $E_{(w,x)} = E^u_{(w,x)} \oplus E^s_{(w,x)}$ which coincides with the Riemannian metric $\langle\ ,\ \rangle$ on $E^u_{(w,x)}$ and $E^s_{(w,x)}$ and which makes $E^u_{(w,x)}$ and $E^s_{(w,x)}$ orthogonal, then the norm $|\cdot|_{(w,x)}$ induced by $\langle\ ,\ \rangle_{(w,x)}$ is equivalent to $|\cdot|$ uniformly for $(w, x) \in \tilde{\Lambda}$. Note also that if $E^\perp_x$ in Proposition III. 4.1 is replaced by a subspace F_x of H which satisfies $\gamma(E_x, F_x) \geq \gamma_0 > 0$ then the conclusion of that proposition holds true as well with the corresponding exponent and constant changing suitably with γ_0. Clearly, the same kind of argument also holds true for Corollary III. 4.1. Then, by (2.3) and Propositions 2.2 and 2.3, one can obtain the desired conclusion from an argument analogous to the proof of Theorem III. 4.1. □

Proposition 2.4 implies that the family of embedded discs $\{W^s_\delta(w, z)\}_{z \in W^u_\delta(w,x)}$ is absolutely continuous for each $0 < \delta \leq \delta_0$ and $(w, x) \in \tilde{\Lambda}$ if $\mathcal{U}(f) \subset \mathcal{U}_6(f)$ and $B \geq B_{f,U}$. To be precise, we first explain the idea of absolutely continuous family of C^1 embedded k-dimensional discs in M. Let $\Delta \subset M$ be a set and let $\{D_x\}_{x \in \Delta}$ be a continuous family of C^1 embedded k-dimensional ($1 \leq k < m_0$, where $m_0 = \dim M$) discs in M such that $D_y \cap D_z = \emptyset$ if $y, z \in \Delta$ and $y \neq z$. Let $x_0 \in \Delta$ and $p, q \in D_{x_0}$, and let W_p, W_q be two smoothly embedded $(m_0 - k)$-dimensional discs transverse to D_{x_0} at p and q respectively. Then there exist two open submanifolds $\hat{W}_p$ and $\hat{W}_q$ of W_p and W_q respectively such that the so-called Poincaré map

$$P_{\hat{W}_p \hat{W}_q} : \hat{W}_p \cap \left(\bigcup_{x \in \Delta} D_x \right) \to \hat{W}_q \cap \left(\bigcup_{x \in \Delta} D_x \right),$$

$$\hat{W}_p \cap D_x \longmapsto \hat{W}_q \cap D_x$$

is a homeomorphism between $\hat{W}_p \cap (\cup_{x \in \Delta} D_x)$ and $\hat{W}_q \cap (\cup_{x \in \Delta} D_x)$. The family of C^1 embedded discs $\{D_x\}_{x \in \Delta}$ is said to be *absolutely continuous* if each of its

Poincaré maps $P_{\hat{W}_p \hat{W}_q}$ is absolutely continuous (with respect to the Lebesgue measures on W_p and W_q).

Proposition 2.5. *Let $\mathcal{U}(f) \subset \mathcal{U}_6(f)$ and $B \geq B_{f,\mathcal{U}}$. Then for each $(w,x) \in \tilde{\Lambda}$ and $0 < \delta \leq \delta_0$ (see Proposition 2.4), the family of C^1 embedded discs $\{W_\delta^s(w,z)\}_{z \in W_\delta^u(w,x)}$ is absolutely continuous.*

Like in the case of Theorem III. 5.1, the proof of this proposition follows the line of the arguments of Part II of [Kat]. We also omit the details here.

Proposition 2.6. *There is a neighbourhood $\mathcal{U}_7(f)$ of f in $C^1(U,U)$ with $\mathcal{U}_7(f) \subset \mathcal{U}_6(f)$, for which there holds the following conclusion: If $\mathcal{U}(f) = \mathcal{U}_7(f)$ and $B \geq B_{f,\mathcal{U}}$, then there exist an open neighbourhood N of Λ in M, a number $0 < \hat{\alpha}_u \leq \delta_0$ (see Proposition 2.4), a continuous family of C^1 embedded k_u-dimensional discs $\{\hat{W}_\delta^u(w,z)\}_{(w,z) \in \Omega^{\mathbf{Z}} \times N}$ for each $0 < \delta \leq \hat{\alpha}_u$ such that the following holds true:*

1) *$W_{\hat{\alpha}_u}^s(w,x) \subset N$ for each $(w,x) \in \tilde{\Lambda}$;*
2) *$g_0(w)\hat{W}_\delta^u(w,z) \supset \hat{W}_\delta^u(G(w,z))$ if $z \in N$ and $g_0(w)z \in N$;*
3) *If $(w,z) \in \tilde{\Lambda}$, then $\hat{W}_\delta^u(w,z) = W_\delta^u(w,z)$;*
4) *For each $(w,x) \in \tilde{\Lambda}, \bigcup_{z \in W_\delta^s(w,x)} \hat{W}_\delta^u(w,z)$ contains an open neighbourhood of x in M;*
5) *For each $(w,x) \in \tilde{\Lambda}, \bigcup_{z \in W_{\delta/4}^s(w,x)} \hat{W}_{\delta/4}^u(w,z) \subset \bigcup_{z \in W_\delta^u(w,y)} W_\delta^s(w,z)$ if $y \in W_{\delta/4}^s(w,x)$ with $(w,y) \in \tilde{\Lambda}$.*

The idea of the proof of the proposition is the same as that of the proof of [Che] Proposition 2.1 or [Hir]$_3$ Theorem 4.2, which deal with the result in the case of deterministic dynamical systems (for endomorphisms and diffeomorphisms respectively). We refer the reader to those references for an analogous proof.

§3 Proof of the Main Result

In this section we complete the proof of Theorem 1.1. Let $\mathcal{U}_i(f), 1 \leq i \leq 7$ be as introduced in the last section. Put

$$\mathcal{U}_0(f) = \bigcap_{i=1}^{7} \mathcal{U}_i(f) = \mathcal{U}_7(f).$$

Lemma 3.1. *Let $\mathcal{U}(f) \subset \mathcal{U}_0(f)$ and $B \geq B_{f,\mathcal{U}}$, and let $(\hat{w},\hat{x}) \in \tilde{\Lambda}$ be a point such that $\tau^n \hat{w} \neq \hat{w}$ for all $n > 0$ and $W_{\hat{\delta}}^u(\hat{w},\hat{x})$ a local unstable manifold introduced in Proposition 2.2 for some $0 < \hat{\delta} \leq \alpha_u$. If we write $L = W_{\hat{\delta}}^u(\hat{w},\hat{x})$*

and let $\tilde{\lambda}_L$ be the measure on $\{\hat{w}\} \times W^u_{\frac{\delta}{5}}(\hat{w}, \hat{x})$ defined by identifying $\tilde{\lambda}_L$ with the normalized Lebesgue measure λ_L on L, then any limit measure $\tilde{\mu}$ of

$$\frac{1}{n}\sum_{k=0}^{n-1} G^k \tilde{\lambda}_L, \quad n \in \mathbf{N}$$

has absolutely continuous conditional measures on W^u-manifolds.

Proof. We first give some useful estimates. From the definition of $\Omega = \Omega_{\mathcal{U}(f),B}$ and 1) (ii) of Proposition 2.2 it follows that there exists $A > 0$ such that for every $0 < \delta \le \alpha_u$, if $(w, x) \in \tilde{\Lambda}$ and $y, z \in W^u_\delta(w, x)$, then

$$|J^u(w, y) - J^u(w, z)| \le A d^u(y, z)$$

where $J^u(w, y) = |\det(T_y g_0(w)|_{E^u_{(w,y)}})|$ and $J^u(w, z)$ is defined analogously. By this fact and 2), 3) of Proposition 2.2, in a way similar to the proof of Lemma VI. 8.1 one can see that there is $C > 0$ such that for each $(w, x) \in \tilde{\Lambda}$ and $0 < \delta \le \alpha_u$ there holds the estimate

$$\frac{1}{C} \le \prod_{k=1}^{n} \frac{J^u(G^{-k}(w, y))}{J^u(G^{-k}(w, z))} \le C \tag{3.1}$$

for all $y, z \in W^u_\delta(w, x)$ and $n \in \mathbf{N}$, and moreover,

$$\Delta((w, x), (w, y)) \stackrel{\text{def}}{=} \prod_{k=1}^{+\infty} \frac{J^u(G^{-k}(w, x))}{J^u(G^{-k}(w, y))}, \quad y \in W^u_\delta(w, x)$$

is a well-defined function of $y \in W^u_\delta(w, x)$.

Suppose that $\tilde{\mu}$ is a limit measure of $\frac{1}{n}\sum_{k=0}^{n-1} G^k \tilde{\lambda}_L, n \in \mathbf{N}$ and

$$\frac{1}{n_i}\sum_{k=0}^{n_i-1} G^k \tilde{\lambda}_L \to \tilde{\mu}$$

as $i \to +\infty$ for some subsequence $\{n_i\}_{i\ge 0}$. Let $\tilde{V}$ be a Borel subset of $\tilde{\Lambda}$ of positive $\tilde{\mu}$ measure such that it is the disjoint union of $\{w\} \times W^u_\delta(w, x), (w, x) \in \sum$ for some $0 < \delta \le \alpha_u$ and some subset $\sum$ of $\tilde{\Lambda}$. For $(w, y) \in \tilde{V}$, let $\tilde{V}_{(w,y)}$ denote the element of the partition $\{\{w\} \times W^u_\delta(w, x)\}_{(w,x)\in\sum}$ of $\tilde{V}$ that contains (w, y), and let $(\tilde{\mu}|_{\tilde{V}})_{(w,y)}$ denote the conditional probability measure of $\tilde{\mu}|_{\tilde{V}}$ on $\tilde{V}_{(w,y)}$. We may regard $(\tilde{\mu}|_{\tilde{V}})_{(w,y)}$ as a measure on $W^u(w, y)$ by identifying $\{w\} \times (\tilde{V}_{(w,y)})_w$ with $(\tilde{V}_{(w,y)})_w \subset W^u(w, y)$. It is easy to see that, if we can find a finite cover of $\tilde{\Lambda}$ by Borel sets $\tilde{V}_i, i = 1, \cdots, l$ of the type of $\tilde{V}$ such that for each $\tilde{V}_i$ one has

$$(\tilde{\mu}\,|_{\tilde{V}_i})_{(w,y)} \ll \lambda^u_{(w,y)}$$

for $\tilde{\mu}$-a.e. $(w, y) \in \tilde{V}_i$, then $\tilde{\mu}$ has absolutely continuous conditional measures on W^u-manifolds.

We now prove the existence of such a finite cover by constructing certain canonical neighbourhoods of $\tilde{\Lambda}$. Let $(w_0, x_0) \in \tilde{\Lambda}$ and $0 < \delta \leq \min\{\alpha_u, \alpha_s\}$, and let W be an open neighbourhood of w_0 in $\Omega^{\mathbf{Z}}$. Writting $T = W^s_\delta(w_0, x_0)$, we put

$$\Sigma = (W \times T) \cap \tilde{\Lambda}$$

and

$$\tilde{V} = \bigcup_{(w,x)\in\Sigma} \{w\} \times W^u_\delta(w, x).$$

By Propositions 2.2 and 2.3 it is easy to see that we can choose δ and W sufficiently small so that $\tilde{V}$ is the disjoint union of $\{w\} \times W^u_\delta(w, x), (w, x) \in \Sigma$. Let $\tilde{V}$ be a Borel subset of $\tilde{\Lambda}$ thus obtained. It clearly contains an open neighbourhood of (w_0, x_0) in $\tilde{\Lambda}$. We may assume that $\tilde{\mu}(\tilde{V}) > 0$, and we may also assume that $\tilde{\mu}(\partial\tilde{V}) = 0$ ($\partial(\cdot)$ is taken in the topology of $\tilde{\Lambda}$) by shrinking δ or W if necessary.

For each $n \in \mathbf{Z}^+$, let $L_n = \{z \in L : G^n(\hat{w}, z) \in \{w\} \times W^u_\delta(w, x)$ for some $(w, x) \in \sum$ but $G^n(\{\hat{w}\} \times L) \not\supset \{w\} \times W^u_\delta(w, x)\}$. From 3) of Proposition 2.2 it follows that $\lambda_L(L_n) \to 0$ as $n \to +\infty$. So we have

$$\lim_{i\to+\infty} \frac{1}{n_i} \sum_{k=0}^{n_i-1} G^k(\tilde{\lambda}_L |_{\{\hat{w}\}\times(L\backslash L_k)}) = \tilde{\mu}. \tag{3.2}$$

Also, since $\tilde{\mu}(\partial\tilde{V}) = 0$, it holds that

$$\lim_{i\to+\infty} \left(\frac{1}{n_i} \sum_{k=0}^{n_i-1} G^k(\tilde{\lambda}_L |_{\{\hat{w}\}\times(L\backslash L_k)}) \right) (\tilde{V}) = \tilde{\mu}(\tilde{V}). \tag{3.3}$$

Suppose that $G^n(\{\hat{w}\} \times (L\backslash L_n)) \supset \{w\} \times W^u_\delta(w, x)$ for some $(w, x) \in \sum$. Denote by $\tilde{m}_{n,(w,x)}$ the conditional probability measure of $[G^n(\tilde{\lambda}_L |_{\{\hat{w}\}\times(L\backslash L_n)})]|_{\tilde{V}}$ on $\{w\} \times W^u_\delta(w, x)$ and let

$$\rho_n = \frac{d\tilde{m}_{n,(w,x)}}{d\lambda^u_{(w,x)}}$$

where we regard $\lambda^u_{(w,x)}$ as a measure on $\{w\} \times W^u_\delta(w, x)$ by identifying $W^u_\delta(w, x)$ with $\{w\} \times W^u_\delta(w, x)$. Then

$$\begin{aligned} \rho_n(w, y) &= \frac{\prod_{k=1}^{n} \frac{1}{J^u(G^{-k}(w, y))}}{\int_{W^u_\delta(w,x)} \prod_{k=1}^{n} \frac{1}{J^u(G^{-k}(w, y))} d\lambda^u_{(w,x)}(y)} \\ &= \frac{\prod_{k=1}^{n} \frac{J^u(G^{-k}(w, x))}{J^u(G^{-k}(w, y))}}{\int_{W^u_\delta(w,x)} \prod_{k=1}^{n} \frac{J^u(G^{-k}(w, x))}{J^u(G^{-k}(w, y))} d\lambda^u_{(w,x)}(y)} \end{aligned} \tag{3.4}$$

for $(w,y) \in \{w\} \times W^u_\delta(w,x)$.

For each $n \in \mathbf{Z}^+$, let $\rho_n : \tilde{V} \to \mathbf{R}^+$ be defined by (3.4). In view of (3.1), it is not difficult to prove that $\rho_n, n \in \mathbf{Z}^+$ are uniformly continuous functions and ρ_n uniformly converges as $n \to +\infty$ to a continuous function $\rho : \tilde{V} \to \mathbf{R}^+$ which is defined by

$$\rho(w,y) = \frac{\Delta((w,x),(w,y))}{\displaystyle\int_{W^u_\delta(w,x)} \Delta((w,x),(w,y))d\lambda^u_{(w,x)}(y)}$$

if $(w,y) \in \{w\} \times W^u_\delta(w.x)$ and $(w,x) \in \sum$. We now define a Borel measure $\tilde{\nu}$ on $\tilde{V}$ by letting

$$\tilde{\nu}(A) = \int \left[\int_{W^u_\delta(w,x)} \chi_A(w,y)\rho(w,y)d\lambda^u_{(w,x)}(y) \right] d(\tilde{\mu}|_{\tilde{V}})(w,x)$$

for $A \in \mathcal{B}(\tilde{V})$. Then, taking (3.2) and (3.3) into consideration, one can easily show that

$$\lim_{i\to+\infty} \frac{1}{n_i} \sum_{k=0}^{n_i-1} [G^k(\tilde{\lambda}_L \mid_{\{\hat{w}\}\times(L\setminus L_k)})]|_{\tilde{V}} = \tilde{\nu}$$

and hence

$$\tilde{\nu} = \tilde{\mu}|_{\tilde{V}}.$$

Since $\tilde{\Lambda}$ is compact, we can find a finite number of Borel subsets $\tilde{V}_1, \cdots, \tilde{V}_l$ of $\tilde{\Lambda}$ of the type of $\tilde{V}$ constructed above such that they cover $\tilde{\Lambda}$. The proof of this proposition is then completed. □

Proposition 3.1. *Let $\mathcal{U}(f) \subset \mathcal{U}_0(f)$ and $B \geq B_{f,U}$, and let v be a Borel probability measure on $\Omega = \Omega_{\mathcal{U}(f),B}$. Then there exists a G-invariant Borel probability measure $\tilde{\mu}$ on $\Omega^{\mathbf{Z}} \times U$ satisfying:*

1) $P_1\tilde{\mu} = v^{\mathbf{Z}}$;

2) *$\tilde{\mu}$ has absolutely continuous conditional measures on W^u-manifolds.*

Proof. Since $\tau : (\Omega^{\mathbf{Z}}, v^{\mathbf{Z}}) \hookleftarrow$ is ergodic, for $v^{\mathbf{Z}}$-a.e. $w \in \Omega^{\mathbf{Z}}$ one has

$$\frac{1}{n}\sum_{k=0}^{n-1} \delta_{\tau^n w} \to v^{\mathbf{Z}} \tag{3.5}$$

as $n \to +\infty$. Let $L = W^u_{\hat{\delta}}(\hat{w},\hat{x})$ for some $0 < \hat{\delta} \leq \alpha_u$ and some $(\hat{w},\hat{x}) \in \tilde{\Lambda}$ with $\hat{w}$ satisfying (3.5), and let $\tilde{\mu}$ be a measure constructed as in Lemma 3.1. Then $\tilde{\mu}$ clearly satisfies the requirements of the proposition. □

Proposition 3.2. *Let $\mathcal{U}(f) \subset \mathcal{U}_0(f)$ and $B \geq B_{f,U}$, and let v be a Borel probability measure on $\Omega = \Omega_{\mathcal{U}(f),B}$. If $\tilde{\mu}$ is a G-invariant Borel probability* *measure on* $\Omega^{\mathbf{Z}} \times U$ *satisfying* $P_1\tilde{\mu} = v^{\mathbf{Z}}$, *then*

$$h_{\tilde{\mu}}(\mathcal{X}(U,v)) \leq \int \sum_i \lambda^{(i)}(w,x)^+ m_i(w,x) d\tilde{\mu}. \tag{3.6}$$

Proof. By (2.1) we know that

$$\tilde{\Lambda} \subset \Omega^{\mathbf{Z}} \times V.$$

Write

$$d_0 = d(\overline{V}, \partial U) \stackrel{\text{def}}{=} \inf\{d(x,y) : x \in \overline{V}, y \in \partial U\} > 0.$$

Fix two numbers $0 < \rho_0 < d_0$ and $b_0 > 0$ such that for each $x \in U$, the exponential map $\exp_x : \{\xi \in T_x M : |\xi| < \rho_0\} \to B(x, \rho_0)$ is a diffeomorphism and

$$b_0^{-1} d(y,z) \leq |\exp_x^{-1} y - \exp_x^{-1} z| \leq b_0 d(y,z)$$

for any $y, z \in B(x, \rho_0)$.

For $n \in \mathbf{N}$ we define

$$\Omega^n = \{g : g = g_n \circ \cdots \circ g_1, g_i \in \Omega, 1 \leq i \leq n\}.$$

By arguments similar to those in Section II. 1 and by the definition of $\Omega = \Omega_{\mathcal{U}(f),B}$, for each $n \in \mathbf{N}$ there exists $0 < r_n < \rho_0/2$ with the following properties: If $x \in U$ with $d(x,V) < r_n$ and $g \in \Omega^n$, then the map

$$H_{(g,x)} \stackrel{\text{def}}{=} \exp_{g(x)}^{-1} \circ g \circ \exp_x : \{\xi \in T_x M : |\xi| \leq r_n\}$$

$$\to \{\eta \in T_{g(x)} M : |\eta| < \rho_0/2\}$$

is well defined and

$$\sup_{\xi \in T_x M, |\xi| \leq r_n} |T_\xi H_{(g,x)} - T_0 H_{(g,x)}| < a_0$$

where $a_0 = \min\{b_0^{-1}, 1\}$. From this it follows that if $x, y \in U$ satisfy $d(x,V) < r_n$ and $d(x,y) < r_n$ then for any $g \in \Omega^n$ it holds that

$$d(g(y), \exp_{g(x)} \circ T_x g \circ \exp_x^{-1} y) \leq d(x,y). \tag{3.7}$$

Now fix $n \in \mathbf{N}$ arbitrarily. For any given sufficiently small $\varepsilon > 0$, take a maximal ε-separated set E_ε of $\overline{U}$ and then define a measurable partition $\alpha_\varepsilon = \{\alpha_\varepsilon(x) : x \in E_\varepsilon\}$ of U such that $\alpha_\varepsilon(x) \subset \overline{\text{int}(\alpha_\varepsilon(x))}$ and $\text{int}\,(\alpha_\varepsilon(x)) = \{y \in U : d(y,x) < d(y,x_i) \text{ if } x \neq x_i \in E_\varepsilon\}$. By Proposition 1.3 we have

$$\begin{aligned} n h_{\tilde{\mu}}(\mathcal{X}(U,v)) = \ & n h_{\tilde{\mu}}^{\sigma}(G) = h_{\tilde{\mu}}^{\sigma}(G^n) \\ = \ & \lim_{\varepsilon \to 0} h_{\tilde{\mu}}^{\sigma}(G^n, \tilde{\alpha}_\varepsilon) \end{aligned} \tag{3.8}$$

where $\tilde{\alpha}_\varepsilon = \{\Omega^{\mathbf{Z}} \times \alpha_\varepsilon(x) : x \in E_\varepsilon\}$.

Let $0 < \varepsilon < r_n/4(1+b_0)$. In a way analogous to (2.3) of Chapter II one has for all $l \geq 1$

$$H_{\tilde{\mu}}\left(\bigvee_{k=0}^{l-1} G^{-kn}\tilde{\alpha}_\varepsilon \Big| \sigma\right) \tag{3.9}$$
$$\leq \; H_{\tilde{\mu}}(\tilde{\alpha}_\varepsilon|\sigma) + (l-1)H_{\tilde{\mu}}(G^{-n}\tilde{\alpha}_\varepsilon \mid \tilde{\alpha}_\varepsilon \vee \sigma).$$

Define $g(x) = \begin{cases} 0 & \text{if } x = 0 \\ -x\log x & \text{if } x > 0 \end{cases}$, by (3.9) we have

$$\begin{aligned} & h_{\tilde{\mu}}^{\sigma}(G^n, \tilde{\alpha}_\varepsilon) \\ \leq \; & H_{\tilde{\mu}}(G^{-n}\tilde{\alpha}_\varepsilon \mid \tilde{\alpha}_\varepsilon \vee \sigma) \\ = \; & \int H_{\mu_w}((g_w^n)^{-1}\alpha_\varepsilon \mid \alpha_\varepsilon) dv^{\mathbf{Z}}(w) \\ = \; & \int \sum_{x\in E_\varepsilon} \mu_w(\alpha_\varepsilon(x)) \sum_{y \in E_\varepsilon} g\left(\frac{\mu_w((g_w^n)^{-1}\alpha_\varepsilon(y) \cap \alpha_\varepsilon(x))}{\mu_w(\alpha_\varepsilon(x))}\right) dv^{\mathbf{Z}}(w) \\ \leq \; & \int \sum_{x \in E_\varepsilon} \mu_w(\alpha_\varepsilon(x)) \log N_x^n(w) dv^{\mathbf{Z}}(w), \end{aligned} \tag{3.10}$$

where $N_x^n(w)$ is the number of elements of α_ε which intersect $g_w^n\alpha_\varepsilon(x)$. Now suppose that $\alpha_\varepsilon(x) \cap V \neq \emptyset$. Then, by (3.7),

$$g_w^n\alpha_\varepsilon(x) \subset g_w^n \exp_x \overline{B(0,\varepsilon)} \subset \exp_{g_w^n x} B(T_x g_w^n B(0,\varepsilon), b_0\varepsilon)$$

where $B(Q,\delta)$ is the δ-neighbourhood of $Q \subset T_{g_w^n x}M$ in $T_{g_w^n x}M$. If

$$\alpha_\varepsilon(y) \cap g_w^n \alpha_\varepsilon(x) \neq \emptyset,$$

we have

$$B(y, b_0^{-2}\frac{\varepsilon}{2}) \cap \exp_{g_w^n x} B(T_x g_w^n B(0,\varepsilon), 2b_0\varepsilon) \neq \emptyset$$

and then

$$B(\exp_{g_w^n x}^{-1} y, b_0^{-1}\frac{\varepsilon}{2}) \cap B(T_x g_w^n B(0,\varepsilon), 2b_0\varepsilon) \neq \emptyset.$$

Since unitary operators preserve distances, it is easy to see that the number of disjoint balls which intersect $B(T_x g_w^n B(0,\varepsilon), 2b_0\varepsilon)$ and whose diameters are $b_0^{-1}\varepsilon$ does not exceed $C\prod_{i=0}^{m_0} \max\{\delta_i(T_x g_w^n), 1\}$ where C is a constant depending only on b_0 and $m_0 (m_0 = \dim M)$ and $\delta_i(T_x g_w^n), 1 \leq i \leq m_0$ are as introduced at the beginning of Section II. 2. Note that $\tilde{\mu}(\tilde{\Lambda}) = 1$ and $\tilde{\Lambda} \subset \Omega^{\mathbf{Z}} \times V$. Thus, by

(3.10),

$$h^{\sigma}_{\tilde{\mu}}(G^n, \tilde{\alpha}_\varepsilon)$$

$$\leq \int \sum_{\substack{x \in E_\varepsilon \\ \alpha_\varepsilon(x) \cap V \neq \emptyset}} \mu_w(\alpha_\varepsilon(x)) \log \left[C \prod_{i=1}^{m_0} \max\{\delta_i(T_x g^n_w), 1\} \right] dv^{\mathbf{Z}}(w)$$

$$= \log C + \sum_{i=1}^{m_0} \int H^{(i)}_\varepsilon(w, y) d\tilde{\mu}$$

where $H^{(i)}_\varepsilon(w, y) = \log^+ \delta_i(T_x g^n_w)$ if $y \in \alpha_\varepsilon(x)$. In a way similar to the proof of (2.5) in Chapter II, we have

$$\begin{aligned} \lim_{\varepsilon \to 0} \sum_{i=1}^{m_0} \int H^{(i)}_\varepsilon(w, y) d\tilde{\mu} &= \sum_{i=1}^{m_0} \int \log^+ \delta_i(T_y g^n_w) d\tilde{\mu} \\ &\leq \int \log |(T_y g^n_w)^\wedge| d\tilde{\mu} \\ &= n \int \sum_i \lambda^{(i)}(w, y)^+ m_i(w, y) d\tilde{\mu}. \end{aligned}$$

Hence, by (3.8), for all $n \in \mathbf{N}$

$$h_{\tilde{\mu}}(\mathcal{X}(U, v)) \leq \frac{1}{n} \log C + \int \sum_i \lambda^{(i)}(w, y)^+ m_i(w, y) d\tilde{\mu}$$

which implies (3.6) by letting $n \to +\infty$. □

Proposition 3.3. *Let $\mathcal{U}(f) \subset \mathcal{U}_0(f)$ and $B \geq B_{f,U}$, and let v be a Borel probability measure on $\Omega = \Omega_{\mathcal{U}(f),B}$. If $\tilde{\mu}$ is a G-invariant Borel probability measure on $\Omega^{\mathbf{Z}} \times U$ with $P_1\tilde{\mu} = v^{\mathbf{Z}}$, then the following two conditions are equivalent:*

1) *$\tilde{\mu}$ has absolutely continuous conditional measures on W^u-manifolds;*
2) *$h_{\tilde{\mu}}(\mathcal{X}(U, v)) = \int \sum_i \lambda^{(i)}(w, x)^+ m_i(w, x) d\tilde{\mu}$.*

As for the proof of this proposition, note that the proof of Theorem VI. 1.1 actually yields the following more general result: Let $\mathcal{X}(M, v)$ and $G : \Omega^{\mathbf{Z}} \times M \hookleftarrow$ be as defined in Section VI. 1 and let $\tilde{\mu}$ be a G-invariant Borel probability measure on $\Omega^{\mathbf{Z}} \times M$ satisfying $P_1\tilde{\mu} = v^{\mathbf{Z}}$, then $\tilde{\mu}$ has absolutely continuous conditional measures on W^u-manifolds if and only if $h^{\sigma}_{\tilde{\mu}}(G) = \int \sum_i \lambda^{(i)}(w, x)^+ m_i(w, x) d\tilde{\mu}$, where $\lambda^{(i)}(w, x), 1 \leq i \leq r(w, x)$ are the Lyapunov exponents of $G : (\Omega^{\mathbf{Z}} \times M, \tilde{\mu}) \hookleftarrow$ at (w, x) and $m_i(w, x)$ is the multiplicity of $\lambda^{(i)}(w, x)$. It is then easy to see that Proposition 3.4 can be proved by a simplification of the arguments of Sections VI. 2–8, the simplification being due to the non-existence of zero

Lyapunov exponent for the present system $G : (\Omega^{\mathbf{Z}} \times U, \tilde{\mu}) \hookleftarrow$. We leave the details of the proof to the reader.

Analogous to Corollary VI. 8.2, there also follows the following result:

Corollary 3.1. *In the circumstances of Proposition 3.3, let $\tilde{\mu}$ be a G-invariant Borel probability measure on $\Omega^{\mathbf{Z}} \times U$ with $P_1\tilde{\mu} = v^{\mathbf{Z}}$ such that 2) of Proposition 3.3 holds true. If η is a partition of $\tilde{\Lambda}$ subordinate to W^u-manifolds with respect to $\tilde{\mu}$ and let ρ be the density of $\tilde{\mu}^{\eta}_{(w,x)}$ with respect to $\lambda^u_{(w,x)}$, then for $\tilde{\mu}$-a.e. (w,x), there exist a countable number of disjoint open subsets $U_n(w,x), n \in \mathbf{N}$ of $W^u(w,x)$ such that $\cup_{n\in\mathbf{N}} U_n(w,x) \subset \eta_w(x), \lambda^u_{(w,x)}(\eta_w(x) \setminus \cup_{n\in\mathbf{N}} U_n(w,x)) = 0$ and on each $U_n(w,x)$ ρ is a strictly positive function satisfying*

$$\frac{\rho(w,y)}{\rho(w,z)} = \prod_{k=1}^{+\infty} \frac{J^u(G^{-k}(w,z))}{J^u(G^{-k}(w,y))}, \quad y,z \in U_n(w,x),$$

in particular, log ρ restricted to each $U_n(w,x)$ is Lipschitz along $W^u(w,x)$.

Proposition 3.4. *Let $B \geq B_{f,U}$ be given. If $\mathcal{U}(f) \subset \mathcal{U}_0(f)$ is sufficiently small, v is a Borel probability measure on $\Omega = \Omega_{\mathcal{U}(f),B}$ and $\tilde{\mu}$ is a G-invariant Borel probability measure on $\Omega^{\mathbf{Z}} \times U$ satisfying 1) and 2) of Proposition 3.1, then*

$$\frac{1}{n}\sum_{k=0}^{n-1} \delta_{G^k(w,y)} \to \tilde{\mu}$$

as $n \to +\infty$ for $v^{\mathbf{Z}}\times$ Leb.-a.e. $(w,y) \in \Omega^{\mathbf{Z}} \times U$.

Proof. First consider $f|_\Lambda$. By the spectral decomposition theorem (see Section 3.B of [Bow]$_1$), every hyperbolic attractor can be decomposed into finitely many components $\Lambda_1, \cdots, \Lambda_{l_0}$ such that $f\Lambda_i = \Lambda_{i+1}, 1 \leq i \leq l_0$ (where $\Lambda_{l_0+1} = \Lambda_1$) and for each $1 \leq i \leq l_0, f^{l_0}|_{\Lambda_i}$ is topologically mixing, i.e. for any two subsets V, W of Λ_i, open in the topology of Λ_i, there is a positive integer K such that $f^{-l_0k}V \cap W \neq \emptyset$ for all $k \geq K$. The basin of attraction U of Λ can also be written as the union of open sets $U_1, \cdots, U_{l_0}$ which satisfy $U_i \supset \Lambda_i$ and $f\overline{U}_i \subset U_{i+1}$ for each $1 \leq i \leq l_0$ (where $U_{l_0+1} = U_1$).

Correspondingly, if $\mathcal{U}(f) \subset \mathcal{U}_0(f)$ is sufficiently small and $B \geq B_{f,U}$, then $G(\Omega^{\mathbf{Z}} \times U_i) \subset \Omega^{\mathbf{Z}} \times U_{i+1}, 1 \leq i \leq l_0$ and there exist disjoint compact sets $\tilde{\Lambda}_1, \cdots, \tilde{\Lambda}_{l_0}$ such that $\tilde{\Lambda} = \cup_{i=0}^{l_0} \tilde{\Lambda}_i, \tilde{\Lambda}_i \subset \Omega^{\mathbf{Z}} \times U_i$ and $G\tilde{\Lambda}_i = \tilde{\Lambda}_{i+1}$ for all $1 \leq i \leq l_0$ (where $\tilde{\Lambda}_{l_0+1} = \tilde{\Lambda}_1$). In this case, if $\tilde{\mu}$ is a G-invariant Borel probability measure on $\Omega^{\mathbf{Z}} \times U$ with $P_1\tilde{\mu} = v^{\mathbf{Z}}$, then $\tilde{\mu}$ has the expression

$$\tilde{\mu} = \frac{1}{l_0}\sum_{i=1}^{l_0} \tilde{\mu}_i$$

with $\tilde{\mu}_i$ being a Borel probability measure on $\Omega^{\mathbf{Z}} \times U_i$ which has support in $\tilde{\Lambda}_i$ and satisfies $G\tilde{\mu}_i = \tilde{\mu}_{i+1}$ and $P_1\tilde{\mu}_i = v^{\mathbf{Z}}, 1 \le i \le l_0$ (where $\tilde{\mu}_{l_0+1} = \tilde{\mu}_1$).

To prove the conclusion of the proposition, we first consider the case $l_0 = 1$. Now let $B \ge B_{f,U}$ be given and let $\mathcal{U}(f) \subset \mathcal{U}_0(f)$. Fix $0 < \delta \le \hat{\alpha}_u/2$ (see Proposition 2.6) and let $\{\hat{W}^u_\delta(w,z)\}_{(w,z)\in\Omega^{\mathbf{Z}}\times N}$ be a continuous family of C^1 embedded k_u-dimensional discs in U given by Proposition 2.6. For each $(w,x) \in \tilde{\Lambda}$, put $N_\delta(w,x) = \bigcup_{z\in W^s_\delta(w,x)} \hat{W}^u_\delta(w,z)$ and $\tilde{N}_\delta(w,x) = N_\delta(w,x) \cap \tilde{\Lambda}_w$, where $\tilde{\Lambda}_w = \{x : (w,x) \in \tilde{\Lambda}\}$. According to Proposition 2.6, $N_\delta(w,x)$ contains an open neighbourhood of x in U.

Denote by w_f the point $(\cdots, f, f, f, \cdots)$ in $\Omega^{\mathbf{Z}}$. Since Λ is compact, there exist $x_1, \cdots, x_l \in \Lambda$ such that $\bigcup_{i=1}^l N_{\delta/8}(w_f, x_i)$ constitutes a neighbourhood of Λ in U. By the topological mixing property of $f|_\Lambda$ and 3) of Proposition 2.2, there exists $K > 0$ such that for all $1 \le i,j \le l$, $f^K\tilde{N}_{\delta/8}(w_f,x_i)$ intersects $\tilde{N}_{\delta/8}(w_f,x_j)$ with $f^K\tilde{N}_{\delta/8}(w_f,x_i) \supset W^u_{2\delta}(w_f,z)$ for some $z \in W^s_{\delta/8}(w_f,x_j) \cap \Lambda$. From this and Proposition 2.6 it follows that for the given $B \ge B_{f,U}$ and $\delta > 0$, if $\mathcal{U}(f) \subset \mathcal{U}_0(f)$ is sufficiently small then for each $w \in \Omega^{\mathbf{Z}}$ there exist $y_1, \cdots, y_l \in \tilde{\Lambda}_w$ and $z_1, \cdots, z_l \in \tilde{\Lambda}_{\tau^K w}$ such that $\bigcup_{i=1}^l N_{\delta/4}(w,y_i)$ is a neighbourhood of $\tilde{\Lambda}_w$, $\bigcup_{i=1}^l N_{\delta/4}(\tau^K w, z_i)$ is a neighbourhood of $\tilde{\Lambda}_{\tau^K w}$ and for all $1 \le i,j \le l$, $g^K_w \tilde{N}_{\delta/4}(w,y_i)$ intersects $\tilde{N}_\delta(\tau^K w, z_j)$ with $g^K_w\tilde{N}_{\delta/4}(w,y_i) \supset W^u_\delta(\tau^K w, z)$ for some $z \in W^s_{\delta/4}(\tau^K w, z_j) \cap \tilde{\Lambda}_{\tau^K w}$.

Now let $\tilde{\mu}'$ be an ergodic component of $\tilde{\mu}$. For a reason analogous to Proposition VI.2.1, $\tilde{\mu}'$ also has absolutely continuous conditional measures on W^u-manifolds. Moreover, by (3.5) we may assume that $P_1\tilde{\mu}' = v^{\mathbf{Z}}$.

Consider $w \in \Omega^{\mathbf{Z}}$. The point w satisfies (3.5) and will be subjected to a finite number of $v^{\mathbf{Z}}$-a.e. assumptions. Let μ'_w be the conditional probability measure of $\tilde{\mu}'$ on $\tilde{\Lambda}_w$ (identified with $\{w\} \times \tilde{\Lambda}_w$) and let i be such that $\mu'_w([\bigcup_{z\in W^s_{\delta/4}(w,y_i)} \hat{W}^u_\delta(w,z)] \cap \tilde{\Lambda}_w) > 0$. From Corollary 3.1 one can see that there exists a local unstable manifold $W^u_\delta(w,p), p \in W^s_{\delta/4}(w,y_i)$ with the property that $\lambda^u_{(w,p)}$-a.e. $y \in W^u_\delta(w,p)$ satisfies

$$\frac{1}{n}\sum_{k=0}^{n-1} \delta_{G^k(w,y)} \to \tilde{\mu}' \quad \text{as } n \to +\infty. \tag{3.11}$$

The absolute continuity property of the embedded discs $\{W^s_\delta(w,z)\}_{z\in W^u_\delta(w,p)}$ (see Proposition 2.5) together with 3) of Proposition 2.3 and 5) of Proposition 2.6 implies that $\lambda^u_{(w,z)}$-a.e. $y \in \hat{W}^u_{\delta/4}(w,z)$ satisfies (3.11) for each $z \in W^s_{\delta/4}(w,y_i)$. From this and our choice of the y_i's and z_j's it follows that for each z_j there exists a local unstable manifold $W^u_\delta(\tau^K w, p_j), p_j \in W^s_{\delta/4}(\tau^K w, z_j)$ with the property that for $\lambda^u_{(\tau^K w, p_j)}$-a.e. $y \in W^u_\delta(\tau^K w, p_j)$

$$\frac{1}{n}\sum_{k=0}^{n-1} \delta_{G^k(\tau^K w,y)} \to \tilde{\mu}' \tag{3.12}$$

as $n \to +\infty$. For each $1 \le j \le l$, the absolute continuity property of the family of embedded discs $\{W^s_\delta(\tau^K w, z)\}_{z \in W^u_\delta(\tau^K w, p_j)}$ together with 3) of Proposition 2.3 implies that Leb.-a.e. $y \in \cup_{z \in W^u_\delta(\tau^K w, p_j)} W^s_\delta(\tau^K w, z)$ satisfies (3.12), and hence, by 5) of Proposition 2.6, Leb.-a.e. $y \in N_{\delta/4}(\tau^K w, z_j)$ satisfies (3.12).

It is easy to see that for the given $B \ge B_{f,U}$ and $\delta > 0$, if $\mathcal{U}(f) \subset \mathcal{U}_0(f)$ is sufficiently small then there exists an open subset W of U such that $W \subset \cup_{j=1}^{l} N_{\delta/4}(w, z_j)$ for $v^{\mathbf{Z}}$-a.e. $w \in \Omega^{\mathbf{Z}}$ and $\Omega^{\mathbf{Z}} \times W \supset \tilde{\Lambda}$. In this case it holds clearly that $v^{\mathbf{Z}} \times$ Leb.-a.e. $(w, y) \in \Omega^{\mathbf{Z}} \times W$ satisfies (3.11). The same is true for $v^{\mathbf{Z}} \times$ Leb.-a.e. $(w, y) \in \Omega^{\mathbf{Z}} \times U$ since $G^n(\Omega^{\mathbf{Z}} \times U) \subset \Omega^{\mathbf{Z}} \times W$ for all n greater than some n_0. This also clearly implies that $\tilde{\mu}' = \tilde{\mu}$. The conclusion of the proposition is then proved for the case $l_0 = 1$.

We now prove the proposition for the case $l_0 > 1$. Let $B \ge B_{f,U}$ be given. Consider $f^{l_0} : U_i \hookleftarrow$ and $G^{l_0} : \Omega^{\mathbf{Z}} \times U_i \hookleftarrow$, $1 \le i \le l_0$ for $\mathcal{U}(f) \subset \mathcal{U}_0(f)$. From the arguments above we see that, if $\mathcal{U}(f)$ is sufficiently small, v is a Borel probability measure on $\Omega = \Omega_{\mathcal{U}(f),B}$ and $\tilde{\mu}$ is a G-invariant Borel probability measure on $\Omega^{\mathbf{Z}} \times U$ satisfying 1) and 2) of Proposition 3.1, then for each $1 \le i \le l_0, v^{\mathbf{Z}} \times$ Leb.-a.e. $(w, y) \in \Omega^{\mathbf{Z}} \times U_i$ satisfies

$$\lim_{n \to +\infty} \frac{1}{n} \sum_{k=0}^{n-1} \delta_{G^{l_0 k}(w,y)} = \tilde{\mu}_i$$

and thus satisfies

$$\lim_{n \to +\infty} \frac{1}{n l_0} \sum_{k=0}^{n l_0 - 1} \delta_{G^k(w,y)} = \frac{1}{l_0} \sum_{i=1}^{l_0} \tilde{\mu}_i = \tilde{\mu}$$

which implies

$$\lim_{n \to +\infty} \frac{1}{n} \sum_{k=0}^{n-1} \delta_{G^k(w,y)} = \tilde{\mu}.$$

This completes the proof of the proposition. □

Theorem 1.1 follows clearly from Propositions 3.1–3.4.

Appendix

A Margulis-Ruelle Inequality for Random Dynamical Systems

As is described in the Introduction and Remark II.2.1, in this appendix we adopt with some modifications J. Bahnmüller and T. Bogenschütz's argument [Bah] about Ruelle's inequality. Their argument is carried out within a more general (than the "i. i. d." case) framework, due to the point of view of L. Arnold, of "stationary" random dynamical systems (see [Arn] for an introduction to this subject).

§1 Notions and Preliminary Results

Let $(\Omega, \mathcal{F}, P)$ be a probability space, θ: $(\Omega, \mathcal{F}, P) \hookleftarrow$ an invertible and ergodic measure-preserving transformation. We call $(\Omega, \mathcal{F}, P, \theta)$ a *Polish system* if Ω is a Polish space and $\mathcal{F}$ is it's Borel σ-algebra. Let $(X, \mathcal{B})$ be a probability space. We call a map

$$\phi : \mathbf{Z}^+ \times \Omega \times X \longrightarrow X, \qquad (n, \omega, x) \mapsto \phi(n, \omega, x)$$

a (measurable) *random dynamical system* (RDS) on $(X, \mathcal{B})$ over $(\Omega, \mathcal{F}, P, \theta)$ if the following hold true:

1) ϕ is measurable;

2) Define $\phi(n, \omega) : X \to X, x \mapsto \phi(n, \omega, x)$ for $n \in \mathbf{Z}^+$ and $\omega \in \Omega$. Then $\phi(n, \cdot)$, $n \in \mathbf{Z}^+$ is a cocycle over θ, i.e. $\phi(0, \omega) = \mathrm{id}$ and

$$\phi(n + m, \omega) = \phi(n, \theta^m \omega) \circ \phi(m, \omega)$$

for all $n, m \in \mathbf{Z}^+$ and all $\omega \in \Omega$.

Let ϕ be an RDS defined above. It is said to be continuous if X is a topological space (with it's Borel σ-algebra) and if $\phi(n, \omega) : X \to X$ is continuous for all $n \in \mathbf{Z}^+$ and all $\omega \in \Omega$. It is said to be C^r $(r \geq 1)$ if X is a C^r manifold and $\phi(n, \omega)$ is C^r differentiable for all $n \in \mathbf{Z}^+$ and all $\omega \in \Omega$. In the sequel we often omit mentioning $(\Omega, \mathcal{F}, P, \theta)$ when speaking of an RDS ϕ.

Let ϕ be an RDS on X. Define

$$\Theta : \Omega \times X \longrightarrow \Omega \times X, \qquad (\omega, x) \mapsto (\theta\omega, \phi(1, \omega)x)$$

and call it the *skew product transformation* induced by ϕ. A probability measure μ on $(\Omega \times X, \mathcal{F} \times \mathcal{B})$ is said to be *ϕ-invariant* if it is invariant under Θ and if it has marginal P on Ω. Invariant measures always exist for continuous RDS on a compact metric space X (which is in complete analogy with deterministic dynamical systems).

Denote by $Pr(X)$ the space of probability measures on $(X, \mathcal{B})$, endowed with the smallest σ-algebra making the maps $Pr(X) \to \mathbf{R}$, $\nu \mapsto \int_X h d\nu$ measurable with h varying over the bounded measurable functions on X. Given a probability measure μ on $\Omega \times X$ with marginal P on Ω, we will call a measurable map $\mu_\cdot$: $\Omega \to Pr(X)$, $\omega \mapsto \mu_\omega$ a *disintegration* of μ (with respect to P) if

$$\mu(A) = \int_\Omega \int_X \chi_A(\omega, x)\, d\mu_\omega(x)\, dP(\omega)$$

for all $A \in \mathcal{F} \times \mathcal{B}$. Disintegrations exist and are unique (P–a.e.) in almost all interesting cases, e. g., if X is a Polish space. In such a situation the ϕ-invariance of μ is equivalent to the validity of the equation

$$\phi(n, \omega)\mu_\omega = \mu_{\theta^n \omega}, \qquad n \in \mathbf{Z}^+, P - \text{a.e.}\omega. \tag{1.1}$$

Henceforth in this appendix we will always assume existence and uniqueness of a disintegration.

Let μ be a ϕ-invariant measure. For every finite measurable partition ξ of X, one can show that the limit

$$h_\mu(\phi, \xi) \stackrel{\text{def}}{=} \lim_{n \to \infty} \frac{1}{n} H_{\mu_\omega}\Big(\bigvee_{k=0}^{n-1} \phi(k, \omega)^{-1}\xi\Big)$$

exists and is constant for P-a.e. ω (see [Bog] Theorem 2.2). The number

$$h_\mu(\phi) \stackrel{\text{def}}{=} \sup\{h_\mu(\phi, \xi) : \xi \text{ is a finite partition of } X\}$$

is called the (measure-theoretic) *entropy* of ϕ with respect to μ. It is shown that the entropy $h_\mu(\phi)$ coincides with the conditional entropy of Θ with respect to $\pi^{-1}\mathcal{F}$, i.e.

$$h_\mu(\phi) = h_\mu^{\pi^{-1}\mathcal{F}}(\Theta) \tag{1.2}$$

where π: $\Omega \times X \to \Omega$ is the natural projection on the first factor (see [Kif]$_1$ Theorem II.1.4 or [Bog] Theorem 3.1).

Lemma 1.1. *Let ϕ be an RDS on $(X, \mathcal{F})$ over $(\Omega, \mathcal{F}, P, \theta)$, and μ a ϕ-invariant measure.*

1) Define ϕ^k by $\phi^k(n, \omega) = \phi(kn, \omega), n \in \mathbf{Z}^+, \omega \in \Omega$ for $k \in \mathbf{N}$. Then for all $k \in \mathbf{N}$ one has

$$h_\mu(\phi^k) = k h_\mu(\phi).$$

2) Assume that X is a compact metric space and $(\Omega, \mathcal{F}, P, \theta)$ is a Polish system. If $\{\xi_k\}_{k=1}^{+\infty}$ is a sequence of finite partitions of X with $\lim_{k \to +\infty}$ diam $\xi_k = 0$, then

$$h_\mu(\phi) = \lim_{k \to +\infty} h_\mu(\phi, \xi_k).$$

Proof: The proof of 1) is analogous to that of Theorem 0.4.3. The proof of 2) is analogous to that of Theorem I.2.5 □

Now let ϕ be a C^1 RDS on a C^∞ Riemannian manifold X and μ a ϕ-invariant measure. If

$$\int_{\Omega\times X} \log^+ |T_x\phi(1,\omega)|\, d\mu < +\infty \tag{1.3}$$

holds true, by Oseledec multiplicative ergodic theorem we know that for μ-a.e. (ω, x) there exist measurable (in (ω, x)) numbers

$$\lambda^{(1)}(\omega,x) < \lambda^{(2)}(\omega,x) < \cdots < \lambda^{(r(\omega,x))}(\omega,x)$$

($\lambda^{(1)}(\omega,x)$ may be $-\infty$) and an associated measurable (in (ω,x)) filtration by linear subspaces of T_xM

$$\{0\} = V^{(0)}_{(\omega,x)} \subset V^{(1)}_{(\omega,x)} \subset \cdots \subset V^{(r(\omega,x))}_{(\omega,x)} = T_xM$$

such that

$$\lim_{n\to\infty} \frac{1}{n} \log |T_x\phi(n,\omega)\xi| = \lambda^{(i)}(\omega,x)$$

if $\xi \in V^{(i)}_{(\omega,x)} \setminus V^{(i-1)}_{(\omega,x)}$, $1 \le i \le r(\omega,x)$. The numbers $\lambda^{(i)}(\omega,x)$, $1 \le i \le r(\omega,x)$ are called the *Lyapunov exponents* of ϕ at point (ω,x) and $m_i(\omega,x) = \dim V^{(i)}_{(\omega,x)} - \dim V^{(i-1)}_{(\omega,x)}$ is called the *multiplicity* of $\lambda^{(i)}(\omega,x)$.

§2 The Main Result and It's Proof

The main result of this appendix is the following random version of the Margulis-Ruelle inequality.

Theorem 2.1 (Margulis-Ruelle inequality for RDS). *Let X be a d-dimensional compact C^∞ Riemannian manifold without boundary. Let ϕ be a C^1 RDS on X over $(\Omega,\mathcal{F},P,\theta)$ and μ a ϕ-invariant measure. Assume that*

$$\int_\Omega \log^+ |\phi(1,\omega)|_{C^1}\, dP(\omega) < +\infty \tag{2.1}$$

where $|\phi(1,\omega)|_{C^1} = \sup_{x\in X} |T_x\phi(1,\omega)|$. Then we have

$$h_\mu(\phi) \le \int_{\Omega\times X} \sum_i \lambda^{(i)}(\omega,x)^+ m_i(\omega,x)\, d\mu. \tag{2.2}$$

Corollary 2.1. *Let* $\mathcal{X}^+(M,v,\mu)$ *be as defined in Section I.1 with* $\mathrm{Diff}^2(M)$ *being replaced by* $\mathrm{Diff}^1(M)$. *Assume that* v *satisfies*

$$\int_{\mathrm{Diff}^1(M)} \log^+ |f|_{C^1}\, dv(f) < +\infty.$$

Then

$$h_\mu(\mathcal{X}^+(M,v)) \leq \int_M \sum_i \lambda^{(i)}(x)^+ m_i(x) d\mu.$$

Proof: Put $W = \mathrm{Diff}^1(M)$. Define an RDS ϕ on M over $(W^{\mathbf{Z}}, \mathcal{B}(W)^{\mathbf{Z}}, v^{\mathbf{Z}}, \tau)$ by

$$\phi : \mathbf{Z}^+ \times W^{\mathbf{Z}} \times M \longrightarrow M, \qquad (n,w,x) \mapsto f_w^n x,$$

where τ and f_w^n is as introduced in Section VI.1.A. By Proposition I.1.2, μ^* is a ϕ-invariant measure. Applying Theorem 2.1 to ϕ with invariant measure μ^*, we prove the corollary by (VI.1.2) and (VI.1.5). □

Remark 2.1. Clearly, Corollary 2.1 confirms in part what we have said in the Introduction and Remark II.2.1 about Ruelle's inequality.

Remark 2.2. Noting that the random maps $\phi(n,\omega)$, $n \in \mathbf{Z}^+, \omega \in \Omega$ in Theorem 2.1 are assumed to be only C^1 maps, one can show that the conclusion of Corollary 2.1 also holds true if $\mathrm{Diff}^1(M)$ is replaced by $C^1(M,M)$. Proof of this fact is analogous to that of Corollary 2.1.

Now we proceed to prove Theorem 2.1. We first reduce the problem to the case when $(\Omega, \mathcal{F}, P, \theta)$ is a Polish system. Let ϕ and μ be as given in the formulation of Theorem 2.1. We now introduce a new, Polish, invertible and ergodic measure-preserving system $(\hat{\Omega}, \hat{\mathcal{F}}, \hat{P}, \hat{\theta})$ by defining

$$\begin{aligned}
\hat{\Omega} &= C^1(X,X)^{\mathbf{Z}} \text{ endowed with the product topology,}\\
\hat{\mathcal{F}} &= \text{the Borel } \sigma\text{-algebra of } \hat{\Omega},\\
\hat{P} &= \hat{\pi} P \text{ where } \hat{\pi} : (\Omega, \mathcal{F}) \to (\hat{\Omega}, \hat{\mathcal{F}}) \text{ is the measurable map}\\
&\quad\ \text{defined by } \hat{\pi}\omega = (\cdots, \phi(1,\theta^{-1}\omega), \phi(1,\omega), \phi(1,\theta\omega), \cdots),\\
\hat{\theta} &= \text{the left shift operator on } \hat{\Omega}.
\end{aligned}$$

Then we may define a C^1 RDS $\hat{\phi}$ on X over the Polish system $(\hat{\Omega}, \hat{\mathcal{F}}, \hat{P}, \hat{\theta})$ by

$$\hat{\phi} : \mathbf{Z}^+ \times \hat{\Omega} \times X \to X, \quad (n, \hat{\omega}, x) \mapsto f_{\hat{\omega}}^n x$$

where

$$f_{\hat{\omega}}^n = \begin{cases} id & \text{if } n = 0 \\ f_{n-1}(\hat{\omega}) \circ \cdots \circ f_0(\hat{\omega}) & \text{if } n > 0 \end{cases}$$

for $\hat{\omega} = (\cdots, f_{-1}(\hat{\omega}), f_0(\hat{\omega}), f_1(\hat{\omega}), \cdots) \in \hat{\Omega}$. It is clear that

$$\Sigma \circ \Theta = \hat{\Theta} \circ \Sigma$$

where $\hat{\Theta}$ is the skew product transformation induced by $\hat{\phi}$ and $\Sigma : \Omega \times X \to \hat{\Omega} \times X$ is a map defined by $\Sigma : (\omega, x) \mapsto (\hat{\pi}\omega, x)$. Put

$$\hat{\mu} = \Sigma\mu.$$

It is easy to see that $\hat{\mu}$ is a $\hat{\phi}$-invariant measure and that, by (1.2)

$$h_{\hat{\mu}}(\hat{\phi}) = h_{\hat{\mu}}^{\pi^{-1}\hat{\mathcal{F}}}(\hat{\Theta}) \geq h_{\mu}^{\pi^{-1}\mathcal{F}}(\Theta) = h_{\mu}(\phi).$$

Since $\hat{\pi}$: $(\Omega, \mathcal{F}, P) \to (\hat{\Omega}, \hat{\mathcal{F}}, \hat{P})$ is measure-preserving, from (2.1) it follows that

$$\int_{\hat{\Omega}} \log^+ |\hat{\phi}(1, \hat{\omega})|_{C^1}\, d\hat{P}(\hat{\omega}) < +\infty.$$

This ensures the existence of the Lyapunov exponents $\lambda^{(1)}(\hat{\omega}, x) < \cdots < \lambda^{(r(\hat{\omega},x))}(\hat{\omega}, x)$ of $\hat{\phi}$ at $\hat{\mu}$-a.e. $(\hat{\omega}, x)$. Clearly,

$$\int \sum_i \lambda^{(i)}(\omega, x)^+ m_i(\omega, x)\, d\mu = \int \sum_i \lambda^{(i)}(\hat{\omega}, x)^+ m_i(\hat{\omega}, x)\, d\hat{\mu}.$$

Hence, in order to prove Theorem 2.1 it is sufficient to consider the case when $(\Omega, \mathcal{F}, P, \theta)$ is Polish.

Next, we will prove the Margulis-Ruelle inequality for a C^1 RDS ϕ on a compact subset of $\mathbf{R}^d$ and then deduce our main result by extending ϕ to a tubular neighbourhood of X. Considerations of this type are standard in differential geometry. In this appendix $\|\cdot\|$ will denote the usual Euclidean norm and $B(X, \varepsilon)$ the closed ε-neighbourhood of X in $\mathbf{R}^d$ if $X \subset \mathbf{R}^d$.

Theorem 2.2. *Let X be a compact subset of $\mathbf{R}^d$, U an open neighbourhood of X, and ϕ a C^1 RDS on U over a Polish system $(\Omega, \mathcal{F}, P, \theta)$ with $\phi(1,\omega)X \subset X$ for all $\omega \in \Omega$. Let μ be a ϕ-invariant measure with support in $\Omega \times X$ and assume that there is an $\varepsilon_0 > 0$ with $B(X, \varepsilon_0) \subset U$ such that $\phi(1,\omega)B(X, \varepsilon_0) \subset B(X, \varepsilon_0)$ for all $\omega \in \Omega$ and*

$$\int_{\Omega} \log^+ \sup_{x \in B(X, \varepsilon_0)} \|T_x\phi(1, \omega)\|\, dP(\omega) < +\infty. \tag{2.3}$$

Then it holds that

$$h_{\mu}(\phi) \leq \int_{\Omega \times X} \sum_i \lambda^{(i)}(\omega, x)^+ m_i(\omega, x)\, d\mu. \tag{2.4}$$

Proof: Fix $n \in \mathbf{N}$ arbitrarily. Let $k \in \mathbf{N}$. We write $\omega \in \Omega_k$ if for all x, $y \in B(X, \varepsilon_0)$ with $\|x - y\| \leq \varepsilon_0/k$ we have

$$\|\phi(n,\omega)y - \phi(n,\omega)x - T_x\phi(n,\omega)(y-x)\| \leq \|y - x\| \tag{2.5}$$

and

$$|\delta_i(T_x\phi(n,\omega)) - \delta_i(T_y\phi(n,\omega))| \leq \frac{1}{2}, \quad 1 \leq i \leq d$$

which clearly implies

$$\frac{1}{2} \leq \frac{\max\{1, \delta_i(T_x\phi(n,\omega))\}}{\max\{1, \delta_i(T_y\phi(n,\omega))\}} \leq 2, \quad 1 \leq i \leq d, \tag{2.6}$$

where $\delta_i(A)$, $1 \leq i \leq d$ are as introduced at the beginning of Section II.2. One can check that each Ω_k is measurable and, since every $\phi(n,\omega)$ is C^1 and $B(X, \varepsilon_0)$ is compact, for every ω there exists $k \in \mathbf{N}$ such that $\omega \in \Omega_k$. Consequently, $\lim_{k\to+\infty} P(\Omega_k) = 1$ since $\Omega_k \subset \Omega_{k+1}$ for all $k \in \mathbf{N}$.

For each $k \in \mathbf{N}$, take a maximal ε_0/k-seperated set E_k of X. We then define a measurable partition $\xi_k = \{\xi_k(x) : x \in E_k\}$ of X such that, with respect to the topology of X as a subspace of $\mathbf{R}^d$, $\xi_k(x) \subset \overline{\mathrm{int}\,(\xi_k(x))}$ and $\mathrm{int}\,(\xi_k(x)) = \{y \in X : \|y - x\| < \|y - x_i\| \text{ if } x \neq x_i \in E_k\}$ for every $x \in E_k$. Clearly $\xi_k(x) \subset B(x, \varepsilon_0/k)$ for all $x \in E_k$ and $\mathrm{diam}\,\xi_k \leq 2\varepsilon_0/k$. Then by Lemma 1.1

$$nh_\mu(\phi) = h_\mu(\phi^n) = \lim_{k\to+\infty} h_\mu(\phi^n, \xi_k). \tag{2.7}$$

Note that for each finite partition ξ of X one has

$$\begin{aligned}
& h_\mu(\phi, \xi) \\
= \;& \lim_{m\to\infty} \frac{1}{m} \int H_{\mu_\omega}\left(\bigvee_{k=0}^{m-1} \phi(k,\omega)^{-1}\xi\right) dP(\omega) \\
\leq \;& \lim_{m\to\infty} \frac{1}{m} \int \left[\sum_{k=1}^{m-1} H_{\mu_\omega}(\phi(k,\omega)^{-1}\xi | \phi(k-1,\omega)^{-1}\xi) + H_{\mu_\omega}(\xi)\right] dP(\omega) \\
= \;& \lim_{m\to\infty} \frac{1}{m} \int \left[\sum_{k=1}^{m-1} H_{\mu_{\theta^{k-1}\omega}}(\phi(1,\theta^{k-1}\omega)^{-1}\xi | \xi) + H_{\mu_\omega}(\xi)\right] dP(\omega) \\
= \;& \lim_{m\to\infty} \frac{1}{m} \left[(m-1) \int H_{\mu_\omega}(\phi(1,\omega)^{-1}\xi|\xi)\, dP(\omega) + \int H_{\mu_\omega}(\xi)\, d(\omega)\right] \\
= \;& \int H_{\mu_\omega}(\phi(1,\omega)^{-1}\xi|\xi)\, dP(\omega).
\end{aligned}$$

Thus from (2.7) it follows that

$$\begin{aligned}
nh_\mu(\phi) \;&\leq\; \overline{\lim_{k\to\infty}} \int H_{\mu_\omega}(\phi(n,\omega)^{-1}\xi_k|\xi_k)\, dP(\omega) \\
&\leq\; \overline{\lim_{k\to\infty}} \int_{\Omega_k} H_{\mu_\omega}(\phi(n,\omega)^{-1}\xi_k|\xi_k)\, dP(\omega)
\end{aligned}$$

$$+\overline{\lim_{k\to\infty}}\int_{\Omega\setminus\Omega_k} H_{\mu_\omega}(\phi(n,\omega)^{-1}\xi_k|\xi_k)\,dP(\omega)$$

$$\stackrel{\text{def}}{=} \overline{\lim_{k\to\infty}} M_k + \overline{\lim_{k\to\infty}} m_k. \tag{2.8}$$

In what follows we estimate M_k and m_k.

Let $\omega \in \Omega_k$. We first estimate the number, written $N_{n,k}(\omega,x)$, of elements of ξ_k which intersect $\phi(n,\omega)\xi_k(x)$ for $x \in E_k$. By (2.5),

$$\begin{aligned}\phi(n,\omega)\xi_k(x) &\subset \phi(n,\omega)B(x,\varepsilon_0/k)\\ &\subset B\left(\phi(n,\omega)x + T_x\phi(n,\omega)B(0,\varepsilon_0/k),\ \ \varepsilon_0/k\right)\\ &= \phi(n,\omega)x + B\left(T_x\phi(n,\omega)B(0,\varepsilon_0/k),\varepsilon_0/k\right).\end{aligned}$$

Hence,

$$\xi_k(x')\cap\phi(n,\omega)\xi_k(x)\neq\emptyset$$

implies

$$B(x',\varepsilon_0/2k)\cap[\phi(n,\omega)x + B(T_x\phi(n,\omega)B(0,\varepsilon_0/k),2\varepsilon_0/k)]\neq\emptyset.$$

Since $B(x',\varepsilon_0/2k)$, $x'\in E_k$ are disjoint, we know that the number $N_{n,k}(\omega,x)$ can not exceed

$$K_n(\omega,x)\stackrel{\text{def}}{=} C_1(d)\prod_{i=1}^{d}\max\{\delta_i(T_x\phi(n,\omega)),1\}$$

where $C_1(d)$ is a constant depending only on d. Therefore,

$$\begin{aligned}&H_{\mu_\omega}(\phi(n,\omega)^{-1}\xi_k|\xi_k)\\ \le\ &\sum_{x\in E_k}\mu_\omega(\xi_k(x))\log K_n(\omega,x)\\ =\ &\sum_{x\in E_k}\int_{\xi_k(x)}\log K_n(\omega,x)\,d\mu_\omega(y).\end{aligned}$$

By (2.6) we have for $y\in\xi_k(x)$

$$\log^+\delta_i(T_x\phi(n,\omega))\le\log 2+\log^+\delta_i(T_y\phi(n,\omega))$$

and hence

$$\log K_n(\omega,x)\le\log C_1(d)+d\log 2+\sum_{i=1}^{d}\log^+\delta_i(T_y\phi(n,\omega)).$$

Consequently, we obtain

$$\begin{aligned}&H_{\mu_\omega}(\phi(n,\omega)^{-1}\xi_k|\xi_k)\\ \le\ &\log C_1(d)+d\log 2+\int_X\sum_{i=1}^{d}\log^+\delta_i(T_y\phi(n,\omega))\,d\mu_\omega(y)\end{aligned}$$

and hence

$$M_k \le \log C_1(d) + d\log 2 + \int_\Omega \int_X \sum_{i=1}^d \log^+ \delta_i(T_y\phi(n,\omega))\, d\mu_\omega(y)\, dP(\omega).$$

Now let $\omega \in \Omega \setminus \Omega_k$. Again we estimate the number $N_{n,k}(\omega, x)$ of elements of ξ_k which intersect $\phi(n,\omega)\xi_k(x)$ for $x \in E_k$. We can not make use of (2.5) in the present case, but we can apply the mean value theorem. For this purpose we define for $k \ge 2$

$$L_{n,k}(\omega) = \sup_{z\in B(X,2\varepsilon_0/k)} \|T_z\phi(n,\omega)\|.$$

Let now $k \ge 2$. By the mean value theorem we know that for all $y, z \in B(X, \varepsilon_0/k)$ with $\|y - z\| \le \varepsilon_0/k$

$$\|\phi(n,\omega)y - \phi(n,\omega)z\| \le L_{n,k}(\omega)\|y - z\|.$$

From this it follows that

$$\begin{aligned}\phi(n,\omega)\xi_k(x) &\subset \phi(n,\omega)B(x,\varepsilon_0/k)\\ &\subset B(\phi(n,\omega)x, L_{n,k}(\omega)\varepsilon_0/k).\end{aligned}$$

Thus, $N_{n,k}(\omega, x)$ can not exceed the maximal number of disjoint balls with radius $\varepsilon_0/2k$ which intersect $B(\phi(n,\omega)x, L_{n,k}(\omega)\varepsilon_0/k + \varepsilon_0/k)$. Hence

$$N_{n,k}(\omega,x) \le C_2(d)\max\{L_{n,k}(\omega), 1\}^d,$$

where $C_2(d)$ is a constant depending only on d. From this it follows that

$$\begin{aligned}H_{\mu_\omega}(\phi(n,\omega)^{-1}\xi_k|\xi_k) &\le \sum_{x\in E_k} \mu_\omega(\xi_k(x))\log N_{n,k}(\omega,x)\\ &\le \log C_2(d) + d\log^+ L_{n,k}(\omega)\\ &\le \log C_2(d) + d\log^+ L_{n,2}(\omega).\end{aligned}$$

Then we obtain

$$m_k \le \log C_2(d) + d\int_{\Omega\setminus\Omega_k} \log^+ L_{n,2}(\omega)dP(\omega). \tag{2.9}$$

From (2.3) it follows that $\log^+ L_{n,2}(\omega) \in L^1(\Omega, P)$. Then by (2.9) we have

$$\overline{\lim_{k\to\infty}}\, m_k \le \log C_2(d)$$

since $P(\Omega\setminus\Omega_k) \longrightarrow 0$ as $k \longrightarrow +\infty$. This together with the estimate of M_k and (2.8) yields

$$\begin{aligned}nh_\mu(\phi) &\le C + \int_\Omega\int_X \sum_{i=1}^d \log^+ \delta_i(T_x\phi(n,\omega))d\mu_\omega(x)dP(\omega)\\ &\le C + \int_{\Omega\times X} \log|(T_x\phi(n,\omega))^\wedge|\, d\mu \qquad (2.10)\end{aligned}$$

where $C = \log C_1(d) + d\log 2 + \log C_2(d)$. Dividing (2.10) by n and letting $n \to +\infty$, we obtain (2.4). □

Proof of Theorem 2.1. Without loss of generality, we assume that $(\Omega, \mathcal{F}, P, \theta)$ is a Polish system. Let h be a C^∞ embedding of X into $\mathbf{R}^{2d+1}$. We will identify X and T_xX, $x \in X$ respectively with their images under h and Th. Let

$$\nu(X) = \{(x, v) \in X \times \mathbf{R}^{2d+1} : v \perp T_x M\}$$

be the normal bundle to X. The exponential map exp: $T\mathbf{R}^{2d+1} = \mathbf{R}^{2d+1} \times \mathbf{R}^{2d+1} \longrightarrow \mathbf{R}^{2d+1}$ is defined by $\exp_x v = x + v$. Let G be the restriction of exp to $\nu(X)$. Then there is an $\varepsilon > 0$ such that $\{(x, v) \in \nu(X) : \|v\| < \varepsilon\}$ is C^∞ diffeomorphically mapped by G onto $\{y \in \mathbf{R}^{2d+1} : d(y, X) < \varepsilon\} \stackrel{\text{def}}{=} N_\varepsilon(X)$ ($N_\varepsilon(X)$ is called a tubular neighbourhood of X, see [Elw]$_1$). We extend ϕ to $N_\varepsilon(X)$ in the following way: Define $\hat{\phi} = j \circ \phi \circ \pi$, where $\pi : N_\varepsilon(X) \to X$ is the projection onto X and $j : X \to N_\varepsilon(X)$ is the inclusion into $N_\varepsilon(X)$. In particular, we have $\hat{\phi}(1, \omega)x = \phi(1, \omega)x$ for all $x \in X$ and for all $\omega \in \Omega$. Choose a number $0 < \delta_0 < \varepsilon$ and put $\hat{X} = B(X, \delta_0)$. The chain rule and the compactness of X and $\hat{X}$ imply that there is a constant C, which depends only on X, such that

$$\sup_{\hat{x} \in \hat{X}} \|T_{\hat{x}} \hat{\phi}(1, \omega)\| \le C \sup_{x \in X} \|T_x \phi(1, \omega)\|$$

for all $\omega \in \Omega$. Since X is compact, all Riemannian metrics on X are equivalent and therefore the Lyapunov exponents of ϕ (with invariant measure μ) are independent of the choice of the Riemannian metric. Note also that no new Lyapunov exponents other than $-\infty$ are created for $\hat{\phi}$ (with invariant measure μ) since for $x \in X$ and $v \in T_x N_\varepsilon(X)$ with $v \perp T_x M$

$$T_x \hat{\phi}(1, \omega)v = \frac{d}{dt}\left(\hat{\phi}(1, \omega)(x + tv)\right)\Big|_{t=0} = \frac{d}{dt}(\phi(1, \omega)x)\Big|_{t=0} = 0.$$

Thus Theorem 2.1 follows from Theorem 2.2. □

References

[Ano] D. V. Anosov, Ya. G. Sinai, *Certain smooth ergodic systems,* Russ. Math. Surveys, 22 (1967), No. 5, 103-167.

[Arn] L. Arnold, H. Crauel, J. P. Eckmann (Eds.), *Lyapunov Exponents,* Lec. Not. Math., Vol. 1486, Springer-Verlag (1991).

[Bah] J. Bahnmüller, T. Bogenschütz, *A Margulis-Ruelle inequality for random dynamical systems,* Report No. 301, Institut für Dynamische Systeme, Universität Bremen, March 1994.

[Bax]$_1$ P. H. Baxendale, *The Lyapunov spectrum of a stochastic flow of diffeomorphisms,* Lec. Not. Math., Vol. 1186, Springer-Verlag (1986), 322-337.

[Bax]$_2$ P. H. Baxendale, *Brownian motions in the diffeomorphism group I,* Compositio Mathematica 53 (1984), 19-50.

[Bog] T. Bogenschütz, *Entropy, pressure, and a variational principle for random dynamical systems,* Random and Computational Dynamics 1 (1992), 219-227.

[Boo] W. M. Boothby, *An Introduction to Differentiable Manifolds and Riemanian Geometry (Second Edition),* Academic Press, INC, 1986.

[Bow]$_1$ R. Bowen, *Equilibrium States and Ergodic Theory of Anosov Diffeomorphisms,* Lec. Not. Math., Vol. 470, Springer-Verlag (1975).

[Bow]$_2$ R. Bowen, D. Ruelle, *The ergodic theory of Axiom A flows,* Invent. Math. 29 (1975), 181-202.

[Bri] M. Brin, Y. Kifer, *Dynamics of Markov chains and stable manifolds for random diffeomorphisms,* Ergod. Th. and Dynam. Sys. 7 (1987), 351-374.

[Car] A. Carverhill, *Flows of stochastic dynamical systems: Ergodic theory,* Stochastic 14 (1985), 273-317.

[Che] Chen Zao-Ping, Liu Pei-Dong, *Orbit-shift stability of a class of self-covering maps,* Science in China (Series A), Vol. 34, No. 1 (1991), 1-13.

[Coh] D. L. Cohn, *Measure Theory,* Birkhäuser, Boston, 1980.

[Dyn] E. B. Dynkin, *Markov Processes, I, II,* Springer-Verlag, Berlin, 1965.

[Eck] J. P. Eckmann, D. Ruelle, *Ergodic theory of chaos and strange attractors,* Reviews of Modern Physics, Vol. 57, No. 3, Part I (1985).

[Elw]$_1$ K. D. Elworthy, *Geometric aspect of diffusions on manifolds,* Lec. Not. Math., Vol. 1362, Springer-Verlag (1988), 277-425.

[Elw]$_2$ K. D. Elworthy, *Stochastic Differential Equations,* L. M. S. Lecture Notices, Cambrige University Press, 1982.

[Fat] F. Fathi, M. R. Herman, J. C. Yoccoz, *A proof of Pesin's stable manifold theorem,* Lec. Not. Math., Vol. 1007, Springer-Verlag (1983), 177-215.

[Fra] J. Franks, *Manifolds of C^r mappings and applications to differentiable dynamical systems,* Studies in analysis, Advances in Mathematics, Supplementary Series 4 (1979), 271-291.

[Guz] M. De Guzmán, *Differentiation of Integrals in* $\mathbf{R}^n$, Lec. Not. Math., Vol. 481, Springer-Verlag (1975).

[Hir]$_1$ M. W. Hirsch, *Differential Topology,* Berlin, Springer, 1976.

[Hir]$_2$ M. W. Hirsch, C. Pugh, *Stable manifolds and hyperbolic sets,* AMS Proc. Symp. Pure Math., Vol. 14 (1970), 133-164.

[Hir]$_3$ M. W. Hirsch, J. Palis, C. Pugh and M. Shub, *Neighborhoods of hyperbolic sets,* Invent. Math. 9 (1970), 121- 134.

[Ike] N. Ikeda, S. Watanabe, *Stochastic Differential Equations and Diffusion Processes,* Tokyo, 1981.

[Kat] A. Katok, J. M. Strelcyn, *Invariant Manifold, Entropy and Billiards; Smooth Maps with Singularities,* Lec. Not. Math., Vol. 1222, Springer-Verlag (1986).

[Kif]$_1$ Y. Kifer, *Ergodic Theory of Random Transformations,* Birkhäuser, Boston, 1986.

[Kif]$_2$ Y. Kifer, *Random Perturbations of Dynamical Systems,* Birkhäuser, Boston, 1987.

[Kif]$_3$ Y. Kifer, *A note on integrability of C^r-norms of stochastic flows and applications,* Lec. Not. Math., Vol. 1325, Springer-Verlag (1988), 125-131.

[Kif]$_4$ Y. Kifer, *On small random perturbations of some smooth dynamical systems,* Math. USSR-Izv 8 (1974), 1083-1107.

[Kin] J. F. C. Kingman, *The ergodic theory of subadditive stochastic processes,* J. Royal Statist. Soc. B 30 (1968), 499-510.

[Kun]$_1$ H. Kunita, *Stochastic Flows and Stochastic Differential Equations,* Cambridge University Press, 1990.

[Kun]$_2$ H. Kunita, *Stochastic differential equations and stochastic flows of diffeomorphisms,* Lec. Not. Math., Vol. 1097, Springer-Verlag (1984), 144-303.

[Led]$_1$ F. Ledrappier, L. -S. Young, *Entropy formula for random transformations,* Probab. Th. Rel. Fields 80 (1988), 217-240.

[Led]$_2$ F. Ledrappier, L. -S. Young, *The metric entropy of diffeomorphisms Part I: Characterization of measures satisfying Pesin's entropy formula,* Ann. Math. 122 (1985), 509-539.

[Led]$_3$ F. Ledrappier, J. M. Strelcyn, *A proof of the estimation from below in Pesin's entropy formula,* Ergod. Th. and Dynam. Sys. 2 (1982), 203-219.

[Led]$_4$ F. Ledrappier, *Quelques propriétés des exposants caractéristiques,* Lec. Not. Math., Vol. 1097, Springer-Verlag (1984), 305-396.

[Liu]$_1$ Liu Pei-Dong, Qian Min, *Ruelle's inequality for the entropy of diffusion processes,* Acta Mathematica Sinica (New series), Vol. 9, No. 2 (1993), 212-224.

[Liu]$_2$ Liu Pei-Dong, *Pesin's entropy formula for stochastic flows of diffeomorphisms (in Chinese),* Science in China (Series A), Vol. 24, No. 1 (1994), 16-27.

[Liu]$_3$ Liu Pei-Dong, Qian Min, *Pesin's entropy formula and SBR-measures of random diffeomorphisms,* Science in China (Series A), Vol. 36, No. 8 (1993), 940-956.

[Liu]$_4$ Liu Pei-Dong, *R-stability of orbit-spaces of endomorphisms (in Chinese),* Ann. of Chinese Math., No. 4 (1991), 415-421.

[Liu]$_5$ Liu Pei-Dong, *Random perturbations of hyperbolic attractors,* Preprint (to appear).

[Liu]$_6$ Liu Pei-Dong, Qian Min, *Conditional entropies of measure-preserving transformations,* Advances in Math. (China) (to appear).

[Liu]$_7$ Liu Pei-Dong, *Nonautonomous perturbations of hyperbolic sets,* Preprint.

[Man]$_1$ R. Mañé, *Ergodic Theory and Differentiable Dynamics,* Springer-Verlag, 1987.

[Man]$_2$ R. Mañé, *A proof of Pesin's formula,* Ergod. Th. and Dynam. Sys. 1 (1981), 95-102.

[Ose] V. I. Oseledec, *The multiplicative ergodic theorem,* Transl. Mosc. Math. Soc. 19 (1968), 197-231.

[Pes]$_1$ Ya. B. Pesin, *Families of invariant manifolds corresponding to non-zero characteristic exponents,* Math. of the USSR-Izvestija, Vol. 10, No. 6 (1976), 1261-1305.

[Pes]$_2$ Ya. B. Pesin, *Lyapunov characteristic exponents and smooth ergodic theory,* Russ. Math. Surveys 32: 4 (1977), 55-114.

[Roh]$_1$ V. A. Rohlin, *On the fundamental ideas of measure theory,* A. M. S. Transl. (1) 10 (1962), 1-52.

[Roh]$_2$ V. A. Rohlin, *Lectures on the theory of entropy of transformations with invariant measures,* Russ. Math. Surveys 22 : 5 (1967), 1-54.

[Roh]$_3$ V. A. Rohlin, *Selected topics from the metric theory of dynamical systems,* Amer. Math. Soc. Transl. (2) 49 (1966), 171-240.

[Roy] H. L. Royden, *Real Analysis (2nd Ed.),* MacMillan, New York, 1968.

[Rud] W. Rudin, *Real and Complex Analysis,* New York, McGraw-Hill, 1974.

[Rue]$_1$ D. Ruelle, *An inequality for the entropy of differentiable maps,* Bol. Soc. Bras. Math. 9 (1978), 83-87.

[Rue]$_2$ D. Ruelle, *Ergodic theory of differentiable dynamical systems,* Publ. Math. IHES 50 (1979), 27-58.

[Rue]$_3$ D. Ruelle, *A measure associated with Axiom A attractors,* Amer. J. Math. 98 (1976), 619-654.

[Shi] G. E. Shilov, B. L. Gurevich, *Integral, Measure and Derivative: A Unified Approach,* Dover Publ., N. Y., 1977.

[Sin] Ya. G. Sinai, *Gibbs measure in ergodic theory,* Russ. Math. Surveys 27 : 4 (1972), 21-69.

[Wal]$_1$ P. Walters, *An Introduction to Ergodic Theory,* New York, Springer, 1982.

[Wal]$_2$ P. Walters, *Review of "Ergodic Theory of Random Transformations" by Kifer,* Bull. Amer. Math. Soc. (N. S.) 21 (1989), 113-117.

[Yos] K. Yosida, *Functional Analysis (3rd edition),* Springer-Verlag, Heidelberg, Berlin, New York, 1971.

[You] L. -S. Young, *Stochastic stability of hyperbolic attractors,* Ergod. Th. and Dynam. Sys. 6 (1986), 311-319.

[Zha] Zhang Zhu-Sheng, *Principle of Differentiable Dynamical Systems (in Chinese),* Science Press (China), 1987.

Subject Index

Lecture Notes in Mathematics

For information about Vols. 1–1425
please contact your bookseller or Springer-Verlag

Vol. 1426: J. Azéma, P.A. Meyer, M. Yor (Eds.), Séminaire de Probabilités XXIV, 1988/89. V, 490 pages. 1990.

Vol. 1427: A. Ancona, D. Geman, N. Ikeda, École d'Eté de Probabilités de Saint Flour XVIII, 1988. Ed.: P.L. Hennequin. VII, 330 pages. 1990.

Vol. 1428: K. Erdmann, Blocks of Tame Representation Type and Related Algebras. XV. 312 pages. 1990.

Vol. 1429: S. Homer, A. Nerode, R.A. Platek, G.E. Sacks, A. Scedrov, Logic and Computer Science. Seminar, 1988. Editor: P. Odifreddi. V, 162 pages. 1990.

Vol. 1430: W. Bruns, A. Simis (Eds.), Commutative Algebra. Proceedings. 1988. V, 160 pages. 1990.

Vol. 1431: J.G. Heywood, K. Masuda, R. Rautmann, V.A. Solonnikov (Eds.), The Navier-Stokes Equations – Theory and Numerical Methods. Proceedings, 1988. VII, 238 pages. 1990.

Vol. 1432: K. Ambos-Spies, G.H. Müller, G.E. Sacks (Eds.), Recursion Theory Week. Proceedings, 1989. VI, 393 pages. 1990.

Vol. 1433: S. Lang, W. Cherry, Topics in Nevanlinna Theory. II, 174 pages. 1990.

Vol. 1434: K. Nagasaka, E. Fouvry (Eds.), Analytic Number Theory. Proceedings, 1988. VI, 218 pages. 1990.

Vol. 1435: St. Ruscheweyh, E.B. Saff, L.C. Salinas, R.S. Varga (Eds.), Computational Methods and Function Theory. Proceedings, 1989. VI, 211 pages. 1990.

Vol. 1436: S. Xambó-Descamps (Ed.), Enumerative Geometry. Proceedings, 1987. V, 303 pages. 1990.

Vol. 1437: H. Inassaridze (Ed.), K-theory and Homological Algebra. Seminar, 1987–88. V, 313 pages. 1990.

Vol. 1438: P.G. Lemarié (Ed.) Les Ondelettes en 1989. Seminar. IV, 212 pages. 1990.

Vol. 1439: E. Bujalance, J.J. Etayo, J.M. Gamboa, G. Gromadzki. Automorphism Groups of Compact Bordered Klein Surfaces: A Combinatorial Approach. XIII, 201 pages. 1990.

Vol. 1440: P. Latiolais (Ed.), Topology and Combinatorial Groups Theory. Seminar, 1985–1988. VI, 207 pages. 1990.

Vol. 1441: M. Coornaert, T. Delzant, A. Papadopoulos. Géométrie et théorie des groupes. X, 165 pages. 1990.

Vol. 1442: L. Accardi, M. von Waldenfels (Eds.), Quantum Probability and Applications V. Proceedings, 1988. VI, 413 pages. 1990.

Vol. 1443: K.H. Dovermann, R. Schultz, Equivariant Surgery Theories and Their Periodicity Properties. VI, 227 pages. 1990.

Vol. 1444: H. Korezlioglu, A.S. Ustunel (Eds.), Stochastic Analysis and Related Topics VI. Proceedings, 1988. V, 268 pages. 1990.

Vol. 1445: F. Schulz, Regularity Theory for Quasilinear Elliptic Systems and – Monge Ampère Equations in Two Dimensions. XV, 123 pages. 1990.

Vol. 1446: Methods of Nonconvex Analysis. Seminar, 1989. Editor: A. Cellina. V, 206 pages. 1990.

Vol. 1447: J.-G. Labesse, J. Schwermer (Eds), Cohomology of Arithmetic Groups and Automorphic Forms. Proceedings, 1989. V, 358 pages. 1990.

Vol. 1448: S.K. Jain, S.R. López-Permouth (Eds.), Non-Commutative Ring Theory. Proceedings, 1989. V, 166 pages. 1990.

Vol. 1449: W. Odyniec, G. Lewicki, Minimal Projections in Banach Spaces. VIII, 168 pages. 1990.

Vol. 1450: H. Fujita, T. Ikebe, S.T. Kuroda (Eds.), Functional-Analytic Methods for Partial Differential Equations. Proceedings, 1989. VII, 252 pages. 1990.

Vol. 1451: L. Alvarez-Gaumé, E. Arbarello, C. De Concini, N.J. Hitchin, Global Geometry and Mathematical Physics. Montecatini Terme 1988. Seminar. Editors: M. Francaviglia, F. Gherardelli. IX, 197 pages. 1990.

Vol. 1452: E. Hlawka, R.F. Tichy (Eds.), Number-Theoretic Analysis. Seminar, 1988–89. V, 220 pages. 1990.

Vol. 1453: Yu.G. Borisovich, Yu.E. Gliklikh (Eds.), Global Analysis – Studies and Applications IV. V, 320 pages. 1990.

Vol. 1454: F. Baldassari, S. Bosch, B. Dwork (Eds.), p-adic Analysis. Proceedings, 1989. V, 382 pages. 1990.

Vol. 1455: J.-P. Françoise, R. Roussarie (Eds.), Bifurcations of Planar Vector Fields. Proceedings, 1989. VI, 396 pages. 1990.

Vol. 1456: L.G. Kovács (Ed.), Groups – Canberra 1989. Proceedings. XII, 198 pages. 1990.

Vol. 1457: O. Axelsson, L.Yu. Kolotilina (Eds.), Preconditioned Conjugate Gradient Methods. Proceedings, 1989. V, 196 pages. 1990.

Vol. 1458: R. Schaaf, Global Solution Branches of Two Point Boundary Value Problems. XIX, 141 pages. 1990.

Vol. 1459: D. Tiba, Optimal Control of Nonsmooth Distributed Parameter Systems. VII, 159 pages. 1990.

Vol. 1460: G. Toscani, V. Boffi, S. Rionero (Eds.), Mathematical Aspects of Fluid Plasma Dynamics. Proceedings, 1988. V, 221 pages. 1991.

Vol. 1461: R. Gorenflo, S. Vessella, Abel Integral Equations. VII, 215 pages. 1991.

Vol. 1462: D. Mond, J. Montaldi (Eds.), Singularity Theory and its Applications. Warwick 1989, Part I. VIII, 405 pages. 1991.

Vol. 1463: R. Roberts, I. Stewart (Eds.), Singularity Theory and its Applications. Warwick 1989, Part II. VIII, 322 pages. 1991.

Vol. 1464: D. L. Burkholder, E. Pardoux, A. Sznitman, Ecole d'Eté de Probabilités de Saint- Flour XIX-1989. Editor: P. L. Hennequin. VI, 256 pages. 1991.

Vol. 1465: G. David, Wavelets and Singular Integrals on Curves and Surfaces. X, 107 pages. 1991.

Vol. 1466: W. Banaszczyk, Additive Subgroups of Topological Vector Spaces. VII, 178 pages. 1991.

Vol. 1467: W. M. Schmidt, Diophantine Approximations and Diophantine Equations. VIII, 217 pages. 1991.

Vol. 1468: J. Noguchi, T. Ohsawa (Eds.), Prospects in Complex Geometry. Proceedings, 1989. VII, 421 pages. 1991.

Vol. 1469: J. Lindenstrauss, V. D. Milman (Eds.), Geometric Aspects of Functional Analysis. Seminar 1989-90. XI, 191 pages. 1991.

Vol. 1470: E. Odell, H. Rosenthal (Eds.), Functional Analysis. Proceedings, 1987-89. VII, 199 pages. 1991.

Vol. 1471: A. A. Panchishkin, Non-Archimedean L-Functions of Siegel and Hilbert Modular Forms. VII, 157 pages. 1991.

Vol. 1472: T. T. Nielsen, Bose Algebras: The Complex and Real Wave Representations. V, 132 pages. 1991.

Vol. 1473: Y. Hino, S. Murakami, T. Naito, Functional Differential Equations with Infinite Delay. X, 317 pages. 1991.

Vol. 1474: S. Jackowski, B. Oliver, K. Pawałowski (Eds.), Algebraic Topology, Poznań 1989. Proceedings. VIII, 397 pages. 1991.

Vol. 1475: S. Busenberg, M. Martelli (Eds.), Delay Differential Equations and Dynamical Systems. Proceedings, 1990. VIII, 249 pages. 1991.

Vol. 1476: M. Bekkali, Topics in Set Theory. VII, 120 pages. 1991.

Vol. 1477: R. Jajte, Strong Limit Theorems in Noncommutative L_2-Spaces. X, 113 pages. 1991.

Vol. 1478: M.-P. Malliavin (Ed.), Topics in Invariant Theory. Seminar 1989-1990. VI, 272 pages. 1991.

Vol. 1479: S. Bloch, I. Dolgachev, W. Fulton (Eds.), Algebraic Geometry. Proceedings, 1989. VII, 300 pages. 1991.

Vol. 1480: F. Dumortier, R. Roussarie, J. Sotomayor, H. Żoładek, Bifurcations of Planar Vector Fields: Nilpotent Singularities and Abelian Integrals. VIII, 226 pages. 1991.

Vol. 1481: D. Ferus, U. Pinkall, U. Simon, B. Wegner (Eds.), Global Differential Geometry and Global Analysis. Proceedings, 1991. VIII, 283 pages. 1991.

Vol. 1482: J. Chabrowski, The Dirichlet Problem with L^2-Boundary Data for Elliptic Linear Equations. VI, 173 pages. 1991.

Vol. 1483: E. Reithmeier, Periodic Solutions of Nonlinear Dynamical Systems. VI, 171 pages. 1991.

Vol. 1484: H. Delfs, Homology of Locally Semialgebraic Spaces. IX, 136 pages. 1991.

Vol. 1485: J. Azéma, P. A. Meyer, M. Yor (Eds.), Séminaire de Probabilités XXV. VIII, 440 pages. 1991.

Vol. 1486: L. Arnold, H. Crauel, J.-P. Eckmann (Eds.), Lyapunov Exponents. Proceedings, 1990. VIII, 365 pages. 1991.

Vol. 1487: E. Freitag, Singular Modular Forms and Theta Relations. VI, 172 pages. 1991.

Vol. 1488: A. Carboni, M. C. Pedicchio, G. Rosolini (Eds.), Category Theory. Proceedings, 1990. VII, 494 pages. 1991.

Vol. 1489: A. Mielke, Hamiltonian and Lagrangian Flows on Center Manifolds. X, 140 pages. 1991.

Vol. 1490: K. Metsch, Linear Spaces with Few Lines. XIII, 196 pages. 1991.

Vol. 1491: E. Lluis-Puebla, J.-L. Loday, H. Gillet, C. Soulé, V. Snaith, Higher Algebraic K-Theory: an overview. IX, 164 pages. 1992.

Vol. 1492: K. R. Wicks, Fractals and Hyperspaces. VIII, 168 pages. 1991.

Vol. 1493: E. Benoît (Ed.), Dynamic Bifurcations. Proceedings, Luminy 1990. VII, 219 pages. 1991.

Vol. 1494: M.-T. Cheng, X.-W. Zhou, D.-G. Deng (Eds.), Harmonic Analysis. Proceedings, 1988. IX, 226 pages. 1991.

Vol. 1495: J. M. Bony, G. Grubb, L. Hörmander, H. Komatsu, J. Sjöstrand, Microlocal Analysis and Applications. Montecatini Terme, 1989. Editors: L. Cattabriga, L. Rodino. VII, 349 pages. 1991.

Vol. 1496: C. Foias, B. Francis, J. W. Helton, H. Kwakernaak, J. B. Pearson, H_∞-Control Theory. Como, 1990. Editors: E. Mosca, L. Pandolfi. VII, 336 pages. 1991.

Vol. 1497: G. T. Herman, A. K. Louis, F. Natterer (Eds.), Mathematical Methods in Tomography. Proceedings 1990. X, 268 pages. 1991.

Vol. 1498: R. Lang, Spectral Theory of Random Schrödinger Operators. X, 125 pages. 1991.

Vol. 1499: K. Taira, Boundary Value Problems and Markov Processes. IX, 132 pages. 1991.

Vol. 1500: J.-P. Serre, Lie Algebras and Lie Groups. VII, 168 pages. 1992.

Vol. 1501: A. De Masi, E. Presutti, Mathematical Methods for Hydrodynamic Limits. IX, 196 pages. 1991.

Vol. 1502: C. Simpson, Asymptotic Behavior of Monodromy. V, 139 pages. 1991.

Vol. 1503: S. Shokranian, The Selberg-Arthur Trace Formula (Lectures by J. Arthur). VII, 97 pages. 1991.

Vol. 1504: J. Cheeger, M. Gromov, C. Okonek, P. Pansu, Geometric Topology: Recent Developments. Editors: P. de Bartolomeis, F. Tricerri. VII, 197 pages. 1991.

Vol. 1505: K. Kajitani, T. Nishitani, The Hyperbolic Cauchy Problem. VII, 168 pages. 1991.

Vol. 1506: A. Buium, Differential Algebraic Groups of Finite Dimension. XV, 145 pages. 1992.

Vol. 1507: K. Hulek, T. Peternell, M. Schneider, F.-O. Schreyer (Eds.), Complex Algebraic Varieties. Proceedings, 1990. VII, 179 pages. 1992.

Vol. 1508: M. Vuorinen (Ed.), Quasiconformal Space Mappings. A Collection of Surveys 1960-1990. IX, 148 pages. 1992.

Vol. 1509: J. Aguadé, M. Castellet, F. R. Cohen (Eds.), Algebraic Topology - Homotopy and Group Cohomology. Proceedings, 1990. X, 330 pages. 1992.

Vol. 1510: P. P. Kulish (Ed.), Quantum Groups. Proceedings, 1990. XII, 398 pages. 1992.

Vol. 1511: B. S. Yadav, D. Singh (Eds.), Functional Analysis and Operator Theory. Proceedings, 1990. VIII, 223 pages. 1992.

Vol. 1512: L. M. Adleman, M.-D. A. Huang, Primality Testing and Abelian Varieties Over Finite Fields. VII, 142 pages. 1992.

Vol. 1513: L. S. Block, W. A. Coppel, Dynamics in One Dimension. VIII, 249 pages. 1992.

Vol. 1514: U. Krengel, K. Richter, V. Warstat (Eds.), Ergodic Theory and Related Topics III, Proceedings, 1990. VIII, 236 pages. 1992.

Vol. 1515: E. Ballico, F. Catanese, C. Ciliberto (Eds.), Classification of Irregular Varieties. Proceedings, 1990. VII, 149 pages. 1992.

Vol. 1516: R. A. Lorentz, Multivariate Birkhoff Interpolation. IX, 192 pages. 1992.

Vol. 1517: K. Keimel, W. Roth, Ordered Cones and Approximation. VI, 134 pages. 1992.

Vol. 1518: H. Stichtenoth, M. A. Tsfasman (Eds.), Coding Theory and Algebraic Geometry. Proceedings, 1991. VIII, 223 pages. 1992.

Vol. 1519: M. W. Short, The Primitive Soluble Permutation Groups of Degree less than 256. IX, 145 pages. 1992.

Vol. 1520: Yu. G. Borisovich, Yu. E. Gliklikh (Eds.), Global Analysis – Studies and Applications V. VII, 284 pages. 1992.

Vol. 1521: S. Busenberg, B. Forte, H. K. Kuiken, Mathematical Modelling of Industrial Process. Bari, 1990. Editors: V. Capasso, A. Fasano. VII, 162 pages. 1992.

Vol. 1522: J.-M. Delort, F. B. I. Transformation. VII, 101 pages. 1992.

Vol. 1523: W. Xue, Rings with Morita Duality. X, 168 pages. 1992.

Vol. 1524: M. Coste, L. Mahé, M.-F. Roy (Eds.), Real Algebraic Geometry. Proceedings, 1991. VIII, 418 pages. 1992.

Vol. 1525: C. Casacuberta, M. Castellet (Eds.), Mathematical Research Today and Tomorrow. VII, 112 pages. 1992.

Vol. 1526: J. Azéma, P. A. Meyer, M. Yor (Eds.), Séminaire de Probabilités XXVI. X, 633 pages. 1992.

Vol. 1527: M. I. Freidlin, J.-F. Le Gall, Ecole d'Eté de Probabilités de Saint-Flour XX – 1990. Editor: P. L. Hennequin. VIII, 244 pages. 1992.

Vol. 1528: G. Isac, Complementarity Problems. VI, 297 pages. 1992.

Vol. 1529: J. van Neerven, The Adjoint of a Semigroup of Linear Operators. X, 195 pages. 1992.

Vol. 1530: J. G. Heywood, K. Masuda, R. Rautmann, S. A. Solonnikov (Eds.), The Navier-Stokes Equations II – Theory and Numerical Methods. IX, 322 pages. 1992.

Vol. 1531: M. Stoer, Design of Survivable Networks. IV, 206 pages. 1992.

Vol. 1532: J. F. Colombeau, Multiplication of Distributions. X, 184 pages. 1992.

Vol. 1533: P. Jipsen, H. Rose, Varieties of Lattices. X, 162 pages. 1992.

Vol. 1534: C. Greither, Cyclic Galois Extensions of Commutative Rings. X, 145 pages. 1992.

Vol. 1535: A. B. Evans, Orthomorphism Graphs of Groups. VIII, 114 pages. 1992.

Vol. 1536: M. K. Kwong, A. Zettl, Norm Inequalities for Derivatives and Differences. VII, 150 pages. 1992.

Vol. 1537: P. Fitzpatrick, M. Martelli, J. Mawhin, R. Nussbaum, Topological Methods for Ordinary Differential Equations. Montecatini Terme, 1991. Editors: M. Furi, P. Zecca. VII, 218 pages. 1993.

Vol. 1538: P.-A. Meyer, Quantum Probability for Probabilists. X, 287 pages. 1993.

Vol. 1539: M. Coornaert, A. Papadopoulos, Symbolic Dynamics and Hyperbolic Groups. VIII, 138 pages. 1993.

Vol. 1540: H. Komatsu (Ed.), Functional Analysis and Related Topics, 1991. Proceedings. XXI, 413 pages. 1993.

Vol. 1541: D. A. Dawson, B. Maisonneuve, J. Spencer, Ecole d' Eté de Probabilités de Saint-Flour XXI - 1991. Editor: P. L. Hennequin. VIII, 356 pages. 1993.

Vol. 1542: J.Fröhlich, Th.Kerler, Quantum Groups, Quantum Categories and Quantum Field Theory. VII, 431 pages. 1993.

Vol. 1543: A. L. Dontchev, T. Zolezzi, Well-Posed Optimization Problems. XII, 421 pages. 1993.

Vol. 1544: M.Schürmann, White Noise on Bialgebras. VII, 146 pages. 1993.

Vol. 1545: J. Morgan, K. O'Grady, Differential Topology of Complex Surfaces. VIII, 224 pages. 1993.

Vol. 1546: V. V. Kalashnikov, V. M. Zolotarev (Eds.), Stability Problems for Stochastic Models. Proceedings, 1991. VIII, 229 pages. 1993.

Vol. 1547: P. Harmand, D. Werner, W. Werner, M-ideals in Banach Spaces and Banach Algebras. VIII, 387 pages. 1993.

Vol. 1548: T. Urabe, Dynkin Graphs and Quadrilateral Singularities. VI, 233 pages. 1993.

Vol. 1549: G. Vainikko, Multidimensional Weakly Singular Integral Equations. XI, 159 pages. 1993.

Vol. 1550: A. A. Gonchar, E. B. Saff (Eds.), Methods of Approximation Theory in Complex Analysis and Mathematical Physics IV, 222 pages, 1993.

Vol. 1551: L. Arkeryd, P. L. Lions, P.A. Markowich, S.R. S. Varadhan. Nonequilibrium Problems in Many-Particle Systems. Montecatini, 1992. Editors: C. Cercignani, M. Pulvirenti. VII, 158 pages 1993.

Vol. 1552: J. Hilgert, K.-H. Neeb, Lie Semigroups and their Applications. XII, 315 pages. 1993.

Vol. 1553: J.-L- Colliot-Thélène, J. Kato, P. Vojta. Arithmetic Algebraic Geometry. Trento, 1991. Editor: E. Ballico. VII, 223 pages. 1993.

Vol. 1554: A. K. Lenstra, H. W. Lenstra, Jr. (Eds.), The Development of the Number Field Sieve. VIII, 131 pages. 1993.

Vol. 1555: O. Liess, Conical Refraction and Higher Microlocalization. X, 389 pages. 1993.

Vol. 1556: S. B. Kuksin, Nearly Integrable Infinite-Dimensional Hamiltonian Systems. XXVII, 101 pages. 1993.

Vol. 1557: J. Azéma, P. A. Meyer, M. Yor (Eds.), Séminaire de Probabilités XXVII. VI, 327 pages. 1993.

Vol. 1558: T. J. Bridges, J. E. Furter, Singularity Theory and Equivariant Symplectic Maps. VI, 226 pages. 1993.

Vol. 1559: V. G. Sprindžuk, Classical Diophantine Equations. XII, 228 pages. 1993.

Vol. 1560: T. Bartsch, Topological Methods for Variational Problems with Symmetries. X, 152 pages. 1993.

Vol. 1561: I. S. Molchanov, Limit Theorems for Unions of Random Closed Sets. X, 157 pages. 1993.

Vol. 1562: G. Harder, Eisensteinkohomologie und die Konstruktion gemischter Motive. XX, 184 pages. 1993.

Vol. 1563: E. Fabes, M. Fukushima, L. Gross, C. Kenig, M. Röckner, D. W. Stroock, Dirichlet Forms. Varenna, 1992. Editors: G. Dell'Antonio, U. Mosco. VII, 245 pages. 1993.

Vol. 1564: J. Jorgenson, S. Lang, Basic Analysis of Regularized Series and Products. IX, 122 pages. 1993.

Vol. 1565: L. Boutet de Monvel, C. De Concini, C. Procesi, P. Schapira, M. Vergne. D-modules, Representation Theory, and Quantum Groups. Venezia, 1992. Editors: G. Zampieri, A. D'Agnolo. VII, 217 pages. 1993.

Vol. 1566: B. Edixhoven, J.-H. Evertse (Eds.), Diophantine Approximation and Abelian Varieties. XIII, 127 pages. 1993.

Vol. 1567: R. L. Dobrushin, S. Kusuoka, Statistical Mechanics and Fractals. VII, 98 pages. 1993.

Vol. 1568: F. Weisz, Martingale Hardy Spaces and their Application in Fourier Analysis. VIII, 217 pages. 1994.

Vol. 1569: V. Totik, Weighted Approximation with Varying Weight. VI, 117 pages. 1994.

Vol. 1570: R. deLaubenfels, Existence Families, Functional Calculi and Evolution Equations. XV, 234 pages. 1994.

Vol. 1571: S. Yu. Pilyugin, The Space of Dynamical Systems with the C^0-Topology. X, 188 pages. 1994.

Vol. 1572: L. Göttsche, Hilbert Schemes of Zero-Dimensional Subschemes of Smooth Varieties. IX, 196 pages. 1994.

Vol. 1573: V. P. Havin, N. K. Nikolski (Eds.), Linear and Complex Analysis – Problem Book 3 – Part I. XXII, 489 pages. 1994.

Vol. 1574: V. P. Havin, N. K. Nikolski (Eds.), Linear and Complex Analysis – Problem Book 3 – Part II. XXII, 507 pages. 1994.

Vol. 1575: M. Mitrea, Clifford Wavelets, Singular Integrals, and Hardy Spaces. XI, 116 pages. 1994.

Vol. 1576: K. Kitahara, Spaces of Approximating Functions with Haar-Like Conditions. X, 110 pages. 1994.

Vol. 1577: N. Obata, White Noise Calculus and Fock Space. X, 183 pages. 1994.

Vol. 1578: J. Bernstein, V. Lunts, Equivariant Sheaves and Functors. V, 139 pages. 1994.

Vol. 1579: N. Kazamaki, Continuous Exponential Martingales and *BMO*. VII, 91 pages. 1994.

Vol. 1580: M. Milman, Extrapolation and Optimal Decompositions with Applications to Analysis. XI, 161 pages. 1994.

Vol. 1581: D. Bakry, R. D. Gill, S. A. Molchanov, Lectures on Probability Theory. Editor: P. Bernard. VIII, 420 pages. 1994.

Vol. 1582: W. Balser, From Divergent Power Series to Analytic Functions. X, 108 pages. 1994.

Vol. 1583: J. Azéma, P. A. Meyer, M. Yor (Eds.), Séminaire de Probabilités XXVIII. VI, 334 pages. 1994.

Vol. 1584: M. Brokate, N. Kenmochi, I. Müller, J. F. Rodriguez, C. Verdi, Phase Transitions and Hysteresis. Montecatini Terme, 1993. Editor: A. Visintin. VII. 291 pages. 1994.

Vol. 1585: G. Frey (Ed.), On Artin's Conjecture for Odd 2-dimensional Representations. VIII, 148 pages. 1994.

Vol. 1586: R. Nillsen, Difference Spaces and Invariant Linear Forms. XII, 186 pages. 1994.

Vol. 1587: N. Xi, Representations of Affine Hecke Algebras. VIII, 137 pages. 1994.

Vol. 1588: C. Scheiderer, Real and Étale Cohomology. XXIV, 273 pages. 1994.

Vol. 1589: J. Bellissard, M. Degli Esposti, G. Forni, S. Graffi, S. Isola, J. N. Mather, Transition to Chaos in Classical and Quantum Mechanics. Montecatini Terme, 1991. Editor: S. Graffi. VII, 192 pages. 1994.

Vol. 1590: P. M. Soardi, Potential Theory on Infinite Networks. VIII, 187 pages. 1994.

Vol. 1591: M. Abate, G. Patrizio, Finsler Metrics – A Global Approach. IX, 180 pages. 1994.

Vol. 1592: K. W. Breitung, Asymptotic Approximations for Probability Integrals. IX, 146 pages. 1994.

Vol. 1593: J. Jorgenson & S. Lang, D. Goldfeld, Explicit Formulas for Regularized Products and Series. VIII, 154 pages. 1994.

Vol. 1594: M. Green, J. Murre, C. Voisin, Algebraic Cycles and Hodge Theory. Torino, 1993. Editors: A. Albano, F. Bardelli. VII, 275 pages. 1994.

Vol. 1595: R.D.M. Accola, Topics in the Theory of Riemann Surfaces. IX, 105 pages. 1994.

Vol. 1596: L. Heindorf, L. B. Shapiro, Nearly Projective Boolean Algebras. X, 202 pages. 1994.

Vol. 1597: B. Herzog, Kodaira-Spencer Maps in Local Algebra. XVII, 176 pages. 1994.

Vol. 1598: J. Berndt, F. Tricerri, L. Vanhecke, Generalized Heisenberg Groups and Damek-Ricci Harmonic Spaces. VIII, 125 pages. 1995.

Vol. 1599: K. Johannson, Topology and Combinatorics of 3-Manifolds. XVIII, 446 pages. 1995.

Vol. 1600: W. Narkiewicz, Polynomial Mappings. VII, 130 pages. 1995.

Vol. 1601: A. Pott, Finite Geometry and Character Theory. VII, 181 pages. 1995.

Vol. 1602: J. Winkelmann, The Classification of Three-dimensional Homogeneous Complex Manifolds. XI, 230 pages. 1995.

Vol. 1603: V. Ene, Real Functions – Current Topics. XIII, 310 pages. 1995.

Vol. 1604: A. Huber, Mixed Motives and their Realization in Derived Categories. XV, 207 pages. 1995.

Vol. 1605: L. B. Wahlbin, Superconvergence in Galerkin Finite Element Methods. XI, 166 pages. 1995.

Vol. 1606: P.-D. Liu, M. Qian, Smooth Ergodic Theory of Random Dynamical Systems. XI, 221 pages. 1995.